Cold Deserts

Yashwant Singh Rawat · Chandra Prakash Kala
Editors

Cold Deserts

Vulnerability, Climate Change Adaptation and Mitigation

Springer

Editors
Yashwant Singh Rawat
Department of Wood Technology
Management, Faculty of Civil Technology
Federal Technical and Vocational Training
Institute (TVTI)
Addis Ababa, Ethiopia

Chandra Prakash Kala
Indian Institute of Forest Management
Bhopal, Madhya Pradesh, India

ISBN 978-3-032-29118-9 ISBN 978-3-032-29119-6 (eBook)
https://doi.org/10.1007/978-3-032-29119-6

This Springer imprint is published by the registered company Springer Nature Switzerland AG
The registered company address is: Gewerbestrasse 11, 6330 Cham, Switzerland

Preface

The book *Cold Deserts: Vulnerability, Climate Change Adaptation and Mitigation* provides a multidisciplinary exploration of high-altitude alpine ecosystems, which are increasingly recognized as terrestrial sentinels of global change. Cold deserts represent some of the most fragile and distinctive ecosystems on Earth, characterized by extreme aridity, intense solar radiation, extreme temperature fluctuations, and minimal precipitation. These ecosystems, spanning from the high-altitude trans-Himalayan plateaus like Ladakh and Tibet to the icy reaches of Antarctica and the Patagonian Steppe, act as terrestrial sentinels of global change. These landscapes support uniquely adapted sparse vegetation patterns, biodiversity and culturally distinct human populations currently facing unprecedented threats from global warming and anthropogenic pressures. Evidence indicates that the Himalayan region exhibits some of the most visible reflections of these shifts, including rapid glacial retreat, abnormal precipitation patterns, and the degradation of high-altitude permafrost. Such environmental perturbations threaten the delicate equilibrium of unique flora and fauna, while increasing the frequency of natural hazards like glacial lake outburst floods (GLOFs), cloudburst and flash floods. This volume synthesizes current research on ecological vulnerabilities, traditional knowledge systems, hydrogeological, socio-cultural dimensions and technological innovations necessary to secure a sustainable future of cold desert environments.

This book consists of sixteen contemporary valuable contributions from dedicated scholars that provide empirical investigations and findings on culture and traditional knowledge systems, climate change mitigation and adaption, sustainability, sustainable forest management, ecosystem services and biodiversity conservation in the context of global environmental change and sustainability. Each article published in this book is focused to bridge the existing knowledge gaps and contributes to the development of effective and sustainable management strategies for the sustainability of cold desert ecosystems thereby contributing to the sustainable development goals (SDGs).

First Chapter "The Cold Desert Ecosystem of Ladakh: Vulnerabilities, Adaptations and Pathways Towards Sustainable Development" establishes the foundational understanding of the Ladakh cold desert's vulnerability to climate change, detailing its strategic and cultural importance, traditional knowledge repository for biodiversity conservation and sustainability, while highlighting the region's high sensitivity to altered precipitation patterns and rapid glacial retreat that threaten water security and biodiversity. **Second Chapter** "Human Ecology, Vulnerability and Adaptation to Climate Change in Cold Desert Landscape of Johar Valley of Pithoragarh District in Uttarakhand Himalaya" investigates the social-ecological systems of the Johar Valley, examining how historical disruptions, such as the 1962 border closure with Tibet (China) that forced communities to adapt their agro-pastoral systems, cultural heritage and forest management in response to cold desert environmental stressors. The chapter further explores how communities like Bhotiyas navigate socio-economic transformations, from the historical closure of trans-border trade routes to the modern expansion of unregulated tourism. The findings emphasize that community-led forest governance and traditional food preservation remain vital for resilience against retreating glaciers and shifting habitats. **Third Chapter** "Impacts of Climate Change on Agriculture in Indian Cold Deserts—Mitigation and Adaption" analyses the transition of traditional farming systems toward cash crops and protected cultivation to enhance the farm income. It evaluates the trade-offs between increased productivity in greenhouses (polyhouses) and the escalating pressure on limited water resources caused by the shift to water-intensive cash crops. For instance, polyhouses have increased productivity by 3 to 4 times, the shift from resilient crops like barley to cauliflower and cabbage. This study suggests that crop diversification, mixed cropping and crop rotation are important for maintaining agro-biodiversity, sustainability, soil and water conservation. It advocates for the integration of indigenous knowledge (e.g. *ice stupas*) with modern climate-smart technologies to maintain food security in extreme cold desert environments.

Fourth Chapter "Climate Change Impacts on Indian Cold Deserts Vegetation and Conservation Implications" categorizes cold desert vegetation zones and assesses the mounting threat of biological invasions. It further reports that vegetation distribution is primarily regulated by elevation and moisture availability. It also explores the unique role of Buddhist monasteries as critical actors in environmental stewardship, livelihoods and biodiversity conservation. **Fifth Chapter** "Plant Life and Adaptations in High Mountainous Regions" provides a scientific deep-dive into the morphological and physiological adaptations of high-altitude plants, focusing on specialized traits like dwarfism, succulent tissues, and subterranean biomass storage that enable survival in extreme environments. The chapter details the complex seed dormancy mechanisms and the rapid growth cycles triggered by snowmelt, while noting high seedling mortality due to soil drought and environmental severity. **Sixth Chapter** "The Use of Remote Sensing and Geographic Information Systems (GIS) for the Mapping of Cold Deserts' Bio-resources" details the application of geospatial technologies (Remote Sensing and GIS) for monitoring inaccessible landscapes, ecology and biodiversity conservation. It showcases the use of multi-sensor data fusion and habitat suitability modelling to map bio-resources and predict climate-induced range shifts in cold desert environments. Case studies demonstrate the efficacy of habitat suitability modelling for keystone species like *Sabina tibetica* in Tibetan Plateau and monitoring greening trends in the Andes. **Seventh**

Chapter "Effects of Primary and Secondary Umbels in the Germination Behavior of *Heraclium candicans* Wall. ex. DC. and *Selinum tenuifolium* Wall ex C.B. Clarke, the High-Altitude Umbellifers" examines the germination ecology of high-altitude umbellifers, demonstrating how prolonged pre-chilling serves as an essential evolutionary cue for breaking dormancy in species with complex morpho-physiological traits to adapt in cold deserts.

Eighth Chapter "Sustaining Livelihoods and Mitigating Climate Change Impacts Through Cold Desert Agroforestry Systems in the Lahaul Valley of Northwestern Himalaya, India" quantifies the contribution of willow-based agroforestry systems to rural livelihoods in the Lahaul Valley, highlighting their role in providing the majority of household food, fuelwood and fodder requirements. The agroforestry practices have currently turned into horticulture and cash crops in the Lahaul valley centring to ensure economic resilience against market fluctuations and climatic shifts. This study reveals that willow-based agroforestry is a cornerstone of sustainable land management, climate change mitigation and livelihood improvements in the cold deserts of Lahaul valley. **Ninth Chapter** "Community-Led Forestry and Landscape Restoration for Climate Change Mitigation and Adaption in Cold Desert, Northwestern Himalaya, India" explores Community-Led Forestry and Landscape Restoration (FLR), assessing how local communities rehabilitate degraded lands through the strategic use of cold-tolerant native species like willows, poplar and seabuckthorn. This practice is essential for sustaining biodiversity, improving soil health, carbon sequestration and environmental sustainability in the cold desert ecosystems. It identifies that farmers prefer planting on private land where they maintain direct control and economic benefits. This study underscores the community participation, technological innovations, and governance are critical in scaling up of restoration efforts for resilient ecosystems and sustainable landscapes. **Tenth Chapter** "Agroforestry for Land Management, Climate Change Mitigation, and Adaptation in Cold Desert Ecosystems" positions agroforestry as a multidisciplinary solution for achieving global Sustainable Development Goals (SDGs), emphasizing its potential for carbon sequestration, soil management, and poverty reduction in resource-scarce environments. This study highlights that tree-based agroforestry system is one of the viable solutions in meeting the ever-increasing demands of growing population in terms of food, fodder, and fuelwood. It details how tree-based agroforestry systems enhance carbon sequestration, regulate microclimates, and improve soil health through nutrient cycling. The chapter calls for a holistic policy development to overcome financial and technical barriers to adoption in resource-scarce cold deserts.

Eleventh Chapter "Diversity, Traditional Knowledge, and Vulnerability of Medicinal Plants to Climate Change in Cold Desert Areas of Indian Himalaya" documents the vast diversity of medicinal plants in the Trans-Himalayan zone, noting that climate change is causing an upward shift in species distribution and threatening the potency of secondary metabolites used in traditional healing. **Twelfth Chapter** "Ethnobotany and Conservation Challenges in the Cold Desert Ecosystems of Uttarakhand, India" reviews ethnobotanical knowledge in Uttarakhand, identifying significant declines in the availability of high-value medicinal plants due to overharvesting and phenological shifts due to global environmental change. This chapter concludes that strengthening community-based conservation practices and interventions, and documentation of traditional knowledge, is vital for biodiversity conservation, and the sustainability of bio-resources in the fragile cold desert ecosystems. Conservation is accomplished through sacred pastures and regulated harvesting systems embedded in religious beliefs.

Thirteenth Chapter "Biogeochemical Dynamics of Ladakh's High-Altitude Permafrost" analyses the biogeochemical dynamics of high-altitude permafrost, investigating how active layer thawing activates microbial decomposition and risks shifting these landscapes from methane sinks to potential sources. This study implies that there is a need to address the gaps in monitoring deep-layer carbon and other elemental stocks, as well as trace metal flux through field interventions and remote sensing techniques. **Fourteenth Chapter** "Hydrology and Water Governance in Cold Deserts: Glacial Dynamics, Resource Management, and Natural Disaster Resilience" examines hydrology and water governance, detailing the rising frequency of glacial lake outburst floods (GLOFs) and showcasing adaptive innovations like *Ice Stupas* and traditional village-led distribution systems. This study argued that building resilient systems that can sustainably manage scarce water resources while safeguarding communities and ecosystems against climate-induced risks. **Fifteenth Chapter** "Transformation in Pastoral Practices and Changing Climate in Cold Desert Areas of India" tracks the transformation of pastoral practices, reviewing how communities like the Changpa and Bhotiya navigate border conflicts, restricted movement due to protected areas regulations, increased livestock density and competition with wild herbivores. Climate change further jeopardizes livelihoods through increased mass mortality of livestock during extreme snowstorms and advance upward migration. This chapter also deals with changes in weather patterns, impact on the high-altitude pastures and subsequent effects on livestock and livelihoods of pastoral communities. **Sixteenth Chapter** "Building Psychological Resilience for Sustainable Adaptation in Cold Desert Regions" examines the psychological dimensions of adaptation by exploring how communities build mental toughness and resilience through traditional wisdom and social cohesion to cope with 'eco-anxiety' and environmental instability. It identifies stressors like 'winter-over syndrome' and eco-anxiety while highlighting traditional knowledge systems and social bonds as critical psychological buffers. This chapter conceives adaptation in cold deserts as a process of mental and emotional transformation in psychological sustainability that is essential for fostering long-term human resilience and well-being in these ecologically fragile landscapes.

In a nutshell, it is concluded that safeguarding the resilience of cold desert ecosystems hinges on integrating traditional ecological knowledge with scientific innovation to ensure sustainable development for future generations. As these landscapes reach critical climate thresholds, the transition towards resilient agroforestry, digital-twin modelling, and participatory governance is imperative to ensure the long-term survival of both such unique ecosystems and the communities they sustain. Addressing these challenges requires an interdisciplinary approach that integrates scientific innovations with time-tested local strategies. This collection of the work delves into mitigation and adaptation pathways, including willow-based agroforestry, horticulture, cash crops, innovative water management techniques like 'ice stupas', and the application of Remote Sensing and GIS for the systematic monitoring of bio-resources. Furthermore, the volume explores the crucial role of psychological resilience and cultural continuity in ensuring the long-term well-being of human populations in these demanding environments. The book highlights pathways toward sustainability that balance environmental integrity with human well-being. The diverse array of chapters collectively underscores that effective responses to climate change in cold deserts require context-specific approaches, informed by science, technology, policy, and local knowledge systems.

We hope this book will serve as a valuable reference for researchers, practitioners, students, policymakers, and grassroots organizations in safeguarding the sustainability of cold desert landscapes for future generations. This book is expected to stimulate further research and informed action in cold desert landscapes, worldwide. Eventually, sustaining cold deserts is not only critical for the communities who inhabit them, but also for broader regional and global ecological stability in the face of sustainability and global environmental change.

Addis Ababa, Ethiopia — Yashwant Singh Rawat
Bhopal, India — Chandra Prakash Kala

Contents

About the Editors

Yashwant Singh Rawat earned his M.S. and Ph.D. in Forestry from H. N. B. Garhwal University (a Central University), Uttarakhand, India. His doctoral research was conducted at the G. B. Pant National Institute of Himalayan Environment (NIHE), Uttarakhand, India, and its Himachal Regional Centre in Kullu, Himachal Pradesh—an autonomous institute under the Ministry of Environment, Forest and Climate Change, Government of India. Dr. Rawat has served in various academic and research capacities at the Uttaranchal (PG) College of Bio-medical Sciences, Dehradun (affiliated with H. N. B. Garhwal University), NIHE, Nelson Mandela University (South Africa), the University of KwaZulu-Natal (South Africa), and the United Nations University (Japan). With over 19 years of professional experience following his Ph.D., Dr. Rawat has published extensively in international peer-reviewed journals, with more than 60 research papers and scholarly articles. He has contributed significantly to the academic community through editorial and peer-review service for leading journals and publishers such as Elsevier, Springer Nature, Frontiers, PLOS, Wiley, Taylor & Francis and MDPI. Dr. Rawat is a recipient of several prestigious competitive fellowships and research support from India and abroad, including the CSIR Senior Research Fellowship (Government of India) and the United Nations University–Institute for the Advanced Study of Sustainability (UNU-IAS), Japan. He is currently a faculty member at the Federal Technical and Vocational Training Institute (TVTI), Addis Ababa, Ethiopia. His research interests include forestry, ecology, biodiversity and ecosystem services, environmental sustainability, sustainable energy and resource-use.

Chandra Prakash Kala earned his M.S. degree in Botany from H. N. B. Garhwal University, and Ph.D. in Forestry from the Forest Research Institute (Deemed University), Dehradun, India. He worked at the Wildlife Institute of India, G. B. Pant National Institute of Himalayan Environment, and the National Medicinal Plants Board with various capacities. His decade long research study on the 'Valley of Flowers' in India laid the foundation stone for tagging it in the list of World Heritage Sites by UNESCO. With over 32 years of professional experience, he has published >200 research papers and articles having over 8900 citations. His area of interest is plant ecology, biodiversity conservation, natural resource management, and he has contributed significantly to policy initiatives, and capacity building. He is the Chairperson of Management Development Programmes and has imparted trainings to academicians, corporate professionals, and policy makers, including the Indian Forest Service Officers as a Course Director. He is on the editorial and advisory board of many national and international scientific journals, including *Journal of Ethnobiology* and *Ethnomedicine*, *PLoS Sustainability and Transformation*, and *PLoS Ecosystems*. He is an elected member of the National Academy of Sciences in India (NASI), and also the recipient of prestigious ICFRE Award for Excellence in the Forest Conservation (Biodiversity and Ecology). He has been a visiting scholar at the Pennsylvania State University, in the United States and the University of Ljubljana in Slovenia. At present, he is a faculty member at the Indian Institute of Forest Management, Bhopal, and he has been ranked among the top 2% of the world's academicians/scientists by the Stanford University.

The Cold Desert Ecosystem of Ladakh: Vulnerabilities, Adaptations, and Pathways Toward Sustainable Development

Shairq Irtiqa, Zishan Ahmad Wani, Vikram S. Negi, and Javid Ahmad Dar

1 Introduction

Cold deserts are distinctive ecosystems marked by intense solar radiation, reduced air partial pressure, extreme temperature fluctuations, and minimal precipitation (Joshipura, 2025). These environmental conditions support sparse but highly adaptive vegetation, which offers vital resources and services to local communities despite their vulnerability (Dolma & Uniyal, 2025; Negi et al., 2018a). Cold deserts comprise about 2.3% (74,809 km^2) of India's total geographical area, classified as part of the Trans-Himalayan zone, encompassing portions of the Leh and Kargil districts of Ladakh and the Lahaul-Spiti district of Himachal Pradesh, and parts of Kinnaur and Chamba districts (Saxena et al., 2011). Ladakh, meaning "Land of high passes," is located at the confluence of Palearctic and Oriental biogeographic zones, situated within the Trans-Himalayan Mountain range. The region is characterized by high elevation, distinguished by its impressive glaciers, elevated passes, riverine valleys, and profound cultural heritage (Fig. 1). The Ladakh is bordered by the Great Himalayan range to the south and the eastern Karakoram to the north, with China to its northeast and Pakistan to its northwest. The upper Himalaya forms its southern boundary, extending into parts of the Himalayan and Karakoram ranges, including

S. Irtiqa · Z. A. Wani (✉) · J. A. Dar
Terrestrial Ecology and Modelling (TEaM) Lab, Department of Environmental Science and Engineering, SRM University-AP, Amaravati, Andhra Pradesh, India
e-mail: zishanwani786@gmail.com; zishan.w@srmap.edu.in

J. A. Dar
e-mail: javidahmad.d@srmap.edu.in

V. S. Negi
Department of Botany and Microbiology, H.N.B. Garhwal University (HNBGU), Pauri, Uttarakhand, India

J. A. Dar
Centre for Geospatial Technology, SRM University-AP, Amaravati, Andhra Pradesh, India

Y. S. Rawat and C. P. Kala (eds.), *Cold Deserts*,
https://doi.org/10.1007/978-3-032-29119-6_1

the upper Indus River valley (Ali et al., 2025; Namga & Nayak, 2024). The geological characteristics of Ladakh exhibit formations created by tectonic and geomorphic processes, resulting in stunning landforms. This region is among the world's most elevated, with human settlements even at higher altitudes (Chauhan et al., 2020). The region is characterized by its arid topography, severe climate, and substantial temperature fluctuations (Singh et al., 2021; Tamchos & Kaul, 2019). In recent decades, the delicate equilibrium of this cold desert has come under increasing threat from the accelerating impacts of global climate change and anthropogenic pressures (Kanwar & Kuniyal, 2022). Further, the cold desert has received limited attention due to their low species richness, distribution, and community structure, inherent inaccessibility, and remoteness (Rawat et al., 2023). In this context, understanding the vulnerabilities and adaptive responses of Ladakh's cold desert ecosystem is even more crucial. This chapter aims to provide a comprehensive understanding of the cold desert ecosystem of Ladakh, encompassing a wide array of topics, including its strategic, cultural and hydrological importance and environmental concerns. It addresses gaps in the existing vulnerability assessment and proposes adaptation and mitigation approaches and policies for prioritizing conservation management for long-term preservation and sustainable development.

Fig. 1 Geographic layout of the Ladakh cold desert, highlighting its major physiographic features. **a** Leh. **b** Moon valley (Lamayuru). **c** Nubra valley. **d** Khardung-La pass. **e** Turtuk. **f** Pangong Tso

2 Characteristics of the Cold Desert of Ladakh

2.1 Culture and Religion

Ladakh has a confluence of diverse civilizations and populations, crucial in developing Ladakhi culture with 77% Buddhists, 14% Muslims, and 8% Hindus (Haq et al., 2020; Sharif & Singh, 2025). Historically, Ladakh is populated by predominantly Buddhist and Muslim communities, where Leh primarily consists of the Buddhist community, and Kargil is largely inhabited by Shia Muslims, and a small percentage are Christian. Leh is also known as "Little Tibet," as Buddhist monks have found ground with their Tibetan counterparts, who were influenced by the Mahayana and Vajrayana schools. The number of monasteries established by the rulers and kings centuries ago may be found atop the hills in every village, reflecting the area's rich and tangible cultural past (Suri, 2018). Some of the renowned monastic institutions where various manifestations of Buddha, referred to as Bodhisattvas, include Hemis Monastery, Thiksey Monastery, Spitok Monastery, and others. Shanti Stupa is another peace monument and one of the most iconic landmarks in Leh, famous for its religious and panoramic views. Historical records indicate that the mosques located in Shey and Tyaqshi were constructed by individuals affiliated with Mir Syed Ali Hamadani. At the same time, disciples of Meer Shams-ud-Din Iraqi established the Khanqahs of Chuchot Yokma and Buru in Kargil. The architectural design of the mosques represents a distinct fusion of Indo-Islamic and Tibetan influences, providing a syncretic cultural exchange reflecting centuries of intercultural interactions and integration of religious customs into a unique local aesthetic (Gupta, 2021).

In Eastern Ladakh, the Changpas, a nomadic Tibetan-speaking pastoralist community inhabiting the high-altitude Changthang plateau, constitute the predominant social group, characterized by a minimal agricultural presence, but serve as a cornerstone for the region's economy and societal customs (Humbert-Droz, 2024; Morup et al., 2024). Yaks, dzomos, and dzos are essential for milk, meat, and transport, and the Changthangi goat breed (Changra) holds global recognition for its Pashmina wool, which is regarded as a distinctive essential from this region (Singh & Malik, 2025). Conversely, Western and Central Ladakh are primarily inhabited by agropastoral communities and are one of the regions where traditional knowledge remains unaffected. Given the influence of ethnic background and traditional practices on environmental and ecological knowledge, pastoralist communities are likely to have a more extensive understanding of forage plant species than their agricultural counterparts (Angmo et al., 2024; Pandit et al., 2024). However, they have developed varied approaches to optimize their migration behaviors, production systems, and livestock composition because of climatic, socio-economic, policy shocks, and changes. Additionally, demographic factors significantly contribute to variations in traditional knowledge within individual households (Bhasin, 2012).

2.2 Biodiversity and Traditional Knowledge Systems

Despite the challenging conditions, Ladakh boasts floral and faunal diversity, harboring a wide variety of intriguing wild plant and animal species crucial for their socio-economic, medicinal, and ecological significance (Fig. 2). Ladakh is home to over 700 species of annual and perennial herbs, shrubs, and cultivated plants (Shukla & Srivastava, 2020; Tamchos & Dorjey, 2025). Ladakh has distinctive varieties of plants and is affected by altitudinal variations, where extreme climatic conditions and elevated altitudes contribute to a diminished vegetation cover and reduced species richness. Populations thrive between 4000 and 5800 m before declining at higher altitudes (Gaur et al., 2024). The flora of Ladakh is classified as alpine and high alpine, Oasitic or Riparian vegetation, and arid vegetation, with annual and perennial herbs predominating and a few stunted shrubs and bushes (Batool et al., 2022; Hussain et al., 2024a). Sub-alpine and alpine zone ranges from 3000 to 4000 m asl, supporting a unique vegetation zone (Hussain et al., 2023a; Kumar et al., 2011a).

The oasis vegetation, characteristics of inhabited areas across Ladakh, thrives in conditions with a limited moisture, comprising a diverse array of endemic and introduced plant species. These generally grow in human-inhibited areas along the channels, moist areas, and streams with species like willows and poplars, as well as various shrubs and herbs. The dominant genera include *Medicago, Lactuca, Carum, Potentilla, Arabidopsis, Mentha, Myricaria, Hippophae, Rosa, Clematis*, and other tree

Fig. 2 Some of the plants and animals found in cold ecosystems of Ladakh. **a** *Ephedra gerardiana*. **b** *Clematis ladakhiana*. **c** Common Merganser. **d** Black headed gull. **e** Yak

genera such as *Salix*, *Populus*, *Prunus*, *Pyrus*, *Morus*, *Juglans*, *Juniperus*, and *Robinia* (Hussain et al., 2023a). The desert vegetation is characterized by xerophytic plants adapted to the cold arid conditions. These plant species exhibit a variety of specialized adaptations that facilitate water conservation, enabling them to survive in Ladakh's demanding ecosystems. Sparse, resilient plants like grasses, shrubs, and species that produce cushions like *Thylacospermum caespitosum* and *Acantholimon lycopodioides* make up a most of the vegetation. Other drought-resistant species include *Artemisia sp.*, *Ephedra gerardiana*, *Caragana sp.*, *Myricaria germanica*, *Rheum sp.*, *Allium sp.*, *Sedum sp.*, and *Salsola sp* (Gaur et al., 2024; Tewari & Kapoor, 2013). Plant species is primarily composed of herbs, with a few small, heavily branched shrubs or bushes such as *Atriplex hortensis*, *Ephedra gerardiana*, *Hippophae rhamnoides*, *Sophora moorcroftiana*, *Tanacetum tibeticum*, *Rosa webbiana*, *Berberis ulicina*, *Myricaria germanica*, *and Tamarix gallica* (Gaur et al., 2024). Hamid and Raina (2024) recorded 24 endangered species, 27 vulnerable species, and six near threatened species of plant varieties in the Indus valley, Ladakh. Ladakh's remarkable floral diversity includes 50% of plants that possess considerable medicinal properties, which are an integral part of the traditional Amchi system of medicine (Kumar et al., 2011b). Gurmet and Stobgais (2016) documented approximately 162 plant species within Ladakh that are utilized in the Amchi system such as *Juniperus indica*, *Thermopsis barbata*, *Rhodiola tibetica*, *Rheum emodi*, *Clematis tibetana*, *Artemisia annua*, and *Angelica glauca*. Dawa et al. (2021) reported *Corydalis stricta*, *Heracleum pinnatum*, *Cicer microphyllum*, *Rosa webbiana*, *Geranium wallichianum*, *Carum carvi*, *Aconogonum tortuosum*, *Argentina anserina*, *Hippophae rhamnoides*, *Astragalus tibetanus*, *Thalictrum foliolosum*, and *Physcholaina praealta* as the dominant medicinal plant species. Therefore, the region holds a significant potential for the conservation, cultivation of medicinal and aromatic plant species, and it is expected that the medicinal plant market will reach economic value in the near future.

The cold desert ecosystem of Ladakh also supports unique and endemic fauna diversity adapted to its extreme cold, high-altitude environment. Some of the prominent animal species found in the Ladakh region include the snow leopard (schan), blue sheep or bharal (napo), Eurasian lynx (eeh), Ladakh pika (zabra), Ladakh urial (shapo), mountain weasel (lakimo), Asiatic ibex (skyin), red fox (watse), Tibetan antelope (tsos), Tibetan argali (nyan), Tibetan gazelle (gowa), woolly hare or Tibetan hare (ribong), Tibetan wild ass (kiang), Tibetan wolf (shanku), and Himalayan marmot (phiya). This region is also rich in avifauna and serves as a critical bird migratory route between the Palearctic and the Indian subcontinent, with numerous high-altitude migratory birds utilizing the region (Jamwal et al., 2020), such as bar-headed goose (ngangpa), ruddy shelduck (Cha muru), Golden eagle (laknak), Lammergeier (skiaklag), Chukar partridge (srakpa), and Himalayan vulture (thang-kar). Wetlands and rangelands serve as a significant breeding ground for various water bird species, notably the endangered Black-necked crane (Cha thung-thung).

Traditional knowledge forms an integral part of the cultural and spiritual identity of the community in which it is practiced and preserved (Negi et al., 2018b, 2019, 2023; Ridwan et al., 2023; Wani et al., 2021). Traditional knowledge systems of Ladakh represent a compilation of practical understanding developed over many centuries, reflecting the complex interdependence of human societies and their natural surroundings (Angmo et al., 2024). The region's culturally rich tribal communities have a long-standing relationship with the biological resources, relying on them for sustenance, medicine, food, fodder, and timber (Dorjey et al., 2012). The local inhabitants possess a significant traditional knowledge and use different plants and animals to treat various diseases and disorders. The traditional healthcare of the Ladakh, the Amchi Medical system, fosters about 60% of public health of Ladakh (Chauhan et al., 2020). Further, the religious healers locally known as "Akhones" are preferred in Muslim dominating regions that use plants, animal parts, and even minerals to treat diseases (Batool et al., 2023).

Many medical plants have been reported to be used in traditional healthcare systems in the region (Batool et al., 2023; Hadi et al., 2022; Hussain et al., 2023a; Jishtu et al., 2023, 2025). Different plant parts including leaves, fruits, seeds, bark, roots, and even flowers are being used in different forms like decoction, paste, and powder to alleviate the diseases (Table 1). Haq et al. (2020) have reported 21 animal species including Asiatic wild dog, Marmots, Tibetan antelope, Common Otter, Tibetan wolf, Snow leopard, and Brown bear that are being used to cure various diseases and disorders like paralysis, joint pain, inflammation, ulcers, rheumatism, arthritis, and cough. Further, wild edible plants are an important constituent of traditional diets in Ladakh. The wild plants provide the majority of dietary requirement of proteins, sugars, vitamins and minerals. Angmo et al. (2024) have reported 84 wild edible plants from the region. Some of the prominent wild edibles include *Medicago sativa, Ephedra gerardiana, Arnebia euchroma, Morus alba, Bergenia stracheyi, Rheum webbianum,* and *Allium tortosum.* The majority of medicinal and edible plants (65%) are harvested by the locals for their own use, with the remaining 35% being sold in local markets to generate revenue (Haq et al., 2022). Numerous species in Ladakh are either vulnerable or endangered, while many are categorized as critically endangered.

2.3 Cryosphere and Water Systems

Ladakh is one of India's extensively glaciated high-altitude cold desert regions. This region is home to approximately 5,000 glaciers, collectively occupying an area of 3,187 km^2 and possessing a total volume of 815.62 km^3 (Koul et al., 2016). Among these, Siachen Glacier, a compound glacier situated in the eastern reaches of the Karakoram Range, ranks as the world's second-longest glacier outside of polar regions with an altitude of 3400–7400 m above sea level. The Siachen catchment encompasses an area of approximately 2052 km^2, of which roughly 1219 km^2 is glaciated. Siachen meltwater serves as a source for Nubra River, which originates from its snout, converges with the Shyok River near Sumur (3130 m asl) after which it continues as the Shyok before eventually merging with the Indus River in the vicinity of Skardu (Kumar et al., 2020; Negi et al., 2022a, 2022b). Thus, it serves

Table 1 Some of the important medical plants used by the local people of Ladakh

Botanical name	Local name	Family	Habit	Parts used	Diseases/ailments	References
Artemisia brevifolia	Khampa, Phur-nag, and Burtse	Asteraceae	Perennial subshrub	Leaves and inflorescence	Fever, blood purifier, and stomach disorders	Hussain et al. (2024a)
Arnebia euchroma	Demok	Boraginaceae	Perennial herbs	Roots and leaves	Eye disease, cuts and wounds, cough, cold, and fever	Jain et al. (2021)
Ephedra gerardiana	Tsepat, and mTshe-lDum	Ephedraceae	Shrub	Ariel part	Asthma, rheumatism, chronic fever, wounds, cough, excess sweating, urinary problems	Rinchen et al. (2021)
Hippophae rhamnoide	Tser-Mang, shibshu-lulu, sTarbu, Nak-tSer	Elaeaganaceae	Perennial Shrubs	Seeds, leaves	Skin disorders, stomach ulcers, cancer, indigestion, cardiovascular diseases, diarrhea, fever, cough, cold, food poisoning, gynaecological problems, acidity, throat infections	Tamchos and Dorjey (2025)
Dolomiaea costus	Kuth	Asteraceae	Perennial herbs	Roots	Kidney stones, rheumatism, stomachache, headaches, cold, and throat infection	Elnour and Abdurahman (2024)
Aconitum heterophyllum	Karpo-Bova	Ranunculaceae	Biennial or perennial herb	Roots and seeds	Tonsillitis, urine infection, cough, bronchitis, guinea-worms, and hypertension	Shukla et al. (2020)
Dactylorhiza hatagirea	Angulakpa/ Wangbolakpa	Orchidaceae	Perennial herb	Root	Wounds, diabetes, chronic fever, fractures, stomachache, dysentery, diarrhea, and cancer	Ali (2021)
Gentianopsis paludosa	Chukiting	Gentianaceae	Annual or biennial herbs	Root	Cough	Banoo et al. (2021)
Rheum webbianum	'Lachhu' or 'Chu-rtsa	Polygonaceae	Perennial herb	Stem, roots and leaves	Chronic bronchitis, cold, piles and constipation	Tabin et al. (2022)
Verbascum thapsus	Dandashal	Scrophulariaceae	Perennial herb	Whole plant	Migraine, edema, diabetes, inflammatory ailments in the respiratory tract, wounds, and urinary diseases	Dolkar et al. (2021)

(continued)

Table 1 (continued)

Botanical name	Local name	Family	Habit	Parts used	Diseases/ailments	References
Waldheimia tomentosa	Palu	Asteraceae	Perennial herb	Leaves and flowers	Common infections, acidity, arthritis and rheumatism	Bhatnagar et al. (2017)
Urtica hyperborea	Zatsot	Urticaceae	Perennial herb	Whole plant	Gut health, cold and cough, and chronic diseases	Dolma et al. (2025)
Oxytropis microphylla	Stag-sha	Fabaceae	Perennial herb	Root	Inflammation and oxidative stress	Zomba et al. (2025)
Lycium ruthenicum	Khizer	Solanaceae	Perennial shrubs	Leaves	Anemia, lung disorders, early-onset of diabetes, vision problem, and liver and kidney function	Dhar et al. (2011)
Juniperus polycarpo	Shukpa, Chilgi	Cupressaceae	Perennial tree	Bulbs, fruit	Regulate urination and Kidney problems	Dorjey and Maurya (2021)
Peganum harmala	Sepan	Zygophyllacea	Perennial herbs	Whole plant	Rheumatism, kidney stones, jaundice, sciatica, psychosis, asthma, digestive issues	Shabir et al. (2023)
Bunium persicum	Nagpo-zer	Apiaceae	Perennial herb	Seeds	Stomach problems, diarrhea, fever and dyspepsia	Khan et al. (2023)
Podophyllum hexandrum	Tandik	Berberidacea	Perennial herb	Whole plant	Gynecological problems, gastritis, improving lung and blood circulation, diarrhea, cholera, ulcers	Shabir et al. (2023)

as a vital resource for the region, particularly during periods of reduced water availability, decreased stream flow, and drought, shaping the local fluvial networks. Other primary glacier systems include Drang Drung and Pensilungpa glaciers located in the Doda and Suru River valleys in Zanskar Himalaya, which exhibit similar meteorological conditions and climatic zones, with the Drang Drung glacier characterized by its clean ice, and the Pensilungpa glacier distinguished by a substantial accumulation of supraglacial debris (Jasrotia et al., 2023; Kunmar et al., 2025).

Ladakh harbors rock glaciers that are recognized as significant manifestations of ice-rich permafrost. Occurring at elevations between 3190 and 5753 m asl, these hidden water reservoirs are regionally vital for sustaining streamflow and water avalaibility, particularly during the lean season across the Himalayas (Bhat et al., 2024). Bhat et al. (2025) identified 5,807 rock glaciers, comprising 5,292 intact and 515 relict formations collectively covering an area of 712 km^2, with an average individual size of 0.12 km^2. Further, an estimated 13.7 km^3 of water is stored, serving as a crucial hydrological reservoir in the region within the mapped rock glaciers. Among the river systems, besides Shyok and Zanskar, the predominant rivers in the region, the Indus River, also known as Sindhu and locally as Senge Tsangpo, are sacred and the lifeline of Ladakh, starting from Tibet and entering at Demchok Ladakh, serves not only as a source of drinking and irrigation water but also central water supply for hydropower in Ladakh (Ali et al., 2025). The Indus River converges with its major tributary, the Zanskar River, at the Sangam point in Leh. It is formed by the meeting of the Stod and Lugnak rivers, originating in the Zanskar region from the glaciers. Collectively, these rivers sustain agriculture, domestic, and industrial water supply and form the backbone of the hydrological system in Ladakh.

3 Vulnerabilities of the Ladakh Cold Desert Ecosystem

3.1 Climate Vulnerability

A vulnerability indicator has multi-dimensional drivers and is a foundational element in climate change research. Therefore, assessing vulnerability is crucial for mitigating the detrimental impacts of climate variability through implementing adaptive and robust strategies, particularly in areas where populations rely on natural resources (Namgyal et al., 2025; Wani et al., 2023). Mountains are among the most sensitive ecosystems to environmental perturbations and anthropogenic disturbances (Wani et al., 2022). Evidence from various studies indicates that the Himalayan region is among the few regions where reflections of climate change are most evident (Negi et al., 2022a, 2022b). This region is susceptible to both natural and anthropogenic hazards. In the recent decade, Ladakh has experienced a significant increase in temperature and abnormal precipitation patterns (Bhat et al., 2023). Although the region is characterized by arid conditions with an average annual rainfall of merely 100 mm, this region is currently experiencing an increase in rainfall events relative to snowfall, which have disrupted ecological equilibrium and precipitated disasters such as flash floods and landslides (Singh & Bhatla, 2025). For instance, a cloudburst in the Leh region in August 2010 resulted in 356 mm of rain in two hours, which led to numerous debris flows and caused a significant widening of the lower portions of the central tributary valleys. This disaster killed hundreds and damaged numerous properties (Mueller et al., 2019).

The latest incident occurred in August 2025, significant flash floods and landslides caused by heavy rainfall in the Leh and Zanskar regions led to extensive inconvenience and destruction. Moreover, rapid glacier retreat and change in their surfaces driven by climatic shifts lead to new glacial lakes and the enlargement of existing ones, which could potentially trigger glacial lake outburst floods. Additionally, meltwater serves as a crucial water source for the entire cold-arid region, particularly during periods of low water availability or stream flow. However, studies have revealed that many glaciers across the region exhibit modest thinning rates alongside pronounced frontal recession, underscoring their sensitivity to local climatic and topographic variables (Table 2). Considerable stress on the water security will result in reduced agricultural productivity, which is the primary driver of the region's economic growth (Rana et al., 2023; Soheb et al., 2022). Ladakh shows high sensitivity and vulnerability to the change in climate and needs urgent attention, as the characteristics of the changing climate are not well understood due to the limited monitoring facilities in the region (Chevuturi et al., 2018). Further, understanding local people's perception of climate change is a crucial prerequisite for exploring their coping strategies and for developing appropriate adaptive and mitigative measures (Ali et al., 2022).

3.2 *Biodiversity Loss*

The region's unique bio-geographical position, lying at the crossroads of the Indian, Central Asian, and Tibetan realms, makes it a repository of rare and endemic biodiversity. Several plant and animal species in Ladakh are considered threatened (Table 3), primarily due to the fragile ecosystems, limited water availability, harsh climatic conditions, and increasing anthropogenic pressures such as overgrazing, habitat fragmentation, and unregulated tourism. The significant impact on biodiversity has usually been caused by climate change and increased human activities, probably due to the rise in tourists and urbanization. Climate change has induced significant effects on species' habitats and distribution. Global temperature increase threatens Ladakh's endemic flora from adjacent biogeographic regions (Zargar et al., 2024a, 2024b). Bhasin et al., (2024) reported that the Changthang region in eastern Ladakh supports numerous ungulate species. Among them, only an estimated 100 Tibetan gazelles inhabit Ladakh, making them critically endangered and at risk of local extinction due to changing climate scenarios in the region. Chauhan et al. (2020) have documented 13 critically endangered, 24 endangered, 27 vulnerable, and 6 near threatened species from. Besides tourism, species also face significant anthropogenic threats, including

Table 2 Annual mass balance of different glaciers (m.w.e./year, meters of water equivalent per year; m a^{-1} is meter per year/annum)

Glacier type	Region	Retreat	Range/period covered	References
Siachen glacier	East Karakoram	-0.11 ± 0.24 m w.e./year	1986–2018	Negi et al. (2022a, 2022b)
Drang Drung glacier	Zanskar valley	-21 ± 12 m a^{-1}	2015–2023	Kunmar et al. (2025)
Pensilungpa glaciers	Zanskar valley	-10 ± 4 m a^{-1}	2015–2023	Kunmar et al. (2025)
Kangriz/ Parkachik glacier	Suru valley	-20.5 ± 12 m a^{-1}	2015–2021	Rana et al. (2023)
Machoi glacier	Drass valley	($\sim$14 m yr^{-1})	2010–2018	Taloor et al. (2021)
Kolahoi glacier	Liddar valley	-0.83 ± 0.34 m w.e. a^{-1}	2014–2019	Romshoo et al. (2023)
Chhota Shigri glacier	Chandra basin	-0.57 ± 0.41 m w.e./year	2000–2014	Sahu et al. (2024)
Pensilungpa Glacier	Suru Basin	-0.36 ± 0.04 m w.e. a^{-1}	2016–2019	Mehta et al. (2021)
Nun-Kun Group of Glaciers	Suru and Chenab basins	21.2 ± 2.2 m·a^{-1}	2000–2020	Romshoo et al. (2024)
Stok Glacier	Stok Range	-0.39 ± 0.38 m w.e. a^{-1}	2015–2019	Soheb et al. (2020)

the excessive harvesting of underground plant parts, overgrazing, and the unsustainable use of resources for fuel and fodder (Hussain et al., 2023b). Human-wildlife conflict is another threat to high-altitude animals, including the Tibetan wolf, which is often hunted to protect their livestock.

Biological invasion is also considered a major threat to the biodiversity (Zargar et al., 2025a). Generally, alien species are more common at lower elevations due to human introduction and disturbances; however, some have also adapted well to higher elevations (Meyer et al., 2021; Iseli et al., 2023; Zargar et al., 2025b). The altitudinal range of alien species within Ladakh region offers significant insights into their capacity for adaptation and habitat selection at higher elevations (Zargar et al., 2025b). *Oxalis corniculata* L., an invasive weed of crops in Ladakh, may spread rapidly in future (Zargar et al., 2024a). Hussain et al. (2024b) have reported the first report on pest *Pheosia albivertex* infesting *Populus* plantations. This pest destroys the leaves, especially in the middle of summer when they are at their best. However, species adjust their geographical distribution as a primary strategy to cope with the impacts of climate change. The rising temperatures may facilitate the establishment, growth, and survival of plant species migrating from lower altitudes to higher elevations, potentially leading to the displacement of native species and the introduction of invasive alien plants. Zargar et al. (2024a, 2024b) reported for the first time a

Table 3 Threatened flora and fauna of cold desert ecosystems of Ladakh

Species	Common name	Category	References
Grus nigricollis Przh	Black-necked crane	Near threatened	Birdlife International (2020)
Ovis vignei vignei Blyth	Ladakh Urail	Endemic and endangered	Namgail et al. (2010)
Panthera uncial Sch	Snow leopard	Vulnerable	Raina et al. (2025)
Procapra picticaudata Hodgson	Tibetan Gazelle	Near threatened	Namgail et al. (2008)
Pantholo hodgsonii Abel	Tibetan Antelope	Endangered	Haq et al. (2020)
Ephedra gerardiana Wall ex. Stapf	Gerard Jointfir	Endangered	Rinchen et al. (2021)
Primula atrodentata W.W. Sm	Drumstick Primrose	Near threatened	Butt et al. (2021)
Juniperus polycarpos C. Koch	Himalayan Pencil Cedar	Near threatened	Dorjey and Maurya (2021)
Arnebia euchroma I.M. Johnst	Ratanjot	Endangered	Devi et al. (2023)
Dactylorhiza hatagirea (D. Don) Soo	Salam Panja	Critically endangered	Chauhan et al. (2020)
Aconitum heterophyllum Wall. ex Royle	Indian Atees	Endangered	Chauhan et al. (2020)
Aconitum violaceum Jacq. Ex Stapf	Violet Monkshood	Endangered	Hamid and Raina (2024)
Aquilegia fragrans Benth	Fragrant Columbine	Endangered	Hamid and Raina (2024)
Thymus serpyllum L	Breckland thyme	Vulnerable	Hamid and Raina (2024)
Hippophae rhamnoides L	Sea Buckthorn	Vulnerable	Hamid and Raina (2024)
Hyoscyamus niger L	Henbane	Vulnerable	Chauhan et al. (2020)

range shift in three plant species, namely *Bellis perennis* L., *Cannabis sativa* L., and *Portulaca oleracea* L.

The challenges of effective conservation in the region are compounded by the lack of sufficient resources and tailored strategies (Angmo et al., 2025). To safeguard this natural wealth, several wildlife sanctuaries and protected areas have been established that not only conserve the region's unique flora and fauna but also help maintain ecological balance and cultural heritage associated with traditional agropastoral systems. Some of the protected areas in the region are;

Hemis National Park: This ultimate wildlife hotspot, located on the east side of Ladakh, renowned for snow leopard sightings and home to blue sheep, bear, Red fox, Ladakh urial, and Tibetan wolf. It is India's largest national park covering an area of 3500 km^2 with elevational range of 3,099–6,368 m asl (Ghosal et al., 2024).

Karakoram Wildlife Sanctuary: **Karakoram Wildlife Sanctuary** is located at the edge of the Karakoram Range in the eastern part of Ladakh, sharing borders with China and Pakistan. It is home to the endangered species like the snow leopard and the Tibetan antelope. It is also one of the few places in India where this Tibetan antelope (Chiru) migrates. Other important animals of the sanctuary include Wolf, Lynx, Bharal (blue sheep), Tibetan Gazelle, Shapo, and Wild Yak. Bactrian camels can also be seen in the surrounding area. Lemmargeir, Himalayan Griffon, Gadwall, Tufted Duck, Goosander, Himalayan Snow Cock, Grey Plover, Lapwing, Little Gull, Hill and Blue Rock Pigeons, and larks are among the notable birds. The sanctuary spans around 5000 km^2, varies roughly from 4,200 m to more than 7,500 m at Saltoro Kangri's summit, and is traversed by the Nubra and Shyok rivers, allows conservation of not only fauna but also diverse flora such as alpine vegetation, medicinal plants like *Ephedra gerardiana* and *Arnebia euchroma*, and cold desert flora like junipers, and cushion plants (Kala, 2005).

Changthang Cold Desert Wildlife Sanctuary: A high-altitude sanctuary, ranging from 4300 to 5800 m, is located in eastern Ladakh in the Changthang plateau (Pandit et al., 2024). This sanctuary is well known for one of the highest altitude lakes on earth, Tso Moriri (4595 m), and Pangong Tso. Spanning approximately 4000 km^2, this sanctuary provides a habitat for numerous endangered and rare animals, including the Tibetan wild ass (kiang), quite popular in this place, snow leopard, Tibetan wolf, and several bird species like the black-necked crane, ruddy shelduck, and bar-headed goose. The vegetation consists of wild grasses such as *Nepeta longibracteata*, *Rheum rhabarbarum* (Himalayan rhubarb), and *Arnebia euchroma* (https://www.lehladakh tourism.com/changthang-wildlife-sanctuary/ assessed on 9th October, 2025). The steppe comprises *Kobresia* species, *Carex* species, and sedges. It is also home to Changpa nomads, well known for Pashmina goat herding (Morup et al., 2024).

3.3 Socio-economic and Cultural Changes

The rapid demographic growth, reflected in uncontrolled urban expansion, exerts pressure on cold desert ecosystem of Ladakh, leading to the depletion of resources and environmental degradation (Khanday et al., 2024). This accelerated urbanization has reduced the availability of suitable land for development, consequently driving construction activities into environmentally sensitive zones susceptible to inundation, slope instability and land degradation (Bansal et al., 2023). Besides, the expansion of tourism, which is recognized as a crucial part of the economy, is the second most important driver of urbanization. The tourism sector is expanding drastically that surpassing Ladakh's entire population. This surge in tourism has prompted concerns about straining water resources and disrupting traditional water-sharing mechanisms, consequently increasing vulnerability to anthropogenic disturbances in the region (Dame et al., 2019; Dar et al., 2022). Ladakh's pastoral practices are influenced by ongoing socio-economic and development changes due to livelihood diversification, acculturation, a growing reliance on external food sources, the declining importance of traditional practices, and the migration of individuals from rural to urban areas. These shifts are driven by improved transportation networks, increased aspirations for higher education, institutions such as the public distribution systems (PDS), and the growth of tourism sectors (Dame et al., 2019; Ladon et al., 2023).

4 Adaptive and Mitigation Strategies

With the increasing effects of climate change, Ladakh communities, local engineers, and NGO's have developed innovative strategies and employed traditional knowledge systems to overcome various challenges in the region.

4.1 Traditional Adaptation Strategies

Traditional knowledge systems have developed solutions that are tried-and-true and long-standing practices and are locally available, addressing issues and are reexamined as essential climate change adaptation approaches.

4.1.1 Architectural Designs

The vernacular architecture is the region's legacy that is being casually and practically carried over centuries and is mainly known for the use of local resources, labor and culture (Fig. 3). Simple methods, such as using "bukharis" (traditional stoves) with walls of the houses more thermally dense and making openings smaller in architectural typology, have allowed the traditional knowledge of managing harsh climatic conditions through its vernacular architecture, enabling human habitations (Mehta & Gajjar, 2023; Nasir & Kamal, 2021).

4.1.2 Water Harvesting Techniques

To address the region's dry climate and ensure a steady water supply for farming and everyday use, communities collect and store glacier meltwater by a technique involving redirecting glaciers' melt water into artificial ponds known as "Zing" and then water is directed through the networks of *kuhls* irrigation channels to villages and agricultural fields which is supervised by chhurpon who ensures equal distribution of water (Islam, 2024). The deposition of fresh silt and un-weathered minerals contributes to glacial sources of freshwater salts, and the melt water from glaciers, often below 2 °C, offers protection against various crop diseases (Saxena et al., 2021).

Fig. 3 Vernacular architecture of Ladakh that is practically being carried over centuries. **a** Housing design. **b** Traditional ceilings

4.1.3 Agropastoral Practices

In Ladakh's transhumant pastoral system, herds are often split, and livestock assets are distributed among family members who oversee distinct grazing groups headed to various summer mountain pastures. This practice reduces concentrated foraging pressures in grassland during increased climate stress, allowing for better resource utilization and risk management (Gaur et al., 2024). Consequently, agropastoralism enhances agricultural subsistence by producing various animal products (meat, leather, wool, and dairy) while reducing crop losses by free cattle grazing (Pandit et al., 2024; Raj et al., 2020). Cold-tolerant and water-intensive, water-conserving, and high-value cash crops should be introduced and traditional crops such as Grim (Naked Barley) and buckwheat should be replaced (Dolker, 2018). This crop improvement and diversification reflect the transition and balancing of the traditional resilience and modern innovation (Gaur et al., 2024).

4.2 Scientific and Technological Interventions

Communities in Ladakh have developed innovative water management techniques, commonly known as "ice reservoirs" or "ice stupas," with the capacity to store ice ranging from 17,000 to 23,500 m^3, which significantly alleviates water scarcity challenges and promotes sustainable agricultural activities (Nüsser et al., 2019). These ice stupas collect water during autumn and winter, freezing it until spring for the use. The water thaws and flows down to irrigate the agriculture fields, utilizing a previously untapped portion of the annual water flow and enabling its use to compensate for the decreased flow in the following spring (Balasubramanian et al., 2023). Leh-Ladakh has a large availability of solar irradiation, making renewable energy solutions among the most important options for meeting Ladakh's energy requirements. Different solar-based technologies include solar lighting systems, inclined solar dryer, PV winnower cum dryer, rooftop PV system, solar greenhouse, solar air heater for buildings, and solar water heaters (Santra, 2015). Solar PV modules are the most effective for generating electricity from the sun. The Ladakh Renewable Energy Development Agency (LREDA), a nodal agency for the Ministry of New and Renewable Energy, Govt. of India, is actively deploying renewable energy technologies throughout the region. Under this program, a solar water heating system with a 15000 m^2 collecting area is undergoing conversion, sanctioned by the Ministry of New and Renewable Energy (MNRE), Govt. of India in 2010. Additionally, in Drubuk Block, Ladakh, a 100 kVA solar power plant has been installed (https://ladakhenergy.org/ assessed on 24th July, 2026). Solar dryers are other convenient devices to preserve fruits and vegetables, especially apricots, while eliminating infection and contamination to improve crop quality and lower post-harvest losses. These have varying capacities from 10 to 100 kg and can save energy equivalent to 290–300 kWh/m^2 (Santra, 2015). For irrigation, solar water pumping devices under the Kargil Renewable Energy Development Agency (KREDA) have grown in popularity as an environmentally responsible and sustainable option for farmers, specifically in Kargil. These systems operate water pumps with solar energy, offering a dependable and affordable substitute for conventional diesel or electrically powered pumps, thus playing a vital role in maintaining sustainable agriculture methods (Kredakargil.org). For space heating in traditional architecture, fuelwood combustion was used, producing black carbon emissions that are a significant factor in melting glaciers in the High Himalayas. For this, the adoption of passive solar building designs in Ladakh was motivated by the region's abundant high-quality solar radiation, particularly given its cold-arid climate with temperatures plummeting below −35 °C (Bajracharya, 2023). With a set timeline and execution strategy, the Ladakh Renewable Energy Development Agency (LREDA) should draft a thorough action plan and policy for renewable energy, potentially in collaboration with the Power Development Department (PDD) (The Energy and Resources Institute, 2021).

Ladakh has a geothermal, micro hydel, and wind energy potential to be harnessed. The Puga valley is recognized as a primary geothermal zone characterized by numerous hot springs, geysers, and deposits of sulfur and borax with a number of applications including heating and cooling structures with geothermal heat pumps, producing electricity via geothermal power plants estimating to have a total stored potential of 36.5 MW per meter of thickness, assuming it is a homogeneous reservoir and direct utilization for domestic, industrial processes, and snow melting (Jha et al., 2020; Kumar et al., 2025). Currently, over 150 MW of hydroelectricity is generated from major and small-scale hydroelectric power projects situated on the Indus River and its tributaries, the primary hydel power source. The Ministry of Power has assessed the hydroelectric potential of the Indus River basin, indicating that while 40% of its total potential has been used, the remaining 60% is to be utilized. These projects generate carbon credits for regional development and have no carbon footprint, while also creating employment opportunities besides supplying the region with electricity (Mishra & Tripathi, 2022).

Collective operations are performed to distribute and use water effectively for agricultural purposes. Advanced irrigation techniques, such as drip and sprinkler irrigation, are being implemented as a water-efficient solution for agriculture. Government initiatives and training programs such as the Per Drop More Crop scheme, part of the Pradhan Mantri Krishna Sinchayee Yojana, are being used to encourage farmers to use this technology. Another technological breakthrough is the Ladakh greenhouse, which represents the advancement of the traditional Ladakhi polyhouse by promoting naturally ventilated passive solar greenhouses with many low and medium-cost designs. The primary focus has been growing leafy vegetables such as cauliflower, cabbage, broccoli, tomatoes, chilies, and cucumbers in the peak winter months and raising nurseries in spring, which was previously impossible in traditional passive solar greenhouses (Stobdan, 2023).

4.3 Mitigation Approaches

In order to ensure the long-term sustainability of Ladakh's fragile ecosystem, it is essential that the initiation of development programs must be harmonized with efforts to maintain the ecological balance (Wani et al., 2025). The Ladakh region has transitioned from traditional subsistence practices to more modern modes of urbanization and livelihood over the past few decades. While this shift has led to economic advancement, it presents considerable risks to traditional practices. Therefore, building capacity through organizing workshops, implementing training programs and community preparedness becomes the foremost priority (Jaiswal, 2024). This will require a careful assessment of community needs and identification of skill gaps. The prevailing agricultural systems are characterized by an increase in vegetables alongside a diminishing the yield of conventional crops. Local government needs to conserve and promote the traditional cropping system, needs, and economic growth. For irrigation, location-specific scientific methods should be involved, such as micro-irrigation technologies, and the construction of artificial glaciers and ice stupas with the involvement of policymakers and the Ladakh Autonomous Hill Development Council (LAHDC) (Dolker, 2018; Saxena et al., 2021).

Implementing strong animal husbandry techniques, offering proper veterinary care, and guaranteeing the availability of enough fodder are crucial for enhancing food security and protecting the livelihoods of rural inhabitants. The construction of protective embankments, also called dyke or levees, along riverbanks and stream channels, is an engineering approach to safeguard agricultural lands from the damaging effects of flash floods (Namgyal et al., 2025). Given Ladakh's high solar radiation potential, promoting renewable energy resources, specifically solar power, offers a sustainable alternative to fossil fuels. This will reduce carbon emissions and alleviate pressure on the region's fragile climate, limited bio-resources, and ecosystems while contributing to long-term energy security. Though the tourism sector in Ladakh is recognized as a crucial part of the economy, it is also seen as highly susceptible to risks and disasters. The government and policymakers can address this issue by regulating sustainable tourism practices, such as promoting the use of electric vehicles and reducing plastic bottle consumption. Furthermore, they can restrict the construction of large hotels and promote community-based eco-tourism in rural areas to ensure the region's environmental and cultural sustainability over the long term (Jaiswal, 2024). In areas like the Changthang pastoral, the degradation of grasslands resulted from uncontrolled grazing and climate change impacts. To safeguard traditional knowledge, indigenous breeds, and sustain grazing resources while boosting production and climate resilience, promising projects combine traditional techniques with modern technologies like satellite forecasts and mobile veterinarian services, encouraging pastoralism to control grazing methods (Pandit et al., 2024).

Forecasting cloudburst events is important for safeguarding both humans and critical infrastructure. It is crucial to augment meteorological observation infrastructure and establish dependable early warning mechanisms calibrated explicitly for the region's distinct climatic characteristics. Integrating disaster risk reduction strategies into infrastructure projects, site selection in less hazardous locations, and designing structures resilient to severe weather phenomena can substantially diminish the adverse effects of cloudbursts (Pareta & Pareta, 2025). Also, it is essential to treat the active sediment sequences and topographies and costly geotechnical treatments to mitigate the impacts of mass-flow, landslides, and associated hazards (Dubey & Dar, 2017). Protecting habitats, particularly Ramsar sites, which are crucial for species breeding, is essential for biodiversity conservation. A strategy is required to reduce the human-wildlife conflicts through community engagement, including sustainable practices, education and awareness, and effectively designating protective areas and ecological corridors.

5 Conclusion

Ladakh represents a fragile yet resilient high-altitude cold desert landscape where cultural, ecological, and environmental systems are deeply intertwined. Its rich cultural and religious diversity, shaped by centuries of Buddhist and Islamic traditions enriched by indigenous knowledge system such as the Amchi medical system, underscores the region's significance as a crossroads of spiritual, social, and livelihood practices. Equally, Ladakh's unique biodiversity, from sparse alpine flora to rare high-altitude fauna, forms the ecological foundation supporting pastoral and agropastoral livelihoods. The cryosphere, through its glaciers, snow cover, and intricate water systems, remains the lifeline of both ecosystems and communities. Yet, these cultural and natural assets are increasingly vulnerable to climate change, urban expansion, and tourism pressures. The resulting challenges, biodiversity loss, water scarcity, habitat degradation, and disruption of traditional practices, threaten both ecological balance and cultural continuity. Addressing these vulnerabilities requires integrated approaches that combine adaptation and mitigation strategies, long-term conservation policies, and inclusive development planning. Active collaboration between government, civil-society, and non-government organizations, grounded in community participation, will be essential to safeguard Ladakh's fragile ecosystems and cultural heritage while ensuring sustainable livelihoods for future generations.

References

Ali, M. (2021). *Dactylorhiza hatagirea*: A highly valued medicinal plant in cold desert of Ladakh and its conservation. *International Research Journal of Plant Sciences, 12*(4), 1. https://doi.org/10.14303/irjps.2021.22

Ali, S. N., Singh, R., Morthekai, P., Sharma, A., Phartiyal, B., Quamar, M. F., et al. (2022). Perception of climate change from the Himalayan 'cold desert' Ladakh, India. *Journal of Palaeosciences, 71*(1), 89–111. https://doi.org/10.54991/jop.2022.35

Ali, G., Chaudhari, M. P., Syed, S., Rajpurohit, D., Sanyal, M., & Shrivastav, P. S. (2025). Hydrogeochemical investigation and water quality assessment of the Indus River in the semiarid region of Ladakh, India. *Marine Pollution Bulletin, 211*, Article 117413. https://doi.org/10.1016/j.marpolbul.2024.117413

Angmo, K., Adhikari, B. S., & Rawat, G. S. (2025). Prioritizing conservation and participatory mapping of ethnomedicinal plant resources in Western Ladakh, Indian trans-Himalaya. *Frontiers in Forests and Global Change, 8*, 1481219. https://doi.org/10.3389/ffgc.2025.1481219

Angmo, K., Adhikari, B. S., Bussmann, R. W., & Rawat, G. S. (2024). Harmony in nature: Understanding the cultural and ecological aspects of plant use in Ladakh. *Journal of Ethnobiology and Ethnomedicine, 20*(1), 34. https://doi.org/10.1186/s13002-024-00670-3

Bajracharya, S. R. (2023). Technical and social challenges for implementation of sustainable architecture in the trans-Himalayan Region: An 'Intuitive Approach' to solar-heated buildings in Ladakh. In *World congress of architects* (pp. 869–895). Springer International Publishing, Cham. https://doi.org/10.1007/978-3-031-36320-7_54

Balasubramanian, S., Hoelzle, M., & Waser, R. (2023). Fountain scheduling strategies for improving water-use efficiency of artificial ice reservoirs (Ice stupas). *Cold Regions Science and Technology, 205*, Article 103706. https://doi.org/10.1016/j.coldregions.2022.103706

Banoo, S., Gulzar, R., Islam, T., Ganie, A. H., & Khuroo, A. A. (2021). *Gentianopsis paludosa* var. *alpina* (Gentianaceae), a new record for India from Ladakh, Trans-Himalaya. *Rheedea, 31*(2), 71–76. https://doi.org/10.22244/rheedea.2021.31.02.00

Bansal, S., Abid, N., Parmar, S. S., & Sen, J. (2023). Urbanization and resilience in mountain soil ecosystem: Case of outwash fan area of Leh, Ladakh, India. In *Understanding soils of mountainous landscapes* (pp. 281–301). Elsevier. https://doi.org/10.1016/B978-0-323-95925-4.00012-1

Batool, Z., Singh, K., & Gairola, S. (2022). Documenting potential ethnoveterinary knowledge from the indigenous communities of Cold Desert of Ladakh: A trans-Himalayan region of India. *Ethnobotany Research and Applications, 24*, 1–20.

Batool, Z., Singh, K., & Gairola, S. (2023). Medicinal plants traditionally used in the health care practices by the indigenous communities of the Trans-Himalayan region of Ladakh, India. *Journal of Ethnopharmacology, 317*, Article 116837. https://doi.org/10.1016/j.jep.2023.116837

Bhasin, V. (2012). Life on an edge among the Changpas of Changthang, Ladakh. *Journal of Biodiversity, 3*(2), 85–129. https://doi.org/10.1080/09766901.2012.11884738

Bhasin, A., Ghosal, S., Raina, P., & Hore, U. (2024). Climate change impacts on high altitude wildlife distribution: predicting range shifts for four ungulates in Changthang, eastern Ladakh. *Ecological Frontiers, 44*(2), 365–380. https://doi.org/10.1016/j.chnaes.2023.09.005

Bhat, M. S., Khan, A. A., Akbar, M., & Mir, S. (2023). Disaster-development interface and its impact on emerging vulnerability scenario in Ladakh region of northwestern Himalayas. *Journal of*

Environmental Studies and Sciences, 13(2), 253–270. https://doi.org/10.1007/s13412-023-00818-9

Bhat, I. A., Rashid, I., Ramsankaran, R. A. A. J., Banerjee, A., & Vijay, S. (2024). A comprehensive rock glacier inventory for Jammu, Kashmir, and Ladakh, western Himalaya, India–Baseline for the permafrost research. *Earth System Science Data Discussions, 2024*, 1–25. https://doi.org/10.5194/essd-2023-522

Bhat, I. A., Rashid, I., Ramsankaran, R. A. A. J., Banerjee, A., & Vijay, S. (2025). Inventorying rock glaciers in the Western Himalaya, India, and assessing their hydrological significance. *Geomorphology, 471*, Article 109514. https://doi.org/10.1016/j.geomorph.2024.109514

Bhatnagar, M., Avasthi, A. S., Singh, S., & Ghosal, S. (2017). Evaluation of anti-leishmanial and antibacterial activity of *Waldheimia tomentosa* (Asteraceae), and chemical profiling of the most bioactive fraction. *Tropical Journal of Pharmaceutical Research, 16*(9), 2169–2178. https://doi.org/10.4314/tjpr.v16i9.18

BirdLife International. (2020). *Grus nigricollis* The IUCN red list of threatened species 2020. Retrieved August 25, 2025, from https://doi.org/10.2305/IUCN.UK.2020-3.RLTS.T22692162A180030167.en

Butt, T. A., Ganie, A. H., Khuroo, A. A., Ahmad, R., Rasool, N., & Basharat, S. (2021). New distributional record of *Primula atrodentata* WW. Sm. from Ladakh region and assessment of its threat status across the Himalaya. *Acta Ecologica Sinica, 41*(3), 183–188. https://doi.org/10.1016/J.CHNAES.2020.08.002

Chauhan, A., Jishtu, V., Thakur, L., & Dolma, T. (2020). Medicinal plants of the Trans-Himalayan Cold Desert of Ladakh-a review. *International Journal of Science, Environment, 9*(2), 239–253.

Chevuturi, A., Dimri, A. P., & Thayyen, R. J. (2018). Climate change over Leh (Ladakh), India. *Theoretical and Applied Climatology, 131*(1), 531–545. https://doi.org/10.1007/s00704-016-1989-1

Dame, J., Schmidt, S., Müller, J., & Nüsser, M. (2019). Urbanization and socio-ecological challenges in high mountain towns: Insights from Leh (Ladakh), India. *Landscape and Urban Planning, 189*, 189–199. https://doi.org/10.1016/j.landurbplan.2019.04.017

Dar, S. N., Shah, S. A., & Wani, M. A. (2022). Geospatial tourist information system for promoting tourism in Trans-Himalayas: A study of Leh Ladakh India. *GeoJournal, 87*(4), 3249–3263. https://doi.org/10.1007/s10708-021-10431-4

Dawa, S., Gurmet, P., Stobgais, T., & Rinchen, T. (2021). Survey and ethno-botanical study of medicinal plants of some selected villages of singay-lalok region of Leh (UT Ladakh Region). *Asian Journal of Research in Botany, 5*(2), 60–75.

Devi, S., Sharma, P., Sharma, R., & Thakur, M. (2023). Current Status and medicinal prominence of *Arnebia euchroma* (Ratanjot): A critically endangered plant of Trans-Himalayan region. *Recent Patents on Biotechnology, 17*(1), 92–102. https://doi.org/10.2174/1872208316666220718154341

Dhar, P., Amol, T., Basant, B., Chaurasia, O. P., Bhatt, P. R., & Srivastava, R. (2011). *Lycium ruthenicum* Murray: A less-explored but high-value medicinal plant from Trans-Himalayan cold deserts of Ladakh, India. *Plant Archives, 11*, 583–586.

Dolkar, P., Rinchen, T., Dolkar, D., & Gurmet, P. (2021). Cultivation Studies of *Verbascum thapsus* L., the Great Mullein in Trans-Himalayan Ladakh Region. *International Journal of Current Microbiology and Applied Sciences, 10*(2), 404–410. https://doi.org/10.20546/ijcmas.2021.1002.047

Dolker, P. (2018). An overview of transition in traditional agriculture of Ladakh. *Journal of Himalayan Ecology and Sustainable Development, 13*, 26–48.

Dolma, N., Shahar, B., Joshi, B., & Chongtham, N. (2025). Exploring the nutritional and medicinal potential of *Urtica hyperborea* Jacq. ex Wedd. from Ladakh, India: A multi-analytical study. *Food and Humanity, 100664*. https://doi.org/10.1016/j.foohum.2025.100664

Dolma, T., & Uniyal, S. K. (2025). Valuing seabuckthorn (*Hippophae rhamnoides* L.) ecosystem services for sustainable development and livelihood: a case study in Ladakh cold desert. *Forests, Trees and Livelihoods*, 1–13. https://doi.org/10.1080/14728028.2025.2480055

Dorjey, K., & Maurya, A. K. (2021). Ethnobotany of *Juniperus polycarpos* C. Koch (Cupressaceae) in the Himalayan cold desert of union territory of Ladakh, India. *Indian Journal of Traditional Knowledge, 20*(1), 83–90.

Dorjey, K., Tamchos, S., & Kumar, S. (2012). Ethnobotanical observations in Trans-Himalayan region of Ladakh. *Journal of Plant Development Sciences, 4*(4), 459–464.

Dubey, R. K., & Dar, J. A. (2017). Geotechnical attributes of active sediments and its prospective in hazards susceptibility and mitigation measures: A study on soils from Leh Valley, Ladakh Himalaya, India. *National Academy Science Letters, 40*, 279–284. https://doi.org/10.1007/s40009-017-0566-1

Elnour, A. A., & Abdurahman, N. H. (2024). Current and potential future biological uses of *Saussurea costus* (Falc.) Lipsch: A comprehensive review. *Heliyon, 10*(18). https://doi.org/10.1016/j.heliyon.2024.e37790

Gaur, M. K., Goyal, R. K., Singh, N., Choudhary, V., & Chichaghare, A. R. (2024). Prospects and challenges for biodiversity in the anthropogenic-induced cold arid landscape of Ladakh, India. *Japan Journal of Research, 5*, 1–15. https://doi.org/10.33425/2690-8077.1123

Ghosal, S., Bhasin, A., Raina, P., Jhala, Y. V., Patel, D. A., Dolma, K., et al. (2024). First confirmed record of Common Leopard *Panthera pardus* From Hemis National Park, Ladakh, India. *Journal of the Bombay Natural History Society, 121*(3), 234–240.

Gupta, A. (2021). Material culture and art practice in Ladakh: notes from a collaborative art project. In *Arts in the margins of world encounters* (p. 147).

Gurmet, P., & Stobgais, T. (2016). *A handbook on medicinal plants of himalayas used in Sowa-Rigpa* (1st ed.). National Research Institute for Sowa Rigpa (Publisher).

Hadi, A., Singh, S., Ali, S., & Mehdi, M. (2022). Traditional uses of medicinal plants by indigenous tribes of Ladakh union territory. *Research Journal of Agricultural Sciences, 13*, 68–77.

Hamid, A., & Raina, A. K. (2024). Threat and conservation status of traditionally important plant species: A case study of Indus valley, Western Ladakh Trans-Himalaya. *Proceedings of the National Academy of Sciences, India Section B: Biological Sciences, 94*(2), 449–464. https://doi.org/10.1007/s40011-023-01542-8

Haq, S. M., Calixto, E. S., Yaqoob, U., Ahmed, R., Mahmoud, A. H., Bussmann, R. W., et al. (2020). Traditional usage of wild fauna among the local inhabitants of Ladakh, Trans-Himalayan Region. *Animals, 10*(12), 2317. https://doi.org/10.3390/ani10122317

Haq, S. M., Hassan, M., Jan, H. A., Al-Ghamdi, A. A., Ahmad, K., & Abbasi, A. M. (2022). Traditions for future cross-national food security—Food and foraging practices among different native communities in the Western Himalayas. *Biology, 11*(3), 455.

Humbert-Droz, B. (2024). Environmental change in Ladakh's Changthang: A local, regional and global phenomenon. *Environmental Change and Development in Ladakh, Indian Trans-Himalaya, 119–146*. https://doi.org/10.1007/978-3-031-42494-6_9

Hussain, S., Sharma, S., Bhatti, R. C., & Singh, A. N. (2023a). Floral diversity expedition in Ladakh: An insight into the exploration, distribution pattern, ethnobotanical, phytochemical studies and conservation strategies. *Ethnobotany Research and Applications, 26*, 1–32.

Hussain, A., Schmidt, S., & Nüsser, M. (2023b). Dynamics of mountain urbanization: Evidence from the Trans-Himalayan town of Kargil, Ladakh, India. *Land, 12*(4), 920. https://doi.org/10.3390/land12040920

Hussain, M., Kiran, S., Sayed, I., Khazir, J., Maqbool, T., Ibrahim, M., et al. (2024a). Altitude-dependent influence of *Artemisia brevifolia* on its rhizosphere microbiome in Ladakh region of the Western Himalayas. *Rhizosphere, 30*, Article 100918. https://doi.org/10.1016/j.rhisph.2024.100918

Hussain, M., Kacho, N. F., Shashank, P. R., Ali, M., & Mir, A. H. (2024b). First report of an invasive pest, *Pheosia albivertex* (Hampson 1983) (Lepidoptera: Notodontidae) on *Populus alba* (Salicaceae), from Ladakh, India. *Journal of Asia-Pacific Biodiversity, 17*(2), 321–326. https://doi.org/10.1016/j.japb.2023.12.007

Iseli, E., Chisholm, C., Lenoir, J., Haider, S., Seipel, T., Barros, A., et al. (2023). Rapid upwards spread of non-native plants in mountains across continents. *Nature Ecology & Evolution, 7*(3), 405–413. https://doi.org/10.1038/s41559-022-01979-6

Islam, M. (2024). Climate change adaptation through a social innovation lens: A framework analysis of Ladakh, India's community-based approach for sustainable agriculture practice. In *International Conference on Sustainable Development (ICSD-2024).*

Jain, R., Bhardwaj, P., Pandey, S. S., & Kumar, S. (2021). *Arnebia euchroma*, a plant species of cold desert in the Himalayas, harbors beneficial cultivable endophytes in roots and leaves. *Frontiers in Microbiology, 12*, Article 696667. https://doi.org/10.3389/fmicb.2021.696667

Jaiswal, I. (2024). Assessing tourism sustainability: A case study of Ladakh region. In *The need for sustainable tourism in an era of global climate change: Pathway to a greener future* (pp. 151–172). Emerald Publishing Limited. https://doi.org/10.1108/978-1-83608-668-020241023

Jamwal, P. S., Shrotriya, S., & Takpa, J. (2020). The pattern of waterbird diversity of the trans-Himalayan wetlands in Changthang Wildlife Sanctuary, Ladakh, India. *Journal of Threatened Taxa, 12*(1), 15129–15139. https://doi.org/10.11609/jott.5122.12.1.15129-15139

Jasrotia, A. S., Ahmad, S., Thakur, P. K., Ridwan, Q., Wani, Z. A., Alamri, S. A. M., et al. (2023). Long-term geospatial observations of the Drang Drung and Pensilungpa Glaciers, Northwestern Himalaya, India, from 1976 to 2020. *Sustainability, 15*(20), 15067. https://doi.org/10.3390/su152015067

Jha, S. K., Puppala, H., & Kumar, M. S. M. (2020). 3D characterization of thermo-hydro-geological fields and estimation of power potential from Puga geothermal reservoir, Ladakh, India. *Renewable Energy, 146*, 1510–1523. https://doi.org/10.1016/j.renene.2019.07.006

Jishtu, V., Chauhan, A., & Bano, H. (2023). Medicinal Flora of the Trans-Himalayan Cold Desert in Ladakh, India. In *Biosynthesis of bioactive compounds in medicinal and aromatic plants: Manipulation by conventional and biotechnological approaches* (pp. 257–286). Springer Nature Switzerland. https://doi.org/10.1007/978-3-031-35221-8_12

Jishtu, V., Moran, A., Ibrahim, M., & Ahmad, Z. (2025). Quantitative study of the Indian ethnobotanical medicinal plant resources in the remote Zanskar Valley of Ladakh. *Ethnobotany Research and Applications, 31*, 1–28.

Joshipura, V. (2025). *Geomorphology of desert environments.* Educohack Press.

Kala, C. P. (2005). Indigenous uses, population density, and conservation of threatened medicinal plants in protected areas of the Indian Himalayas. *Conservation Biology, 19*(2), 368–378. https://doi.org/10.1111/j.1523-1739.2005.00602.x

Kanwar, N., & Kuniyal, J. C. (2022). Vulnerability assessment of forest ecosystems focusing on climate change, hazards and anthropogenic pressures in the cold desert of Kinnaur district, northwestern Indian Himalaya. *Journal of Earth System Science, 131*(1), 51. https://doi.org/10.1007/s12040-021-01775-z

Khan, M. H., Dar, N. A., Alie, B. A., Mir, G. H., Al-Sadoon, M. K., Hefft, D. I., & Rather, R. A. (2023). Morpho-biochemical characterization of Kalazeera (*Bunium persicum* Boiss. Fedts) germplasm grown in Global temperate ecologies. *Saudi Journal of Biological Sciences, 30*(5), 103633.https://doi.org/10.1016/j.sjbs.2023.103633

Khanday, A. A., Rather, G. M., Kumar, M. A., Lone, A. H. (2024). Environmental sustainability and its impact on public health in Ladakh-India. In *Sustainability and health informatics: A systems approach to address the climate action induced global challenge* (pp. 183–198). Springer Nature Singapore. https://doi.org/10.1007/978-981-97-6706-9_9

Koul, M. N., Bahuguna, I. M., Rajawat, A. S., Ali, S., & Koul, S. (2016). Glacier area change over past 50 years to stable phase in Drass valley, Ladakh Himalaya (India). *American Journal of Climate Change, 5*(01), 88. https://doi.org/10.4236/ajcc.2016.51010

Kumar, G. P., Kumar, R., & Chaurasia, O. P. (2011a). Conservation status of medicinal plants in Ladakh: Cold arid zone of trans-Himalayas. *Research Journal of Medicinal Plants, 5*, 685–694. https://doi.org/10.3923/rjmp.2011.685.694

Kumar, G. P., Kumar, R., Chaurasia, O. P., & Singh, S. B. (2011b). Current status and potential prospects of medicinal plant sector in trans-Himalayan Ladakh. *Journal of Medicinal Plants Research, 5*(14), 2929–2940.

Kumar, A., Negi, H. S., & Kumar, K. (2020). Long-term mass balance modelling (1986–2018) and climate sensitivity of Siachen Glacier, East Karakoram. *Environmental Monitoring and Assessment, 192*(6), 368. https://doi.org/10.1007/s10661-020-08323-0

Kumar, D., Ojha, A. K., Maurya, V. P., Satyanarayanan, M., Dhakate, R., Kumar, P., et al. (2025). New insight into the shallow sub-surface for geothermal prospection of Puga valley, Ladakh, India from electrical resistivity tomography. *Geothermics, 131*, Article 103409.

Kunmar, P., Rana, A. S., Kumar, V., Mehta, M., & Nainwal, H. C. (2025). Glacial landforms and geometric transformations: Tracing the history of Pensilungpa and Durung-Drung glaciers in Suru and Doda River valleys, Western Himalaya, Ladakh. *Environmental Monitoring and Assessment, 197*(3), 272. https://doi.org/10.1007/s10661-025-13727-x

Ladon, P., Nüsser, M., & Garkoti, S. C. (2023). Mountain agropastoralism: Traditional practices, institutions and pressures in the Indian Trans-Himalaya of Ladakh. *Pastoralism, 13*(1), 30. https://doi.org/10.1186/s13570-023-00289-1

Mehta, D., & Gajjar, V. (2023). Analysis of thermal comfort of Vernacular Ladakhi Dwellings of India. In *Future is urban: Livability, resilience & resource conservation* (pp. 136–144). Routledge. https://doi.org/10.4324/9781003342441-16

Mehta, M., Kumar, V., Garg, S., & Shukla, A. (2021). Little ice age glacier extent and temporal changes in annual mass balance (2016–2019) of Pensilungpa Glacier, Zanskar Himalaya. *Regional Environmental Change, 21*(2), 38. https://doi.org/10.1007/s10113-021-01766-2

Meyer, S. E., Callaham Jr, M. A., Stewart, J. E., & Warren, S. D. (2021). Invasive species response to natural and anthropogenic disturbance. In *Invasive species in forests and rangelands of the United States: A comprehensive science synthesis for the United States forest sector* (pp. 85–110). Springer International Publishing. https://doi.org/10.1007/978-3-030-45367-1_5

Mishra, A., & Tripathi, P. (2022). Role and potential of hydro-electrical projects in development in union territory of Ladakh. Policy Paper No. 4. https://www.ioe.du.ac.in/delhi-school-of-public-policy-and-governance/policy-papers/

Morup, T., Namgail, T., & Tsering, T. (2024). Wildlife versus livestock: Conservation Dilemma of the Pastoralists of Changthang. In *Environmental change and development in Ladakh, Indian trans-Himalaya* (pp. 147–161). Springer Nature Switzerland. https://doi.org/10.1007/978-3-031-42494-6_10

Mueller, S., Sammonds, P., Bhat, G. M., Pandita, S., Suri, K., Thusu, B., & Le Masson, V. (2019). Disaster scenario simulation of the 2010 cloudburst in Leh, Ladakh, India. *International Journal of Disaster Risk Reduction, 33*, 485–494. https://doi.org/10.1016/j.ijdrr.2018.09.004

Namga, S., & Nayak, R. (2024). Geological features of Northwestern Himalaya, Ladakh, India: Geosites and their significance in the Ladakh Himalaya. *Geoheritage, 16*(2), 39. https://doi.org/10.1007/s12371-024-00942-0

Namgail, T., Bagchi, S., Mishra, C., & Bhatnagar, Y. V. (2008). Distributional correlates of the Tibetan gazelle *Procapra picticaudata* in Ladakh, northern India: Towards a recovery programme. *Oryx, 42*(1), 107–112. https://doi.org/10.1017/S0030605308000768

Namgail, T., Van Wieren, S. E., Mishra, C., & Prins, H. H. T. (2010). Multi-spatial co-distribution of the endangered Ladakh urial and blue sheep in the arid Trans-Himalayan mountains. *Journal of Arid Environments, 74*(10), 1162–1169. https://doi.org/10.1016/j.jaridenv.2010.04.013

Namgyal, P., Sarkar, S., & Kumar, R. (2025). Vulnerability assessment of rural households to climate change using livelihood vulnerability framework approach in the trans-Hidarmalayan region of Ladakh, India. *Anthropocene, 49*, Article 100467. https://doi.org/10.1016/j.ancene.2025.100467

Nasir, O., & Kamal, M. A. (2021). Vernacular architecture as a design paradigm for sustainability and identity: The case of Ladakh, India. *American Journal of Civil Engineering and Architecture, 9*(6), 219–231. https://doi.org/10.12691/ajcea-9-6-2

Negi, V. S., Joshi, B. C., Pathak, R., Rawal, R. S., & Sekar, K. C. (2018a). Assessment of fuelwood diversity and consumption patterns in cold desert part of Indian Himalaya: Implication for conservation and quality of life. *Journal of Cleaner Production, 196*, 23–31. https://doi.org/10.1016/j.jclepro.2018.05.237

Negi, V. S., Pathak, R., Sekar, K. C., Rawal, R. S., Bhatt, I. D., Nandi, S. K., & Dhyani, P. P. (2018b). Traditional knowledge and biodiversity conservation: A case study from Byans Valley in Kailash Sacred Landscape, India. *Journal of Environmental Planning and Management, 61*(10), 1722–1743. https://doi.org/10.1080/09640568.2017.1371006

Negi, V. S., Maikhuri, R. K., Maletha, A., & Phondani, P. C. (2019). Ethnobotanical knowledge and population density of threatened medicinal plants of Nanda Devi Biosphere Reserve, Western Himalaya, India. *Iranian Journal of Science and Technology, Transactions A: Science, 43*(1), 63–73. https://doi.org/10.1007/s40995-018-0545-5

Negi, H. S., Gusain, H. S., Dhahiya, T., & Kumar, A. (2022a). Hydrological mass balance study of Siachen glacier, East Karakoram. *Journal of Earth System Science, 131*(1), 33. https://doi.org/10.1007/s12040-021-01788-8

Negi, V. S., Tiwari, D. C., Singh, L., Thakur, S., & Bhatt, I. D. (2022b). Review and synthesis of climate change studies in the Himalayan region. *Environment, Development and Sustainability, 24*(9), 10471–10502. https://doi.org/10.1007/s10668-021-01880-5

Negi, V. S., Pathak, R., Thakur, S., Joshi, R. K., Bhatt, I. D., & Rawal, R. S. (2023). Scoping the need of mainstreaming indigenous knowledge for sustainable use of bioresources in the Indian Himalayan region. *Environmental Management, 72*(1), 135–146. https://doi.org/10.1007/s00267-021-01510-w

Nüsser, M., Dame, J., Kraus, B., Baghel, R., & Schmidt, S. (2019). Socio-hydrology of "artificial glaciers" in Ladakh, India: Assessing adaptive strategies in a changing cryosphere. *Regional Environmental Change, 19*(5), 1327–1337. https://doi.org/10.3389/frwa.2023.1100977

Pandit, A., Mir, M., Mir, M., Wani, Y., Bisati, I., Nisa, S. U., et al. (2024). Pastoralism in Changthang, Ladakh: Adaptations, challenges, and pathways for sustainability. *Mountain Research and Development, 44*(1), A1–A7. https://doi.org/10.1659/mrd.2023.00028

Pareta, K., & Pareta, U. (2025). A review of cloudbursts events in the Himalaya region, and 2D hydrodynamic simulation using MIKE models. *Natural Hazards, 1–33*. https://doi.org/10.1007/s11069-025-07173-8

Raina, P., Mungi, N. A., Kumar, U., Rathi, A. D., Khan, N. H., & Patel, D. A. et al. (2025). Comprehensive assessment of snow leopard distribution and population in the Indian Trans-Himalaya, Ladakh: Standardizing methods for evidence-based conservation. *PLoS One, 20*(5), e0322136. https://doi.org/10.1371/journal.pone.0322136

Raj, A., Angchuk, D., & Islam, M. A. (2020). Livelihood dependence on highland pastoralism (Doksa) in Trans-Himalayan region of Zanskar, Ladakh. *International Journal of Environment and Climate Change, 10*(12), 67–76. https://doi.org/10.9734/ijecc/2020/v10i1230285

Rana, A. S., Kunmar, P., Mehta, M., & Kumar, V. (2023). Glacier retreat, dynamics and bed overdeepenings of Parkachik Glacier, Ladakh Himalaya, India. *Annals of Glaciology, 64*(92), 254–267. https://doi.org/10.1017/aog.2023.50

Rawat, Y. S., Negi, V. S., Moussa, I. M., Zaman, W., & Elansary, H. O. (2023). Diversity, distribution and vegetation assessment of woody plant species in the cold desert environment, North–Western Himalaya, India. *Sustainability, 15*(13), 10429. https://doi.org/10.3390/su151310429

Ridwan, Q., Wani, Z. A., Hanief, M., Pant, S., Shah, A. A., Siddiqui, S., & Alamri, S. (2023). Indigenous knowledge and perception of local people towards biodiversity conservation in Rajouri district of Jammu and Kashmir, India. *Sustainability, 15*(4), 3198. https://doi.org/10.3390/su15043198

Rinchen, T., Gurmet, P., Dolker, P., & Stobgais, T. (2021). Cultivation and population status of Ephedra (*Ephedra gerardiana* Wall. Ex Stapf) critically endangered medicinal plant for the conservation in cold desert of Ladakh, India. *Research Square*. https://doi.org/10.21203/rs.3.rs-246584/v1

Romshoo, S. A., Abdullah, T., Murtaza, K. O., & Bhat, M. H. (2023). Direct, geodetic and simulated mass balance studies of the Kolahoi Glacier in the Kashmir Himalaya, India. *Journal of Hydrology, 617*, Article 129019. https://doi.org/10.1016/j.jhydrol.2022.129019

Romshoo, S. A., Ameen, U., Bhat, M. H., & Abdullah, T. (2024). Comprehensive analysis of glacier recession (2000–2020) in the Nun-Kun Group of Glaciers, Northwestern Himalaya. *Journal of Mountain Science, 21*(3), 768–785. https://doi.org/10.1007/s11629-023-8266-4

Sahu, R., Gupta, R. D., Ramanathan, A., Kumar, P., & Eidhammer, T. (2024). Long-term annual and seasonal mass balance reconstruction and sensitivity analysis of Chhota Shigri Glacier in Western Himalaya. *Environmental Science and Pollution Research, 31*(3), 4910–4924. https://doi.org/10.1007/s11356-023-31537-x

Santra, P. (2015). Scope of solar energy in cold arid region of India at Leh Ladakh. *Annals of Arid Zone, 54*(3&4), 109–117.

Saxena, K. G., Liang, L., & Xue, X. (2011). *Global change, biodiversity, and livelihood in the cold desert region of Asia.* Bishen Singh Mahendra Pal Singh. ISBN: 978-81-211-0780-8.

Saxena, A., Raghuvanshi, M. S., & Suna, T. (2021). Traditional water management of subsistence agriculture system in cold arid Ladakh: A review. *International Journal of Scientific & Engineering Research, 12*(4), 1019–1020.

Shabir, M., Banoo, A., & Salim, M. (2023). An Ethnopharmacological review of three notable medicinal plants found in the union territory of Ladakh, India. *Exploratory Animal and Medical Research, 13*, 85–90. https://doi.org/10.52635/eamr/13(S):85-90

Sharif, M., & Singh, R. K. (2025). Contemporary politics in Ladakh: Contestations and identity construction. *Social Identities, 1–14*. https://doi.org/10.1080/13504630.2025.2465366

Shukla, A. N., & Srivastava, S. K. (2020). Flora of Ladakh: An annotated inventory of flowering plants. In *Biodiversity of the Himalaya: Jammu and Kashmir state* (pp. 673–730). Springer Singapore.

Shukla, S., Sharma, A., Kumar, N., Kaur, B., Kumar, S., Patel, M. K., & Saxena, S. (2020). Leh-Ladakh: A new potential habitat for aruna (*Aconitum heterophyllum*). *International Journal of Current Microbiology and Applied Sciences, 11*, 2570–2575.

Singh, N., Angchuk, D., Phour, M., & Chaurasia, O. P. (2021). Studies on physicochemical and sensory attributes of potatoes stored during winter in high-altitude cold desert, Ladakh, India. *Potato Research, 64*(2), 197–212. https://doi.org/10.1007/s11540-020-09472-y

Singh, R., & Bhatla, R. (2025). Assessment of rainfall variability in Ladakh amidst evolving climate. *Natural Hazards, 121*(3), 3073–3097. https://doi.org/10.1007/s11069-024-06922-5

Singh, S. P., & Malik, J. K. (2025). Migratory patterns and migratory routes of pastoral small Ruminants in the Himalayas. In *Migratory small ruminant farming system in the Himalayas-peculiarities, problems, prospects, climate effects, nutrition and health* (pp. 45–66). Springer Nature Singapore.

Soheb, M., Ramanathan, A., Angchuk, T., Mandal, A., Kumar, N., & Lotus, S. (2020). Mass-balance observation, reconstruction and sensitivity of Stok glacier, Ladakh region, India, between 1978 and 2019. *Journal of Glaciology, 66*(258), 627–642. https://doi.org/10.1017/jog.2020.34

Soheb, M., Ramanathan, A., Bhardwaj, A., Coleman, M., Rea, B., Spagnolo, M., et al. (2022). Multitemporal glacier inventory revealing four decades of glacier changes in the Ladakh region. *Earth System Science Data Discussions, 1–26*. https://doi.org/10.5194/essd-14-4171-2022

Stobdan, T. (2023). Passive solar greenhouse for growing vegetables under extreme conditions of Ladakh. In *Agriculture in Ladakh: A step towards sustainable mountain development*, Beeja House.

Suri, K. (2018). Understanding historical, cultural and religious frameworks of mountain communities and disasters in Nubra valley of Ladakh. *International Journal of Disaster Risk Reduction, 31*, 504–513. https://doi.org/10.1016/j.ijdrr.2018.06.004

Tabin, S., Gupta, R. C., Kamili, A. N., & Parray, J. A. (2022). Medical and medicinal importance of Rheum spp. collected from different altitudes of the Kashmir Himalayan range. *Cellular, Molecular and Biomedical Reports, 2*(3), 187–201. https://doi.org/10.55705/cmbr.2022.349901.1050

Taloor, A. K., Kothyari, G. C., Manhas, D. S., Bisht, H., Mehta, P., Sharma, M., et al. (2021). Spatio-temporal changes in the Machoi glacier Zanskar Himalaya India using geospatial technology. *Quaternary Science Advances, 4*, Article 100031. https://doi.org/10.1016/j.qsa.2021.100031

Tamchos, S., & Dorjey, K. (2025). *Hippophae rhamnoides* subsp. *turkestanica*: A Potential Species for Sustainable Propagation in the Cold Desert of Ladakh. *National Academy Science Letters, 48*(1), 89–94. https://doi.org/10.1007/s40009-024-01431-2

Tamchos, S., & Kaul, V. (2019). Seabuckthorn: Opportunities and challenges in Ladakh. *National Academy Science Letters, 42*(2), 175–178. https://doi.org/10.1007/s40009-018-0707-1

Tewari, V. P., & Kapoor, K. S. (2013). Western Himalayan cold deserts: Biodiversity, eco-restoration, ecological concerns and securities. *Annals of Arid Zone, 52*(3 & 4), 225–232.

The Energy and resources institute. (2021). Carbon Neutral and Climate Resilient Ladakh A strategy document-Action plan and Roadmap.

Wani, Z. A., Pant, S., & Singh, B. (2021). Descriptive study of plant resources in context of the ethnomedicinal relevance of indigenous flora; a case study from Rajouri-Poonch region of Himalaya. *Ethnobotany Research and Applications, 21*, 1–22.

Wani, Z. A., Bhat, J. A., Negi, V. S., Satish, K. V., Siddiqui, S., & Pant, S. (2022). Conservation priority index of species, communities, and habitats for biodiversity conservation and their management planning: A case study in Gulmarg Wildlife Sanctuary, Kashmir Himalaya. *Frontiers in Forests and Global Change, 5*, Article 995427. https://doi.org/10.3389/ffgc.2022.995427

Wani, Z. A., Fatima, T., Verma, O., Sharma, V., & Pant, S. (2023). Impacts of climate change on plants with special reference to the Himalayan Region. In *Climate change adaptation, risk management and sustainable practices in the Himalaya* (pp. 237–251). Springer International Publishing. https://doi.org/10.1007/978-3-031-24659-3_11

Wani, M. D., Akhter, S., Shah, S. A., & Dada, Z. A. (2025). Does mountain tourism impact the socio-ecological balance of the Indian Trans Himalayas, Ladakh?. In *Balancing mountain tourism, cultural heritage and environmental stability* (pp. 89–108). IGI Global Scientific Publishing.

Zargar, S. A., Ganie, A. H., Reshi, Z. A., Shah, M. A., Sharma, N., & Khuroo, A. A. (2024a). *Oxalis corniculata* L. (Oxalidaceae), an addition of an alien plant species to the flora of Ladakh, India. *Vegetos, 37*(1), 373–378. https://doi.org/10.1007/s42535-023-00612-6

Zargar, S. A., Khuroo, A. A., Reshi, Z. A., & Ganie, A. H. (2024b). Recent species' range expansion in the Himalayan Highlands: New distribution records of naturalized alien plants from Ladakh, India. *Biology Bulletin, 51*(3), 576–585. https://doi.org/10.1134/S1062359023606079

Zargar, S. A., Reshi, Z. A., Ganie, A. H., Dutt, H. C., Shah, M. A., & Sharma, N. (2025a). Local inventories for effective management of alien species: Insights from the alien flora of Jammu, Kashmir, and Ladakh, India. *Environmental Monitoring and Assessment, 197*(1), 1–18. https://doi.org/10.1007/s10661-024-13479-0

Zargar, S. A., Malik, R. A., Khuroo, A. A., Ganie, A. H., & Reshi, Z. A. (2025b). Alien flora of the Himalayan highlands: Naturalized and invasive plants in the Trans-Himalayan region of Ladakh, India. *Biological Invasions, 27*(4), 1–19. https://doi.org/10.1007/s10530-025-03562-z

Zomba, D., Dar, M. A., Singh, R., Batish, D. R., & Kaur, S. (2025). Exploring the antimicrobial, antioxidant, and in vitro anti-inflammatory potential of *Oxytropis microphylla* (Pall.) DC. plant species of Ladakh region. *Discover Food, 5*(1), 1–14. https://doi.org/10.1007/s44187-025-00553-w

Human Ecology, Vulnerability, and Adaptation to Climate Change in Cold Desert Landscape of Johar Valley of Pithoragarh District in Uttarakhand Himalaya

Shalini Rawat, Pradeep Kumar Sharma, Suman Naithani, and V. P. Uniyal

1 Introduction

Cold desert ecosystems, characterized by low precipitation, high diurnal temperature variation, fragile soils, and unique biodiversity, occupy significant yet often overlooked ecological niches across the globe (Singh & Thadani, 2015). For instance, Ladakh, Lahaul, and Spiti regions in India, situated in the trans-Himalayan zone, receive less than 100 mm of precipitation annually and host a mix of alpine steppe and semi-arid ecosystems (Passang et al., 2022). The Tibetan plateau in Tibet (China) stands out by its significant daily temperature fluctuations and vast permafrost (Qiu, 2008). In South America, the Altiplano, a high Andean plateau spanning Bolivia, Peru, and Chile, experiences high solar radiation, arid soils, and supports endemic vegetation adapted to freezing nights and dry days (Vuille et al., 2018). Similarly, the desert steppes of Mongolia and Central Asia are characterized by a continental climate with frequent droughts and fragile alpine ecosystems. Despite harsh environmental constraints, cold deserts sustain unique biodiversity and culturally adapted human populations (Nüsser & Baghel, 2014). From endemic plant species to nomadic pastoralism, these regions have developed survival strategies deeply intertwined with climatic rhythms and ecological limitations (Dame & Nüsser, 2011; Kala, 2005).

Johar valley presents a distinctive example of a high-altitude cold desert ecosystem situated in the trans-Himalayan rain shadow zone of the Pithoragarh district of Uttarakhand state. Historically this area was a trade route connecting India and

S. Rawat · P. K. Sharma (✉) · S. Naithani
Department of Environmental Science, Graphic Era (Deemed to be) University, Dehradun, Uttarakhand, India
e-mail: pksharma@geu.ac.in

V. P. Uniyal
Department of Biosciences, Graphic Era (Deemed to be) University, Dehradun, Uttarakhand, India

Y. S. Rawat and C. P. Kala (eds.), *Cold Deserts*,
https://doi.org/10.1007/978-3-032-29119-6_2

western Tibet. Johar's landscape and communities have evolved within an intricate socio-ecological matrix shaped by altitudinal zonation, pastoralism, and forest-based resource use. Unlike some other Indian cold deserts such as Ladakh and Spiti, which are characterized by extreme aridity and Buddhist cultural traditions (Dame & Nüsser, 2011), Johar is notable for its mixed agro-pastoral systems and Bhotiya tribal communities, who have traditionally practiced seasonal migration (Kala et al., 2005). Recent shifts in climate patterns due to global change, especially glacial retreat and declining snowfall along with socio-economic transitions such as rural outmigration, have brought new pressures on local forest ecosystems and resource governance (Negi et al., 2021). These changes underscore the urgency of examining Johar valley not only as a representative cold desert system but also as a microcosm for understanding broader challenges of sustainability, resilience, and community-led rangelands resource management under global change (Pandey et al., 2019).

Currently, cold deserts face vulnerabilities in the form of glacial retreat, altered hydrological regimes, biodiversity loss, and socio-economic shifts that challenge the sustainability of traditional livelihoods (Ali et al., 2022). Due to extreme climatic conditions and remoteness communities have developed intricate adaptive practices rooted in indigenous knowledge systems, collective resource management, and seasonal mobility in cold deserts (Mahdavi et al., 2023). Building on existing scholarly literature in climate change resilience, biodiversity conservation, and sustainable forest management, this chapter introduces two under-represented but critical lenses: (1) the historical and cultural context of cold deserts and (2) local tribes' assistance in sustaining forests as ecosystems in midst of global change. Hence, this study focuses on community-based adaptation, traditional knowledge systems, and participatory conservation of forest resources in the Johar valley of Uttarakhand.

2 Materials and Methods

2.1 Study Area

The present study was undertaken in the Johar valley of Pithoragarh district in Kumaon region of Uttarakhand state in India. It represents a unique socio-ecological landscape, predominantly inhabited by the Rung, Shauka, and Barpatiya communities in different parts of Munsyari block of Pithoragarh district. These groups have historically developed intricate adaptation strategies to cope with the challenges of high-altitude living in high cold desert of Johar valley (upper part of the Munsyari block), including climatic extremes, short growing seasons, limited arable land, and restricted access to markets and infrastructure. Their socio-economic practices reflect a deep integration of ecological knowledge and cultural tradition, enabling sustainable use of the region's scarce natural resources. This high-altitude mountain valley is distinguished by pronounced elevational gradients. A prominent chain of five peaks exceeding 6,300 m collectively known as the Panchachuli range forms a natural

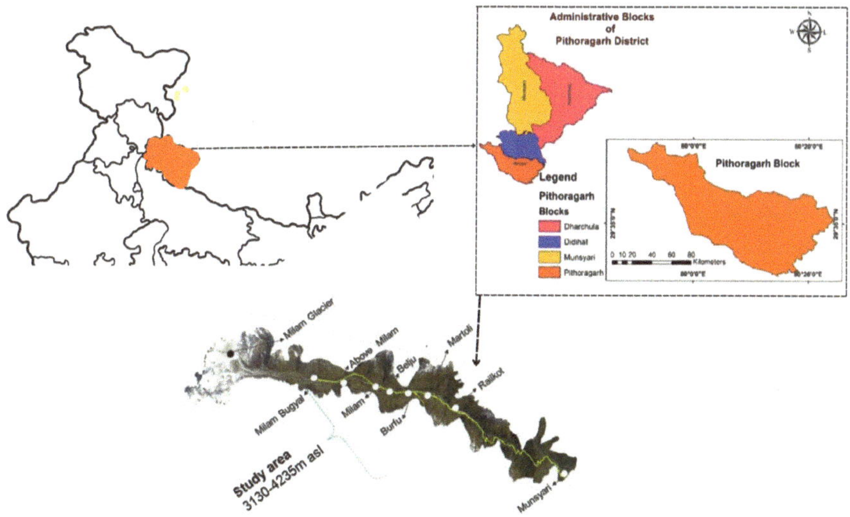

Fig. 1 Map of the study area depicting the transhumance villages

divide between the Gori and Darmaganga river basins. The glacier-fed Gori Ganga River descends from the glaciated upper catchments to elevations below 1,000 m near the town of Munsyari (Fig. 1). A broad spectrum of organisms and plants is supported by the region's significant microclimatic variability, which is triggered by changes in slope orientation and aspect due to its difficult topography.

The Johar valley exemplifies the biophysical characteristics of a cold desert ecosystem. Positioned within the trans-Himalayan rain shadow of the Panchachuli massif and sharing a boundary with the Tibet Autonomous Region, the valley spans an altitudinal range from approximately 2,500 m in its lower reaches to over 6,000 m near its glaciated headwaters (Pandey et al., 2019). The combined effects of high elevation and orographic shielding resulted in minimal summer monsoonal precipitation, while substantial snow accumulation occurs during the winter season. Meltwater from surrounding glaciers and snowfields constitutes the primary source of discharge for the Gori Ganga River, which functions as the principal hydrological artery of the valley (Parihar, 2021). Climatic conditions are characterized by marked seasonality, with winter temperatures frequently dropping below −20 °C and short, cool summers with daytime temperatures rarely exceeding 20 °C.

2.2 *Ecology and Vegetation*

Ecological zonation within the Johar valley is closely aligned with altitudinal gradients. The upper reaches are dominated by glacial landforms such as moraines and periglacial scree fields, interspersed with alpine meadows. Vegetation in these zones is sparse and specialized, consisting mainly of cold-adapted herbaceous species, grasses, and cryptogams like lichens (Rawal et al., 2018). The landscapes of the Johar valley exhibit a high degree of ecological diversity, shaped by altitudinal gradients and the region's distinct climatic conditions. Despite its cold, arid, and high-elevation environment, the valley supports a surprisingly rich biodiversity, including both endemic and highly specialized species adapted to extreme abiotic stresses. The native vegetation comprises hardy woody species, such as *Juniperus* spp. (juniper), *Hippophae rhamnoides* (seabuckthorn), and alpine herbs, well-suited to nutrient-poor, shallow soils and extended snow cover. The regions also support a high diversity of medicinal plants and non-timber forest produce (NTFP), which are essential for local healthcare systems and subsistence economies. Alpine species such as *Rheum australe* (wild rhubarb), *Picrorhiza kurroa* (kutki), and *Aconitum heterophyllum* (Atish) are especially valued for their medicinal properties and are traditionally harvested by local communities for both therapeutic use and economic exchange (Acharya et al., 2012).

Johar valley is equally notable for its faunal richness, providing critical habitat for a number of rare and threatened high-altitude species. Among the most emblematic is *Panthera uncia* (snow leopard), a keystone predator inhabiting the upper rocky and glaciated terrains. Other key mammals include *Capra sibirica* (Himalayan ibex), *Pseudois nayaur* (bharal or blue sheep), and *Vulpes vulpes* (red fox), all of which are well-adapted to the valley's rugged landscapes and climatic extremities (Shrestha & Wegge, 2008). Charismatic fauna such as *Ursus thibetanus* (Himalayan black bear), *Moschus leucogaster* (musk deer), and the elusive snow leopard are found in the upper elevations and serve as indicators of ecosystem health (Koju et al., 2023). The avifaunal diversity includes high-altitude species such as *Lophophorus impejanus* (Himalayan monal), the state bird of Uttarakhand, and *Tetraogallus tibetanus* (Tibetan snowcock), which are common in alpine meadows and scree habitats (Rawat, 2017). Additional avian taxa include various species of pheasants, partridges, and migratory birds that rely on the forest's structural diversity for breeding and foraging.

Integral to the region's cultural landscape is seasonal alpine meadows locally known as *Bugyals*, which form the foundation of traditional transhumant pastoral systems. These high-altitude grasslands become accessible during the summer thaw and serve as essential grazing grounds for livestock such as sheep, goats, and yaks. For generations, local communities have practiced rotational grazing across these meadows, an adaptive livelihood strategy that aligns ecological conservation with subsistence needs (Gaur et al., 2024). These dynamic human–environment interactions exemplify the valley's ecological complexity and the embedded cultural knowledge systems that underpin its sustainable use. With this backdrop, the pronounced ecological gradients in Johar valley provide an ideal natural laboratory for examining the interactive effects of elevation, climate variability, and anthropogenic land use. Consequently, the valley serves as a representative model system for understanding the dynamics of cold desert ecosystems under the pressures of global environmental change.

2.3 *Data Collection, Processing, and Analysis*

The fieldwork conducted between September and November 2023 across 15 villages in the Johar valley. In coordination with *Himal Prakriti*, a local civil society organization, the study gained critical insights into forest governance management, i.e., van-panchayats, traditional ecological knowledge, and socio-cultural values associated with natural-resource use. Around 150 individuals were approached for a series of open-ended interviews which addressed topics such as livelihood practices, traditional practices, community difficulties, and forest management (Fig. 2). The sampling followed a purposive strategy designed to capture a diversity of perspectives from knowledge holders, including forest council leaders (van panchayat members), elders, women, ecotourism professionals, and local environmental NGO personnel (Kar et al., 2025; Kumar, 2022; Small, 2009). Participants included members from General Caste, Scheduled Castes, and Scheduled Tribes, encompassing a range of socio-economic backgrounds. Data collection continued until thematic saturation was reached. Interviews were conducted primarily in Hindi, with occasional use of English phrases and the Kumaoni dialect. All interviews were audio-recorded, translated into English, and transcribed during the fieldwork.

Ten participants were selected from each village, (male or female over the age of 18 years old) from each household, resulted in 92 women and 58 men between the ages of 22 and 84. While most interviews were audio-recorded, detailed notes were also taken during all sessions. The open-ended format allowed for the emergence of unanticipated themes and deeper insights with each interaction. To further contextualize the data, we participated in local heritage museum visits, guided forest walks, and excursions to sacred sites. These immersive experiences enriched the qualitative

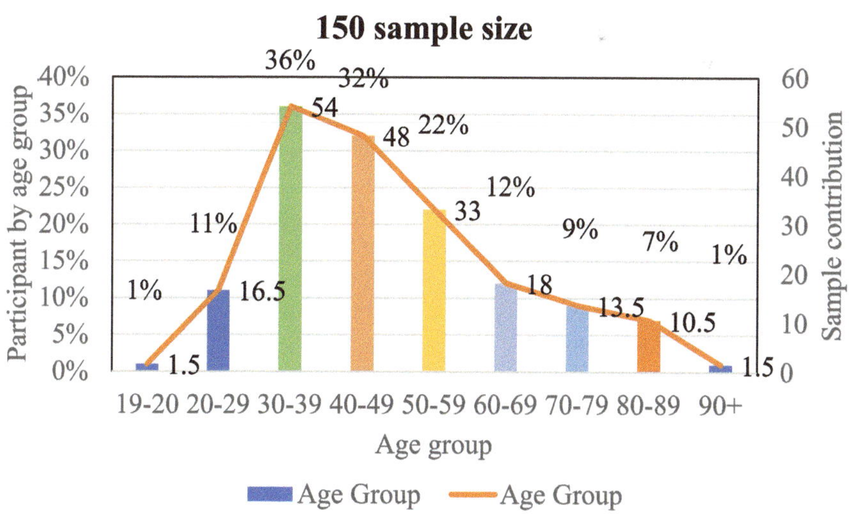

Fig. 2 Age group distribution of participants

dataset and illuminated how local worldviews shape perceptions of natural resource stewardship and conservation.

3 Results and Discussion

3.1 The Historical, Livelihoods, and Socio-cultural Setup of the Johar Valley

The Johar valley holds a prominent place in the history of the trans-Himalayan trade networks, serving as a critical corridor that connected India and western Tibet for centuries. This historical bond was deeply connected with the culture of both countries. From the early medieval period up until the mid-twentieth century, the Bhotiya communities of Johar primarily from villages like Martoli and Milam facilitated the movement of goods such as salt, wool, borax, livestock, grains, and jaggery across the formidable Himalayan terrain (Kumar & Sati, 2023). These trade routes were not merely socio-economic lifelines but also vectors of cultural exchange, linguistic blending, and religious syncretism, fostering long-standing relationships between Indian and Tibetan communities (Singh, 2025). The interdependence between agriculture, pastoralism, forest-based resource use, and artisanal production reflects adaptive strategies developed by local communities over generations to cope with ecological constraints and socio-economic isolation.

Seasonal migration was an essential feature of this trade-based lifestyle. Johari traders had to travel over high mountain passes such as the Unta Dhura and Kingribingri La, spending summers in the Tibetan plateau and returning to the Indian valleys in winter. This pattern of transhumance and trade shaped not only the local economy but also social institutions and cultural practices. Trade caravans facilitated the spread of Buddhist and Hindu religious traditions, artisanal crafts, and knowledge systems that are still embedded in local culture today (Bora, 2025).

The closure of the Indo-Tibetan border following the 1962 Sino-Indian War abruptly ended this centuries old exchange, resulted in major socio-economic transformations. Traditional trade-based livelihoods were disrupted, leading to increased outmigration, reliance on subsistence agriculture, and gradual shifts in cultural identity (Bora, 2025). Nonetheless, the legacy of trans-Himalayan trade continues to shape local memory, place-based identity, and ongoing conversations around heritage and development in Johar valley.

Trade and Economy: The Johar valley's prehistoric trans-Himalayan trade was essential to support the local financial system while providing substantial resources to an area that would otherwise have been hampered by its meager agricultural potential. The trading of commodities like salt, wool, and wheat, and animals sustained an array of livelihood customs, such as seasonal employment, handicrafts, and pastoralism, in addition to providing necessities (Bhatt et al., 2009). This trade economy enabled the Johar communities to maintain a degree of economic resilience, despite the harsh climatic and topographic constraints, fostering social cohesion and cultural vibrancy in the valley.

3.1.1 Culture and Communities of Cold Deserts

The cold desert inhabitants have developed nuanced socio-ecological systems that integrate pastoralism, agriculture, trade, and spiritual practices deeply tied to the landscape (Bora, 2025). These communities have cultivated a deep understanding of climatic variability, seasonal rhythms, and resource limitations, often expressed through rotational grazing, water-sharing customs, and local festivals that mark ecological transitions (Bhatt et al., 2009). The indigenous knowledge systems play a crucial role in guiding adaptive strategies to environmental stressors (Sospeter et al., 2025). These include the use of cold-resistant crop varieties, the practice of transhumance, and the community management of forests and water through traditional governance institutions (Rawat et al., 2025). Such systems are not only ecologically sensitive but are also socially embedded, emphasizing collective responsibility, intergenerational learning, and spiritual reverence for nature.

3.1.2 Livelihood Patterns

The valley's high-altitude environment, characterized by low productivity, climatic variability, and limited infrastructure, has shaped a diversified subsistence-oriented livelihood system. Agriculture is practiced at altitudes ranging from 2,000 to 3,000 m in the Johar valley, relying on hardy crops such as barley, buckwheat, and potatoes, which are suited to the valley's short and intense growing season. Animal husbandry plays a vital role, involving the rearing of sheep, goats, yaks, and cattle, which provide meat, milk, wool, and traction for plowing. Livestock also serve as a critical form of mobile capital, particularly in times of economic uncertainty (Negi, 2007a, 2007b).

In addition, forest-based resource extraction including the collection of fuelwood, fodder, wild edibles, and medicinal plants remains a key component of household economies. These practices are often guided by indigenous ecological knowledge (IEK), which governs seasonal harvesting, species selection, and habitat protection. The sustainable use of these resources has historically ensured both ecological stability and community food and health security.

The data indicates a significant and growing reliance on natural resources in the Johar valley, driven in part by both subsistence needs and limited access to modern alternatives. Fuelwood is the most heavily utilized resource, with 87% of respondents depending on it as a primary source of domestic energy. This high dependency reflects the lack of accessible and affordable alternatives such as LPG or electricity in remote high-altitude areas. Similarly, 63% of households collect leaf litter for composting and 59% harvest grass for livestock fodder activities that are integral to agro-pastoral systems but increasingly intensive due to growing household demands and shrinking natural availability (Fig. 3).

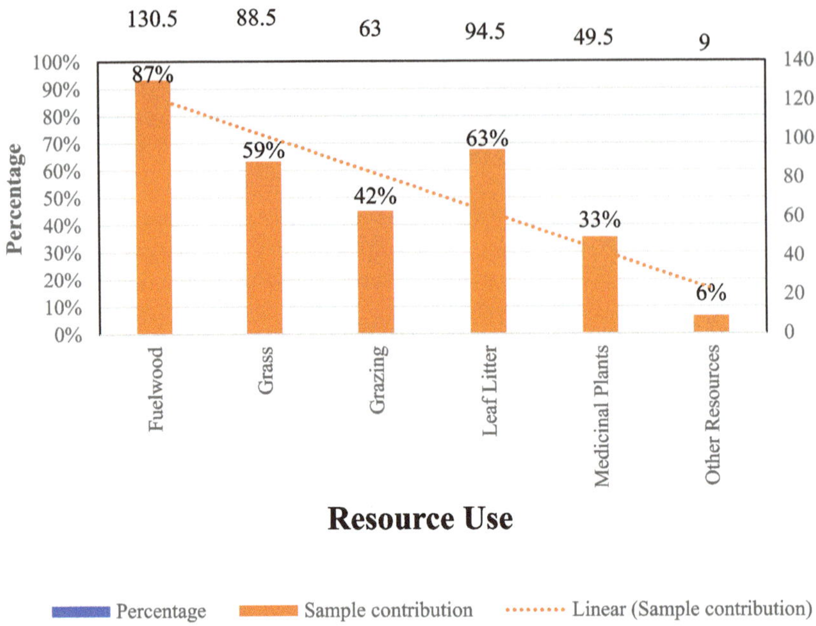

Fig. 3 Forest resources use pattern

Grazing, reported by 42% of respondents, continues to be a vital land-use practice, though it often overlaps with forested and alpine regions that are ecologically sensitive and prone to degradation. The collection of medicinal plants by 33% of respondents further emphasizes the importance of traditional ecological knowledge and the economic value of non-timber forest products (NTFPs). Meanwhile, 6% of respondents indicated the use of other forest resources, suggesting that while resource use is concentrated around a few key categories, broader forest dependence persists.

The increased frequency of natural resource use, particularly in the context of a fragile Himalayan ecosystem, raises concerns about sustainability, forest health, and long-term ecological balance (Bhatt et al., 2024). These trends highlight the urgent need for community-based forest governance, sustainable harvesting practices, and diversification of energy and livelihood sources to reduce pressure on natural ecosystems.

3.1.3 Seasonal Mobility and Transhumance

The communityof Johar valley practice transhumance, which is a seasonal migration of households and livestock between high-altitude summer settlements (*Bugyals*) and lower-altitude winter villages in the foothills and valleys. Key summer settlements, also known as Bugyals or alpine pastures in the Johar valley include villages such as Milam, Burfu, Martoli, Rilkot, Ganghar, and Lwan, which are inhabited during the warmer months (May–October) when snow recedes and grazing lands become accessible. These high-altitude hamlets lie in proximity to the Indo-Tibetan border and are often used for grazing livestock, collection of medicinal plants, and cultivating short-season crops like barley and buckwheat. During the harsh winter months (November–April), these communities descend to Mid Johar and Talla Johar, occupying lower-altitude villages such as Bui, Sarmoli, Suring, Darkot in Mid Johar and Chirkila, Sobla, Dhapa, Sosa, and Balwakot in Talla Johar. These winter settlements offer milder climates and proximity to markets and administrative centers, enabling households to sustain livelihoods through agriculture, animal husbandry, and trade. The cyclical movement between these altitudes forms the basis of vertical transhumance, a practice that not only sustains pastoral and agricultural productivity but also conserves ecological balance by allowing time for natural regeneration in alpine zones.

Animal husbandry is a key livelihood activity that complements subsistence agriculture and mitigates climatic and market risks. The communities rear a combination of goats, sheep, yaks, and cattle, which provide milk, meat, wool, hides, and manure. During summer, livestock is moved to high-altitude alpine pastures (*Bugyals*) toward Malla Johar in 12–14 transhumant villages, while in winter, they are sheltered in lower elevations (Talla Johar, i.e., Munsyari) to avoid harsh weather and snow-covered terrain (Srichandan et al., 2021). This cyclical migration not only optimizes forage availability but also reduces the pressure on local pasturelands, enabling ecological regeneration (Fernández-Guisuraga et al., 2022). However, shifts in pasture phenology due to rising temperatures, the encroachment of invasive plant species, and reduced snowfall have begun to disrupt transhumant patterns (Ingty et al., 2023). Simultaneously, socio-economic changes such as youth outmigration and the declining cultural transmission of herding knowledge have contributed to the gradual erosion of this adaptive system. However, the vertical mobility system enables the optimal utilization of grazing and agricultural resources, minimizes pressure on any single ecological zone, and reduces vulnerability to seasonal climatic extremes. Transhumance is also essential to safeguarding continuity in culture and cohesiveness in society since patterns of migration have become embedded in customs and conventions (Mannia, 2024).

Over the past two decades, transhumance has witnessed a marked decline. Drivers of this shift include changing climatic patterns (e.g., decreased snowfall and glacial melt affecting pasture availability), infrastructural development (e.g., road access reducing the need for seasonal relocation), and socio-economic transformations such as outmigration, education, and employment in urban areas. These changes have led to the semi-sedentarization of many households and a growing dependence on external income sources, which in turn affects the management and regeneration of high-altitude forest and pasture ecosystems (Pasakhala et al., 2021).

3.1.4 Subsistence Agriculture and Agro-Ecological Resilience

Agriculture in the Johar valley is predominantly subsistence-based and employs agro-ecological practices adapted to the region's steep terrain, thin soils, and short growing season. Terraced fields between 2,000 and 3,200 m above sea level are used for growing crops such as barley (*Hordeum vulgare* L.), buckwheat (*Fagopyrum esculentum* Moench), potatoes (*Solanum tuberosum* L.), and legumes like kidney beans (*Phaseolus vulgaris L.*) and peas (*Pisum sativum* L.). These crops are selected for their tolerance to frost, drought, and nutrient-poor soils, thus enhancing the resilience of local food systems (Patel et al., 2024).

Traditional cropping systems often incorporate crop rotation, mixed cropping, and the use of organic fertilizers such as livestock manure, which contribute to soil fertility and erosion control. However, climate-induced changes including erratic snowfall, shorter winters, and increased incidence of extreme weather events are undermining agricultural productivity. Observations from the region suggested delayed sowing periods, reduced crop yields, and increased vulnerability to pest outbreaks, all of which threaten household food security (Kumar et al., 2025). The agricultural system that needs to be adopted is chosen by the transhumant migrants' limited cropping season. Therefore, only commodities that ripe in 3.5–4 months are farmed. The earliest start of a crop season is rarely leveraged by the people since upward migration is controlled by other factors, primarily because of the existence of undamaged trails that allow families to move with their cattle and other essentials (Nyssen et al., 2009).

One estimate places the number of crops or vegetables and medicinal plants currently farmed in these settlements at about 21, all of which are seasonal (Tables 2 and 3). Fields for these crops are usually ploughed in October every year, right before the migration lower to the winter dwellings. This is done mainly to keep the majority of the moisture from the snow melt in the soil and to prevent the growth of weeds and soil-borne pests. When they arrive the next year, the fields are ploughed once again to soften the soil, weeds are pulled, and manure in whatever form is available is spread.

Two high-value, intensively grown crops stand out among the overall number of crops grown: caraway (*Carum carvi*) and jambu (chives, *Allium stracheyi*). An expensive perennial plant, jambu (*Allium spp.)*, costs between 150 and 200 rupees at the Munsiari local market. In Johaar valley, this crop and caraway occupy the majority of the arable ground. These crops can be harvested twice during the extremely short growing season and yield more per unit area.

Allium humile, *A. wallichi*, and *A. stracheyii* are the three species that are cultivated in the valley. Continuous propagation of *Jambu* from the same rootstock ensures that neither quantity nor quality is lost. This simplifies farming, despite the laborious post-harvest processing (Marín et al., 2023). In late May or early June, harvesting takes place, and then again in late September or early October. This valley is the site of the biennial caraway crop. The Foundation for Ecological Security, an NGO based in Munsiari, estimates that the Johar valley alone produces about 1100 kg of caraway seeds a year, which adds 1.76 lacs to agricultural revenue.

Although the crop can only be harvested after two years, people are dedicating a larger percentage of their cultivable land to it because it sells for 250 rupees per kilogram. In these farms, crop rotation is not practiced, and the following seeds are sown right away after each harvest. The earth receives the essential nutrients from the caraway stalks.

3.1.5 Agricultural Evolution and Lifestyle Dynamics

Currently, massoor (*Vigna spp.*), owa or jau (*Hordeum himalayens*), and nappal or durum wheat (*Triticum spp.*) are among the major and traditional crops (Table 1) with very little acreage that are under danger in the area because of a lack of interest. These three products were in high demand during the height of commerce with Tibet; nappal was enormously produced as sattu (flour). Since the villages' populations at the time frequently numbered in the hundreds, Jau was used to make and sell the traditional local beverage, chaktiki. People used to arrive to their summer residences at the end of March and stay until late October, so they had time to grow, harvest, and trade nappal, which was the main reason for its cultivation.

3.1.6 Forest Resources and Ethnobotanical Knowledge System

Communities in Johar valley possess extensive ethnobotanical knowledge while utilizing a diverse range of alpine and subalpine plant species for health, food, and ritual purposes. Species such as *Artemisia maritima, Picrorhiza kurroa*, and *Nardostachys jatamasi* are used in traditional medicine and have cultural as well as economic value (Bisht et al., 2021; Kaur, 2020) (Tables 2 and 3). The harvesting practices have traditionally been governed by local norms and panchayat regulations, with rules designed to ensure seasonal restrictions, rotational harvesting, and non-destructive collection methods. However, climate-induced changes in species distribution, reduced plant regeneration rates, and increased commercialization of non-timber forest products (NTFPs) are putting pressure on these fragile ecosystems. Additionally, forest degradation linked to increased fuelwood extraction and reduced community monitoring, due to depopulation and weakened customary institutions, threatens the sustainability of these resources.

3.1.7 Crafts, Trade, and Emerging Economies

Beyond subsistence, artisanal production and trade provide critical avenues for income generation. The Shauka and Rung subsets of the Bhotiya community have a rich tradition of wool-based handicrafts, including the production of carpets, shawls, and woolen garments using yak and sheep wool. These crafts are deeply embedded in the region's cultural heritage and are traditionally produced during the winter

Table 1 Major crops in the Johar valley after Negi (2007a)

Crop species	Vernacular name	Area (ha) 2002	% of total area	Production (kg) 2001
Fagopyrum esculentum Moench	Oggal/palthi	0.020	0.09	18
F. tataricum	Phaphar	4.380	21.03	2975
Hordeum himalayens	Owa-jau	0.166	0.80	60
Solanum tuberosum	Alu	3.950	18.97	18,556
Brassica campestris	Sarson/mustard	6.960	33.42	3683
Triticum spp.	Nappal/durum wheat	0.218	1.04	40
Vigna spp.	Massoor	0.086	0.41	19
Pisum spp.	Kalon	0.004	0.02	7
	Other vegetables	0.012	0.06	–
	Total food crops	15.795	75.85	
Allium spp.	Jambu/chives	3.499	16.80	2431
Carum carvi	Thoya/caraway	1.328	6.38	1121
	Total		4.827	23.18
Angelica glauca	Gandarayan	0.137	0.66	15
Aconitum heterophyllum	Atis	0.024	0.12	5
Rheum emodi	Dolu	0.005	0.02	6
Sassurea costus	Koot	0.030	0.15	–
Picrorhiza scrophulariiflora	Kutki	0.005	0.02	–
	Total	20.823	100.00	

months when agricultural and pastoral activities are limited. With the decline of cross-border trade post-1962 and increased market integration, many households have shifted toward the commercial sale of handicrafts through cooperatives and seasonal tourism. Furthermore, the collection of high-value NTFPs particularly *Ophiocordyceps sinensis* (commonly known as *Keeda jadi* or caterpillar fungus) has emerged as a significant, though ecologically contentious, income source (Laha et al., 2018). This fungus, prized in international markets for its purported medicinal properties, is harvested from alpine meadows during the pre-monsoon season.

While *Ophiocordyceps* collection has boosted local cash incomes, it has also led to ecological stress due to overharvesting, habitat trampling, and intensified competition. Climate change has further affected its availability, with earlier melting of snow and shifting phenological windows impacting its life cycle. Without clear regulatory frameworks and sustainable harvest guidelines, there is a risk of resource depletion and social conflict over access rights.

Table 2 Medicinal values of some traditional crops Negi (2007a)

Scientific name	Local name	Chemical constituents	Medicinal uses
Avena sativa	Jau	Saponin, hordenine, glucosides	Seeds are an antispasmodic, stimulant, and nerve tonic. Beneficial to the kidneys and heart. Also employed in the economic exploitation of the fungus ergot
Fagopyrum esculentum	Oggal/palthi	Rutin, fagopyrin, flavanones, glycosides	Headaches are treated with leaf paste, while fever is treated with fried leaves in ghee. Grains for stomach problems, choleric diarrhea, and colic
Fagopyrum tataricum	Phaphar	Same as above (Rutin-rich)	Bark used as an astringent; known as a good source of rutin
Macrotyloma uniflorum	Gahat	Prolamine, gluteline, niacin, riboflavin	Grain soup traditionally used to treat kidney stones
Panicum miliaceum	Cheena	Essential oil, sulfur	Traditional Baluchistani medicine uses grains cooked with curd to treat gonorrhea and jaundice
Allium stracheyi	Jimbu	Organic compounds (unspecified)	Used as a condiment. Ground with Saussurea costus root and fried in ghee for relief from stomach aches
Carum carvi	Thoya/kala jeera	Umbelliferone, scopoletine, glycosides, essential oil	For stomach ache, consume seeds that have been seasoned with salt. It is also frequently used as a spice and a therapeutic condiment

Insights from the livelihood data reveal that the economy of the Johar valley remains predominantly agrarian, with 95% of households relying on agriculture as a primary source of subsistence and income. This is followed by animal husbandry, which supports 71% of respondents and serves as an essential complementary activity to farming (Fig. 4), particularly in high-altitude regions where livestock are integral to both food security and manure-based soil fertility. The collection and trade of medicinal plants constitute a significant non-timber forest product (NTFP)-based livelihood for 29% of the population, reflecting both ecological knowledge and dependency on local biodiversity.

3.1.8 Food Preservation Practices in the Johar Valley

The Johar valley poses significant challenges to agricultural practices and food storage likewise other high-altitude areas (Roy et al., 2004). Seasonal migration and transhumance are historically tied to these constraints; however, for those remaining in the valley during winter, food preservation is essential. Traditional

Table 3 Medicinal plants are used for domestic and commercial purposes

Scientific name	Local name	Part used	Folk medicinal uses
Saussurea costus (Falc) Lipsch	Kut	Root	For toothaches, jaundice, snakebite, and skin conditions, apply root paste
Cannabis sativa L	Bhaang	Leaves	For cuts, piles, and skin ulcers, apply leaf paste Use of seed oil for muscle soreness and burns
Rhododendron arboreum sm	Burans	Flower	To treat bleeding nose, use flower paste. Use warm leaves in oil to treat arthritis. Seed paste for wounds and boils Jaundice is treated with leaf juice, butter, and pepper. Flower juice is a heart tonic and blood pressure medication
Swertia chirayita *(Roxb.)*	Chirayata	Whole	Fever, hyperglycemia, and dyspepsia are treated with fresh leaf and stem juice
Aesculus indica (Wall. ex Cambess.)	Pangar	Fruits	As a galactagogue, goats are given ground fruits External fruit paste is used to treat sore throats Oil and seed paste are administered to rheumatism pain
Podophyllum hexandrum Royle	Bankakari	Fruits, root	Ripe fruit was consumed to treat dyspnea. Use of root paste for wounds and cuts
Rheum emodi Wall. ex Meisn	Dolu	Root	Mustard oil and dried leaf powder are combined and administered externally to cuts, boils, and wounds
Dactylorhiza hatagirea (D. Don) Soo	Hatajadi	Tuber	Tuber paste is used to treat fever and diarrhea and applied to cuts and wounds
Aconitum balfourii Stapf	Meetha Atish	Tuber	Snakebite is treated with tuber paste
Aconitum heterophyllum Wall. ex Royle	Atis	Root	For headaches, root paste is utilized. Milk and root powder are used to treat fever and stomachaches
Zanthoxylum armatum DC	Timoor	Bark, fruits	Toothaches are treated with fruit powder and bark paste As a toothbrush, branches are utilized. Seeds for seasoning
Picrorhiza kurroo Royle ex Benth	Kutki	Root	Typhoid, jaundice, and stomachaches can be treated with sugar and dried roots

ecological knowledge embedded in the food systems of the Bhotiya tribal communities offers insight into adaptive strategies for surviving in extreme environments. Sun drying is a primary preservation method employed during the short summer and autumn harvest period. Leafy greens (e.g., *Brassica juncea*), tubers (*Colocasia esculenta*), and flowers (e.g., *Rhododendron arboreum*) are pre-treated via blanching and then dehydrated under solar radiation for 10–20 days (Kaushal et al., 2013). This significantly reduces water activity, thus inhibiting microbial growth and enzymatic spoilage. Preserved and fermented foods in Johar valley are not merely utilitarian

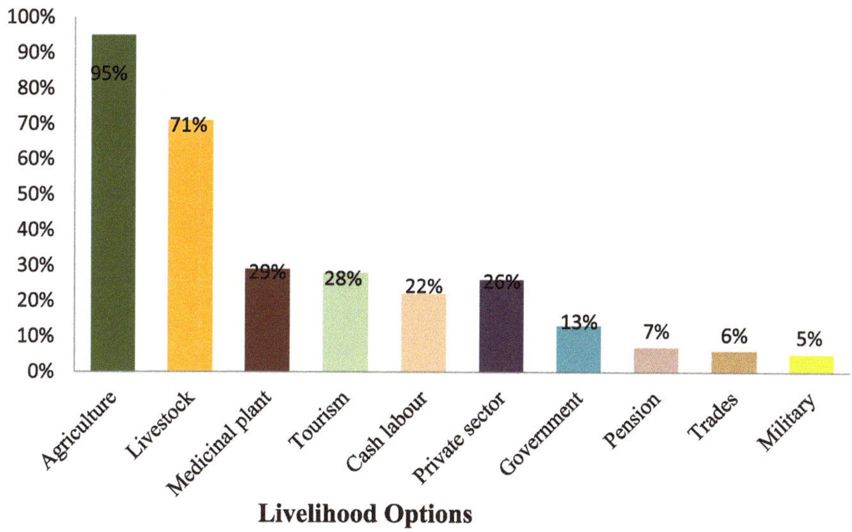

Fig. 4 Respondent's livelihood options

but are interwoven with cultural identity, rituals, and community resilience. They ensure dietary diversity during food-scarce winters, offer probiotic benefits, and reduce dependency on external food systems. These practices reflect in an intricate understanding of local ecology and microbial dynamics (Dwivedi et al., 2024). Food preservation in the Johar valley represents a sophisticated adaptive response to environmental extremities. Traditional knowledge systems, particularly regarding alcohol fermentation and meat preservation, provide sustainable, low-energy models of food security that may inform broader cold-climate adaptation strategies under climate change scenarios.

3.1.9 Meat Preservation Techniques

The traditional meat preservation is deeply rooted in the region's high-altitude, Tibetan-influenced heritage. Animal proteins, primarily from goat and sheep, are preserved through a combination of salting, oiling, and drying. Strips of meat are coated in turmeric (*Curcuma longa L.*), rock salt, and mustard oil (*Brassica juncea L.*), and suspended over the kitchen hearth. Fresh goat or sheep meat is cut into cubes, smeared with turmeric, salt, and mustard oil, and hung in cold, dry air to air-dry for several days (Roy et al., 2004). Partially dried meat is then cold-smoked over a wood fire (often in the kitchen), infusing flavor and adding antimicrobial protection. The combination of drying spicing draws moisture out, slows bacterial growth, and maintains hygiene in off-grid homes. This yields sukha meat (dry meat), which can be stored for months and later rehydrated and cooked into stews with local vegetables. In colder valleys like Johar, raw meat strips are coated in turmeric and salt, threaded on strings, and hung indoors, where smoke further enhances preservation and taste. This process not only prevents spoilage but imparts distinct organoleptic properties suited for cold climates.

3.1.10 Alcoholic Fermentation for Self-use

Traditional fermented beverages, such as *Jaan* (mild cereal beer) and *kacchi* (distilled liquor), are culturally and nutritionally significant. *Jaan* typically contains 3–5% alcohol by volume (ABV), while *kacchi*, distilled using indigenous apparatus, can exceed 15% ABV (Tamang, 2010) (Table 4). These beverages are rich in B-complex vitamins and organic acids and serve ritualistic as well as thermogenic functions during winter months (Majumder & Bhattacharya, 2024). The Bhotiya community has mastered the skill of manufacturing distinctive alcohol-based beverages that symbolize multiple aspects of their cultural identity while simultaneously functioning as social mixing equipment. These kinds of drinks are the most appropriate refreshments to have during liturgical or hedonistic rituals because of their high nutritional value and delicious flavor (Tamang, 2010). Often, the choice of raw materials is based on what is easily accessible in the region, including fruits, grains, or herbs. The type of vessel used, the length of the fermenting process, and the use of local herbs or spices are further characteristics that set Bhotiya traditional alcohol apart. The tribe's harmonious connection with the land and their profound awareness of their natural environment is apparent at every stage of the process. Wine consumption and serving blossomed became a familial tradition that was used for both rehabilitative and savory purposes (Mekita, 2018).

3.2 The Ecological and Climatic Dynamics Characterization of the Johar Valley

The Johar valley is undergoing profound ecological and socio-economic transformations driven by climate change. One of the most visible changes is the accelerating retreat of glaciers, such as the Milam and Panchachuli glaciers, which directly threatens long-term water security for downstream communities (Bolch et al., 2012). These glaciers are the primary sources of the Gori Ganga River, and their retreat has reduced the volume and reliability of melt water during the crucial pre-monsoon period, impacting irrigation, drinking water, and hydrological balance.

Simultaneously, reduced snowfall and increasingly erratic precipitation are disrupting traditional seasonal cycles, essential for both winter preparedness and summer crop cultivation. The unpredictability of snowfall limits the ability of communities to store water or plan sowing seasons, and in years of below-average snow accumulation, soil moisture levels are insufficient to support even short-season

Table 4 List of traditional fermented alcoholic beverages in Johar valley with their dialect name (Kishor et al., 2013)

S. No	Name of beverage (in dialect)	Region of Uttarakhand/ community	Fermented medium	Entrepreneurial culture	Sort of beverage
1	Jann	Kumaun Himalayas/ Johar valley	Cereals comprise grains such as rice (*Oryza sativa*), grains such as wheat (*Triticum aestivum*), and barley. Millets of various kinds feature the grains proso millet (*Panicum miliaceum*) with the foxtail millet (*Setaria italica*)	Balam/Balma	A mild beverage similar to beer
2	Daru	Kumaun Himalayas/ Johar valley	Cereal: jaggery, rice, jaggery for mild sweet flavor	Balam	Alcoholic beverage with a strong distillation
3	Kacchi	Kumaun Himalayas/ Johar valley/ Dharchula, Garhwal Himalayas	Cereal: seeds of barley	Balam	Hard alcohol that has been distilled
4	Chyakti	Kumaun Himalayas/ Johar valley/ Dharchula	Cereal: sugar from wheat or barley seeds	Balam/Balma	Without distillation

crops (Liang et al., 2020). This is particularly concerning for subsistence farmers who rely on single-season agriculture at high altitudes.

Furthermore, wildlife habitats are shifting upward due to rising temperatures, leading to increased encounters between humans and wild species such as snow leopards, black bears, and bharal (blue sheep). These shifts have intensified human–wildlife conflict, particularly in areas where pastoralists and wild herbivores compete for shrinking grazing lands (Acharya et al., 2017). Such ecological changes underscore the intricate interdependencies in cold desert ecosystems and highlight the urgency of developing community-centered, climate-resilient adaptation strategies.

3.2.1 The Johar Valley's Climate Change Manifestations

Mountain dwellers are particularly vulnerable to the effects of climate change, which include altered precipitation and temperature patterns, a rise in the frequency and intensity of natural disasters (such as flooding), and ongoing obstacles to adaptation (Adler et al., 2023; McDowell et al., 2021; Palomo et al., 2021; Tiwari et al., 2020). At the same time, considerable social instability is being brought on by trends of urbanization, outmigration, and resource commercialization in mountain areas, including the HKH (Gagné, 2019; Naudiyal et al., 2019). According to the respondents' opinions and graphical depictions of how climate change is influencing the local community's nutrition and means of subsistence, 95% of the community's primary source of income comes from agriculture (Table 5). Table 5 also shows the respondents' considerations of changing farming practices due to the following proportion. While 96% of respondents observed a gradual decline in snowfall and a change in snowfall patterns that impacted agricultural productivity, all respondents (100%) viewed a drop in snowfall as a direct consequence of their comprehension of environmental and climatic change.

3.2.2 Erratic Snowfall and Precipitation Patterns

One of the most pronounced indicators of climate change in the Johar valley is the increasing irregularity of snowfall and rainfall patterns. Historically, snowfall during the winter months served as a natural reservoir, gradually melting during spring and summer to sustain agricultural irrigation and groundwater recharge. However, over the past two decades, communities have observed a marked delay in snowfall onset, reductions in total accumulation, and heightened year-to-year variability (Malik & Ford, 2025). In parallel, monsoonal and pre-monsoonal rainfall events have become more sporadic and intense, often leading to flash flooding and soil erosion while failing to provide consistent agricultural support (Chuphal et al., 2025).

These hydro-meteorological changes undermine the temporal predictability of water availability, which is critical for crop selection, sowing schedules, and pasture management (Sati, 2021). Crops that depend on predictable snowmelt such as barley and pulses are particularly affected, leading to yield reductions and growing food insecurity in isolated hamlets. In essence, the climatic stability that once underpinned the valley's agro-pastoral systems is rapidly eroding.

3.2.3 Degradation of Alpine Pastures

Rising mean temperatures, shifting precipitation regimes, and changes in snowmelt timing have altered the phenology, productivity, and species composition of the alpine pastures (Kumar & Pathak, 2025). Warmer conditions favor the encroachment of non-palatable, woody, or invasive species, such as *Artemisia* and *Polygonum*, displacing traditional fodder species like *Festuca* and *Poa*. The net result is a decline in pasture quality and biomass availability, which negatively impacts livestock health, milk yields, and overall herd productivity. Additionally, reduced pasture regeneration rates due to altered snowpack dynamics limit the capacity for traditional rotational grazing to restore ecosystem balance (Oyinbo et al., 2019).

Table 5 Distribution of respondents according to their knowledge about the climate change and changes in environment ($n = 150$)

Observed changes by respondents	Statements	Respondent percentage (%)	No. of respondents
Rainfall	Unsettled rainfall timing	45	67
	Escalating rainfall	50	75
	In contrast to the season	3	5
	A situation similar to flooding	2	3
	No changes	0	0
Snowfall rate	Increasing	0	0
	Decreasing	100	150
Water resources	Increasing	3	5
	Decreasing	95	142
	Stayed the same	2	3
Water condition	In good condition as previously	82	123
	When it rains, insects are present	14	21
	The worst state	4	6
Cloud burst like activities	Increasing	59	88
	Stayed the same	21	31
	Decreasing	2	3
	Never heard of this term before	5	8
	We have heard this term lately, and it is occurring	8	12
	Exclusively occurs in Talla Johar	1	2
	Nothing ever occurs in Malla Johar	4	6
Natural disasters	Increasing	82	123
	Decreasing	2	3
	Stayed the same	16	24

3.2.4 Emergence of Agricultural Pests and Diseases

Climatic warming in the Johar valley is expanding the altitudinal range of crop pests and plant pathogens that were historically restricted to lower elevations. Warmer winters and increased humidity during the growing season have facilitated the proliferation of aphids, fungal pathogens (such as *Puccinia* spp.), and root rot organisms affecting cereals and legumes (Singh et al., 2023). These outbreaks are particularly harmful to barley and buckwheat, which form the backbone of local food security and are not typically bred for pest resistance. Farmers in the region generally lack access to chemical or organic pest control technologies due to remoteness, cost, and limited extension services. As a result, outbreaks often lead to significant crop losses, triggering cascading impacts on household food stocks, income generation, and nutrition. Moreover, the disruption of historical climatic cues has complicated farmers' ability to anticipate and adapt to pest cycles, further increasing vulnerability (Hossain et al., 2024).

3.2.5 Glacial Retreat and Hydrological Dynamics

Glacial retreat is one of the most visually and ecologically consequential indicators of climate change in the Johar valley. Key glaciers such as Milam, Uttari Balati, and Panchachuli have experienced accelerated mass loss in recent decades, with retreat rates documented at approximately 20–30 m per year in some locations (Mal et al., 2019). This retreat is directly attributed to rising mean annual temperatures and reduced accumulation zones due to diminished winter snowfall. These glaciers form the headwaters of the Gori Ganga River, a crucial water source for both upstream and downstream communities. As glaciers shrink, the immediate effect is on seasonal water availability, particularly during the pre-monsoon months when meltwater historically sustained agricultural irrigation and daily consumption needs in lower-altitude villages (Wood et al., 2020). The diminishing glacier mass reduces long-term water security and alters the hydrological cycle, creating uncertainty for traditional livelihoods.

Beyond water scarcity, glacial retreat contributes to increased landslides, slope destabilization, and soil erosion, particularly during intense or erratic rainfall events. The loss of glacial cover exposes loose moraine and sediment, which becomes highly susceptible to movement, especially in steep terrain. The fragile geological makeup of the region, combined with increased water runoff, heightens the frequency and severity of natural hazards, threatening infrastructure, trails, and human settlements (Bolch et al., 2012). These physical changes are compounded by socio-economic vulnerabilities, especially for communities whose livelihoods depend on predictable water flow for high-altitude farming and transhumant pastoralism. Reduced pasture productivity and disrupted cropping patterns have already been reported in parts of Munsyari and nearby valleys, prompting concern over long-term sustainability and migration pressures along with tangible effects on agriculture, livestock, and domestic water supply (Thapa et al., 2025). Thus, the retreat of the Milam and

Panchachuli glaciers is not merely a biophysical phenomenon but a complex socio-ecological crisis that jeopardizes both natural systems and cultural resilience in the Johar valley (Rawat, 2023).

3.2.6 Increased Frequency of Natural Calamities

The destabilization of permafrost, increased freeze thaw cycles, and changing precipitation patterns have contributed to a higher frequency of landslides and slope failures in the region. These geohazards are particularly prevalent along newly constructed or expanded roads and near human settlements located on or below unstable terrain (Lennartz, 2013). Landslides not only result in the direct loss of arable land and infrastructure but also sever critical transportation routes, affecting trade, emergency services, and tourism flows (Winter et al., 2019).

Perceptions gathered from local respondents in the Johar valley strongly indicate a growing awareness and concern regarding the increased frequency of natural calamities. According to the data presented in Table 3, 59% of respondents reported a noticeable increase in cloudburst-like activities, while 21% stated that the frequency of such events has remained the same, and 2% observed a decrease. Interestingly, 5% of respondents admitted they had never heard of the term "cloudburst," while 8% noted that they had only recently become aware of it, coinciding with its emergence as a local phenomenon. Additionally, 1% perceived that such events occur exclusively in Talla Johar, whereas 4% believed that nothing of this nature happens in Malla Johar.

Broader perceptions regarding natural disasters reflect an even more significant trend: 82% of respondents believed that the frequency of natural disasters is increasing, while only 16% felt it had remained the same, and a minimal 2% observed a decline. These insights not only emphasize the community's lived experiences of environmental change but also align with scientific projections of increased geohazard risks in fragile Himalayan ecosystems due to climate change, glacial retreat, and shifting precipitation patterns (Shrestha, 2025).The observed perceptions thus provide a valuable lens into localized climate vulnerability and highlight the urgency for adaptive and community-based disaster risk management strategies (Malik & Ford, 2025).

4 Conclusion

The Johar valley is a prime example of a high-altitude socio-ecological system that is severely impacted by socio-economic changes and climatic change. Traditional transhumance techniques are being disrupted, alpine meadows are changing, and feed supply is decreasing due to altered snowfall patterns, retreating glaciers, and rising temperatures. Together with youth movement and the decline in indigenous knowledge transfer, these environmental changes pose a challenge to the long-term viability of forest-based livelihoods, livestock rearing, and subsistence farming. At the same time, increased resource extraction, excessive non-timber forest product harvesting, and invasive plant species invasion exacerbate ecological stress, threatening biodiversity and cultural legacy.

The study emphasizes that maintaining food security, livestock productivity, and ecosystem resilience still depends on adaptive practices including community-based forest governance, rotational grazing, and agro-ecological crop management. However, changes in the socio-economic landscape, climatic variability, and infrastructure measures are all limiting the efficiency of these strategies. The valley can become more adapted by integrating traditional ecological knowledge with contemporary resource management, encouraging climate-resilient agricultural and pastoral activities, and fortifying local institutions. In order to maintain the vibrancy of the Johar valley's natural and cultural landscapes, it is imperative that specific policy interventions and community-led adaptation measures be implemented that balance ecological sustainability with livelihood security.

References

Acharya, S., Charan, G., Singh, N., & Srivastava, R. B. (2012). Soil organic carbon sequestration of cold desert Ladakh. *Range Management and Agroforestry, 33*(1), 79–82.

Acharya, K. P., Paudel, P. K., Jnawali, S. R., Neupane, P. R., & Köhl, M. (2017). Can forest fragmentation and configuration work as indicators of human–wildlife conflict? Evidences from human death and injury by wildlife attacks in Nepal. *Ecological Indicators, 80*, 74–83.

Adler, C., Wester, P., Bhatt, I., Huggel, C., Insarov, G., Morecroft, M., & Werners, S. (2023). Cross-chapter paper 5: mountains. Climate change 2022: Impacts, adaptation, and vulnerability. https://doi.org/10.1017/9781009325844.022

Ali, S. N., Singh, R., Morthekai, P., Sharma, A., Phartiyal, B., Quamar, M. F., & Arora, P. (2022). Perception of climate change from the Himalayan 'cold desert'Ladakh, India. *Journal of Palaeosciences, 71*(1), 89–111.

Bisht, D., Kumar, D., Kumar, D., Dua, K., & Chellappan, D. K. (2021). Phytochemistry and pharmacological activity of the genus artemisia. *Archives of Pharmacal Research, 44*(5), 439–474. https://doi.org/10.3390/molecules17032542

Bora, A. (2025). The Bhotiya Tribe of Uttarakhand: Identity, struggles and resilience. In A. Kumar, D. V. Bhosale, & S. S. Ramajayam (Eds.), *Nomads of India: History and the struggle for recognition* (pp. 56–71). Bharti Publications.

Bhatt, D., Joshi, G. C., & Tiwari, L. M. (2009). Culture, habitat and ethno-medicinal practices by Bhotia Tribe people of Dharchula Region of Pithoragarh District in Kumaun Himalaya, Uttarakhand. *Ethnobotanical Leaflets, 2009*(8), 2.

Bhatt, H., Verma, D., Jugran, H. P., Kumar, V., & Pandey, R. (2024). Community perception about ecosystem services assessment from forests managed under different regimes in Hindu Kush Himalaya. *Environmental and Sustainability Indicators, 24*, Article 100463. https://doi.org/10.1016/j.indic.2024.100463

Bolch, T., Kulkarni, A., Kääb, A., Huggel, C., Paul, F., Cogley, J. G., & Stoffel, M. (2012). The state and fate of Himalayan glaciers. *Science, 336*(6079), 310–314. https://doi.org/10.1126/science.1215828

Chuphal, D. S., Malik, I., Singh, R., Vangala, G., Niranjannaik, M., Vegad, U., & Mishra, V. (2025). Multi-day extreme precipitation caused major floods in India during summer monsoon of 2024. *Earth's Future, 13*(5), e2024EF005497. https://doi.org/10.1029/2024EF005497

Dame, J., & Nüsser, M. (2011). Food security in high mountain regions: Agricultural production and the impact of food subsidies in Ladakh, Northern India. *Food Security, 3*(2), 179–194. https://doi.org/10.1007/s12571-011-0127-2

Dwivedi, S., Singh, V., Mahra, K., Sharma, K., Baunthiyal, M., & Shin, J. H. (2024). Functional foods in the northwestern Himalayan Region of India and their significance: a healthy dietary tradition of Uttarakhand and Himachal Pradesh. *Joural of Ethnic Foods, 11*(1), 20. https://link.springer.com/article/10.1186/s42779-024-00236-4

Dolma, T., & Uniyal, S.K. (2025). Valuing seabuckthorn (*Hippophae rhamnoides* L.) ecosystem services for sustainable development and livelihood: a case study in Ladakh cold desert. *Forests, Trees and Livelihoods,* 1–13.

Fernández-Guisuraga, J. M., Fernández-García, V., Tárrega, R., Marcos, E., Valbuena, L., Pinto, R., & Calvo, L. (2022). Transhumant sheep grazing enhances ecosystem multifunctionality in productive mountain grasslands: A case study in the Cantabrian Mountains. *Frontiers in Ecology and Evolution, 10,* Article 861611. https://doi.org/10.3389/fevo.2022.861611

Gagné, K. (2019). *Caring for glaciers: Land, animals, and humanity in the Himalayas.* University of Washington Press.

Gaur, M. K., Goyal, R. K., Choudhary, V., Kanwar, M. S., Singh, N., Chichaghare, A. R., & Raghuvanshi, M. S. (2024). Traditional ecological knowledge and natural resource management in the cold desert of Leh-Ladakh, India. *Ecofeminism and Climate Change, 5*(2), 115–122. https://doi.org/10.26480/efcc.02.2024.115.122

Hossain, M. M., Sultana, F., Mostafa, M., Ferdus, H., Rahman, M., Rana, J. A., & Al Sabbir, M. A. (2024). Plant disease dynamics in a changing climate: impacts, molecular mechanisms, and climate-informed strategies for sustainable management. *Discover Agriculture, 2*(1), 132. https://link.springer.com/article/10.1007/s44279-024-00144-w

Ingty, T., Erb, A., Zhang, X., Schaaf, C., & Bawa, K. S. (2023). Climate change is leading to rapid shifts in seasonality in the Himalaya. *International Journal of Biometeorology, 67*(5), 913–925. https://link.springer.com/article/10.1007/s00484-023-02465-9

Kaur, H. (2020). *Nardostachys jatamansi* (D. Don) DC: An invaluable and endangered medicinal plant of the Himalayas. *Journal of Medicinal Plants, 12*(2), 45–52. https://doi.org/10.1016/j.jmp.2020.07.002

Kaushal, M., Sharma, K. D., & Attri, S. (2013). Effect of blanching on nutritional quality of dehydrated colocasia, *Colocasia esculenta* (L.) Schott leaves. *Indian Journal of Natural Products and Resources, 4*(2), 161–164

Kala, C. P. (2005). Indigenous uses, population density, and conservation of threatened medicinal plants in protected areas of the Indian Himalayas. *Conservation Biology, 19*(2), 368–378. https://doi.org/10.1111/j.1523-1739.2005.00602.x

Kala, C. P., Farooquee, N. A., & Dhar, U. (2005). Traditional uses and conservation of timur (*Zanthoxylum armatum* DC.) through social institutions in Uttaranchal Himalaya, India. *Conservation and Society, 3*(1), 224–230

Kar, S., Subedi, M. R., Kadam, P., & Dwivedi, P. (2025). Assessing the impact of customary ownership of trees and socioeconomic factors on the participatory forest management in Jharkhand, India. *Environmental Management, 75*(5), 1201–1215.

Kishor, K., Tewari, L. M., Tewari, A., & Upreti, B. M. (2013). Traditionally fermented alcoholic beverages of high altitude Bhotiya tribal community in Kumaun Himalaya. *India. African J of Food Science, 7*(12), 508–510.

Koju, N. P., Gosai, K. R., Bashyal, B., Byanju, R., Shrestha, A., Buzzard, P., & Khanal, L. (2023). Seasonal Prey Abundance and food plasticity of the vulnerable snow leopard (*Panthera uncia*) in the Lapchi valley, Nepal Himalayas. *Animals, 13*(20), 3182.

Kumar, S. (2022). *Use of traditional knowledge in sustainable forest management and provisioning of ecosystem services in Jharkhand.*

Kumar, S., & Sati, V. P. (2023). Highland-lowland interaction in mountain regions: A study of changing patterns among Bhotiya Tribes of Garhwal Himalaya. *Human Geographies, 17*(1), 87–105. https://doi.org/10.5719/hgeo.2023.171.5

Kumar, S., & Pathak, B. (2025). Observed changes in rainfall extremes and current trends in Gandhinagar district, Gujarat. *Theoretical and Applied Climatology, 156*(10), 1–19.

Kumar, P., Sarda, R., Yadav, A., Ashwani, & Gonencgil, B., Rai, A. (2025). Farmer's perception of climate change and factors determining the adaptation strategies to ensure sustainable agriculture in the cold desert region of Himachal Himalayas, India. *Sustainability, 17*(6), 2548. https://doi.org/10.3390/su17062548

Laha, A., Badola, R., & Hussain, S. A. (2018). Earning a livelihood from Himalayan caterpillar fungus in Kumaon Himalaya: Opportunities, uncertainties, and implications. *Mountain Research and Development, 38*(4), 323–331. https://doi.org/10.1659/MRD-JOURNAL-D-17-00063.1

Liang, S., Li, X., Zheng, X., Jiang, T., Li, X., & Qiao, D. (2020). Effects of winter snow cover on spring soil moisture based on remote sensing data product over farmland in northeast China. *Remote Sensing, 12*(17), 2716. https://doi.org/10.3390/rs12172716

Lennartz, T. (2013). Constructing roads—Constructing risks? Settlement decisions in view of landslide risk and economic opportunities in Western Nepal. *Mountain Research and Development, 33*(4), 364–371. https://doi.org/10.1659/MRD-JOURNAL-D-13-00048.1

Majumder, S., & Bhattacharya, M. (2024). Himalayan fermented beverages and their therapeutic properties with scientific validations: A comprehensive review. *Journal of Ethnic Foods, 11*(1), 44.

Malik, I. H., & Ford, J. D. (2025). Monitoring climate change vulnerability in the Himalayas. *Ambio, 54*(1), 1–19.

Mal, S., Mehta, M., Singh, R. B., Schickhoff, U., & Bisht, M. P. S. (2019). Recession and morphological changes of the debris-covered Milam Glacier in Gori Ganga valley, Central Himalaya, India, derived from satellite data. *Frontiers in Environmental Science, 7*, 42. https://doi.org/10.3389/fenvs.2019.00042

Mahdavi, S. K., Shahraki, M., & Sharafatmandrad, M. (2023). The mechanism of knowledge-based behavior of pastoralists for rangeland management: Exploitation, restoration and conservation. *Scientific Reports, 13*(1), 16296.

Mannia, S. (2024). Transhumance in the time of UNESCO: Political narratives, rhetorical representations, enhancement practices. Martor. *Revue d'anthropologie du Musée du Paysan Roumain, 29*, 87–100.

Marín, J. A., et al. (2023). Assessing effect of rootstock micropropagation on field performance. *PMC*. https://doi.org/10.3390/plants12030674

McDowell, G., Stevens, M., Lesnikowski, A., Huggel, C., Harden, A., & DiBella, J. (2021). Global Adaptation Mapping Initiative. Closing the adaptation gap in mountains. *Mountain Research and Development, 41*(3), A1. https://doi.org/10.1659/MRD-JOURNAL-D-21-00033.1

Mekita. (2018). The long, winding path of wine as medicine. Wine Enthusiast. https://www.wineenthusiast.com/culture/wine/history-wine-medicine-health/

Naudiyal, N., Arunachalam, K., & Kumar, U. (2019). The future of mountain agriculture amidst continual farm-exit, livelihood diversification and outmigration in the Central Himalayan villages. *Journal of Mountain Science, 16*, 755–768.

Negi, C. S. (2007a). Changing face of polyculture in the Darma and Johaar valleys, Pithoragarh, Kumaun Himalayas. *The International J of Sustainable Development & World Ecology, 14*(4), 428–436. https://doi.org/10.1080/13504500709469743

Negi, C. S. (2007b). Declining transhumance and subtle changes in livelihood patterns and biodiversity in the Kumaon Himalaya. *Mountain Research and Development, 27*(2), 114–118. https://doi.org/10.1659/mrd.0818

Negi, V. S., Tiwari, D. C., Singh, L., & Bhatt, I. D. (2021). Review and synthesis of climate change studies in the Himalayan region. *Environmental Development and Sustainability, 23*(3), 453–472.

Nüsser, M., Baghel, R. (2014). The emergence of the cryoscape: Contested narratives of Himalayan glacier dynamics and climate change. In *Environmental and climate change in South and Southeast Asia: How are local cultures coping* (pp. 138–156).

Nyssen, J., Descheemaeker, K., Zenebe, A., Poesen, J., Deckers, J., & Haile, M. (2009). Transhumance in the Tigray highlands (Ethiopia). *Mountain Research and Development, 29*(3), 255–264. https://doi.org/10.1659/mrd.00033

Oyinbo, O., Chamberlin, J., Vanlauwe, B., Vranken, L., Kamara, Y. A., Craufurd, P., & Maertens, M. (2019). Farmers' preferences for high-input agriculture supported by site-specific extension services: Evidence from a choice experiment in Nigeria. *Agricultural Systems, 173*, 12–26. https://doi.org/10.1016/j.agsy.2019.02.003

Pasakhala, B., Ghate, R., Phuntsho, K., Gentle, P., Gurung, J., Shrestha, A., & Thapa, S. (2021). Against the tide: The future of transhumant herders in the Kailash sacred landscape of Nepal. *Mountain Research and Development, 41*(4), R8–R15. https://doi.org/10.1659/MRD-JOURNAL-D-20-00073.1

Passang, S., Schmidt, S., & Nüsser, M. (2022). Topographical impact on snow cover distribution in the trans-Himalayan region of Ladakh. *India. Geosciences, 12*(8), 311. https://doi.org/10.3390/geosciences12080311

Parihar, D. S. (2021). Physiography and geology of Gori Ganga watershed, Kumaun Higher Himalaya. *International Journal of Science and Research, 10*(2), 591–595.

Palomo, I., Locatelli, B., Otero, I., Colloff, M., Crouzat, E., Cuni-Sanchez, A., Gómez-Baggethun, E., González-García, A., Grêt-Regamey, A., Jiménez-Aceituno, A., Martín-López, B., Pascual, U., Zafra-Calvo, N., Bruley, E., Fischborn, M., Metz, R., & Lavorel, S. (2021). Assessing nature-based solutions for transformative change. *One Earth, 4*(5), 730–741.

Pandey, A., Chandra Sekar, K., Joshi, B., & Rawal, R. S. (2019). Threat assessment of high-value medicinal plants of cold desert areas in Johar valley, Kailash sacred landscape, India. *Plant Biosystems-an International Journal of Dealing with All Aspects of Plant Biology, 153*(1), 39–47. https://doi.org/10.1080/11263504.2018.1448010

Patel, S. K., Sharma, A., Barla, A., & Singh, G. S. (2024). Traditional farming in the Indian Himalayan region: Preserving agro-biodiversity and future food security. In *The resilience of traditional knowledge systems for a sustainable future: A focus on agriculture and food practices in the Himalayas* (pp. 27–57).

Qiu, J. (2008). China: The third pole. *Nature, 454*(7203), 393–396. https://doi.org/10.1038/454393a

Rawal, R. S., Rawal, R., Rawat, B., Negi, V. S., & Pathak, R. (2018). Plant species diversity and rarity patterns along altitude range covering treeline ecotone in Uttarakhand: Conservation implications. *Tropical Ecology, 59*(2), 225–239.

Rawat, G. S. (2017). The Himalayan vegetation along horizontal and vertical gradients. In H. H. T. Prins & Namgail (Eds.), *Bird migration across the Himalayas: Wetland functioning amidst mountains and glaciers* (pp. 189–204).

Rawat, D. (2023). Promoting sustainable livelihoods and climate change adaptation through Ecotourism: A case study of Darma Valley in the Indian Himalayan Region (Doctoral dissertation, Staats-und Universitätsbibliothek Hamburg Carl von Ossietzky). https://ediss.sub.uni-hamburg.de/handle/ediss/10981

Rawat, S., Sharma, P. K., Naithani, S., & Uniyal, V. P. (2025). Consequences of climate change on medicinal plants in the Western Himalayas with special reference to Johar valley, Uttarakhand: A systematic review. *Environment Conservation Journal, 26*(3), 1116–1126. https://doi.org/10.36953/ECJ.31513027

Roy, B., Kala, C. P., Farooquee, N. A., & Majila, B. S. (2004). Indigenous fermented food and beverages: A potential for economic development of the high-altitude societies in Uttaranchal. *Journal of Human Ecology, 15*(1), 45–49.

Sati, V. P. (2021). *Resilience and adaptation strategies in the Indian Himalayan Region*. Springer Nature.

Shrestha, R., & Wegge, P. (2008). Habitat relationships between wild and domestic ungulates in Nepalese trans-Himalaya. *Journal of Arid Environments, 72*(6), 914–925. https://doi.org/10.1016/j.jaridenv.2007.12.002

Shrestha, P. (2025). Hydropower Governance in Nepal: A social-ecological systems assessment (Doctoral dissertation, The University of Arizona). https://www.proquest.com/openview/6196a6a24109f42eb408d506ec207a24/1?pq-origsite=gscholar&cbl=18750&diss=y

Srichandan, S., Pasupuleti, R. S., & Mishra, A. J. (2021). The transhumance route of Pithoragarh: A cultural route? *Environmental Challenges, 5*, Article 100291. https://doi.org/10.1016/j.envc.2021.100291

Singh, S. P., & Thadani, R. (2015). Complexities and controversies in Himalayan research: A call for collaboration and rigor for better data. *Mountain Research and Development, 35*(4), 401–409.

Singh, B. K., Delgado-Baquerizo, M., Egidi, E., Guirado, E., Leach, J. E., Liu, H., & Trivedi, P. (2023). Climate change impacts on plant pathogens, food security and paths forward. *Nature Reviews Microbiology, 21*(10), 640–656.

Singh, S. (2025). Historic trade routes and economic networks of Uttarakhand Himalayas-evolution, decline, and revival. Authorea Preprints.

Small, M. L. (2009). 'How many cases do I need?': On science and the logic of case selection in field-based research. *Ethnography, 10*(1), 5–38. https://doi.org/10.1177/1466138108099586

Sospeter, M. E., Mtega, W. P., & Malekani, A. (2025). Indigenous knowledge for climate change adaptation among Tanzanian smallholder farmers: A systematic review. *IFLA Journal*, 03400352251317997. https://doi.org/10.1177/03400352251317997

Tamang, J. P. (2010). *Himalayan fermented foods: Microbiology, nutrition, and ethnic values*. CRC Press. https://doi.org/10.1201/9781420093254

Thapa, S., Rawat, A., & Rajput, S. (2025). The impact of climate change on livestock in the Uttarakhand Himalayan region. *HE AGBIZ TODAY E-Magazine, 2*(3), 067–071.

Tiwari, K. R., Sitaula, B. K., Bajracharya, R. M., Raut, N., Bhusal, P., & Sengel, M. (2020). Vulnerability of pastoralism: A case study from the High Mountains of Nepal. *Sustainability, 12*(7), 2737.

Vuille, M., Carey, M., Huggel, C., Buytaert, W., Rabatel, A., Jacobsen, D., & Villacis, M. (2018). Rapid decline of snow and ice in the tropical Andes-impacts, uncertainties and challenges ahead. *Earth-Science Reviews, 176*, 195–213. https://doi.org/10.1016/j.earscirev.2017.09.019

Winter, M. G., Peeling, D., Palmer, D., & Peeling, J. (2019). Economic impacts of landslides and floods on a road network. *Auc Geographica, 54*(2), 207–220. https://karolinum.cz/data/clanek/7542/Geogr_54_2_0207.pdf

Wood, L. R., Neumann, K., Nicholson, K. N., Bird, B. W., Dowling, C. B., & Sharma, S. (2020). Melting Himalayan glaciers threaten domestic water resources in the Mount Everest Region, Nepal. *Frontiers in Earth Science, 8*, 128.

Impacts of Climate Change on Agriculture in Indian Cold Deserts—Mitigation and Adaptation

Chandra Prakash Kala and Yashwant S. Rawat

1 Introduction

The cold deserts of India, such as those in Ladakh, Lahaul-Spiti, a portion of district Kinnaur and Chamba and parts of Sikkim and Uttarakhand states, are characterized by extreme temperature, low precipitation, and harsh dry winds. These conditions have led to the evolution of specialized plant communities, many of which are endemic and not found elsewhere (Kala, 2003; Kala & Mathur, 2002). Agriculture in these areas is challenging due to shallow soils, labor intensity, and remoteness from markets, yet it supports the majority of the local population (Spehia et al., 2024). Over the period, local communities have developed and selected some unique agricultural techniques, irrigation systems, and crop varieties suited to the cold desert environment (Singh, 2009). Traditional farming systems in these regions are highly adapted to the environment, using glacial meltwater for irrigation and organic practices to enrich the soil (Saxena et al., 2021; Singh, 2009; Vishvakarma et al., 2005).

Living in a small, isolated, sparsely populated settlements, the indigenous communities in cold deserts of India have evolved a range of farming systems for raising crops, despite the short-growing period, extremely low rainfall, and harsh climate (Shaheen et al., 2013; Verma et al., 2025). In Ladakh region of India, it is achieved through a complex network of irrigation canals and improving soil fertility by adding farm yard manure. The farmers in Indian cold desert have evolved creativity and adaptability to the prevailing harsh climatic conditions by practicing all the available techniques and use of resources for agriculture, which include human labor, animal power, and handmade tools (Richards, 1985; Singh, 2009). The suitable season for

C. P. Kala (✉)
Indian Institute of Forest Management, Nehru Nagar, Bhopal, Madhya Pradesh, India
e-mail: cpkala@yahoo.co.uk

Y. S. Rawat
Department of Wood Technology Management, Faculty of Civil Technology, Federal Technical and Vocational Training Institute (TVTI), Addis Ababa, Ethiopia

Y. S. Rawat and C. P. Kala (eds.), *Cold Deserts*,
https://doi.org/10.1007/978-3-032-29119-6_3

raising crops is restricted to spring and summer (April–October), hence rabi (such as wheat, barley, mustard, peas along with vegetable crops, potatoes, and onions) and some kharif (maize, and vegetables) crops are grown simultaneously within such limited period (Verma et al., 2025).

The cold deserts are not untouched from the global climate change impacts rather such impacts are crucial and more prominent in such unique biome (Kala & Silori, 2012; Shrestha et al., 2012). Warming in the cold deserts of the Indian Himalaya is driving rapid environmental and socio-economic changes. The region faces a complex interplay of glacial melt, shifting precipitation, land degradation, and ecosystem disruption, all of which threaten the sustainability of indigenous communities and their fragile environment (Shaheen et al., 2013; Sharma et al., 2024). Though the Indian cold desert spans over a large geographical area, the land under agriculture is extremely low. For instance, the Spiti block in Himachal Pradesh has 2,033 ha under cultivation out of 711,900 ha, accounting for 0.28% of the land used for agriculture by approximately 2,583 farming families (Spehia et al., 2024).

The limited agriculture land within cold desert regions of India is under tremendous stress due to global warming and deglaciation (Banerji & Basu, 2010; Kala & Silori, 2012). Maintaining and conservation of cold deserts agrobiodiversity is a need of hour in view of ensuring food security to such marginalized communities, especially during the harsh winter months when access to fresh produce is limited. Working on agriculture ecosystem of cold deserts not only helps to suggest the conservation priorities but also support the livelihood, traditional knowledge, and the cultural heritage associated with it. In this context, this chapter aims to examine the impacts of climate change on agriculture ecosystem along with the possible adaptation and mitigation strategies of climate change.

2 Crop Diversity in Indian Cold Deserts

Traditional cropping systems that have evolved in the cold deserts of the Indian Himalaya are particularly well-suited to withstand climate variability, including drought, erratic rainfall, and poor soils. The most resilient among these are millets. Millets are highly drought-tolerant and require much less water than rice or wheat, thriving even in rainfed and marginal conditions typical of cold deserts (Kheya et al., 2023). Barley (*Hordeum vulgare* L.) and grim barley (*Hordeum aegiceras* Nees ex Royle) are the traditional crops in high-altitude Himalayan regions, known for their ability to grow in cold, dry climates and poor soils. Buckwheat (*Fagopyrum tataricum* (L.) Gaertn. and *F. esculentum* Moench) is adapted to high altitudes and short-growing seasons, making them suitable for the cold deserts. The other crops traditionally grown in the cold deserts are *Triticum aestivum* L., *Panicum miliaceum* L., and *Avena sativa* L. Amaranth is cultivated for its grains and leaves, both of which are nutritious and can withstand dry spells. Mustard is another resilient oilseed crop, often grown alongside millets and pulses in traditional agriculture systems. Chickpea (*Cicer arietinum* L.), fenugreek (*Trigonella foenum-graecum*), and alfalfa (*Medicago *sativa* L.) are also grown in the cold deserts. Among vegetables and spices, the prominent crops grown in cold desert of India are *Pisum sativum* L., *Solanum tuberosum* L., *Allium cepa* L., *Allium stracheyi* Baker, *Coriandrum sativum* L., and *Carum carvi* L. At present, the farmers have started growing cauliflower (*Brassica oleracea* var. botrytis L.), cabbage (*Brassica oleracea* var. capitata (L.) DC.), turnip (*Brassica rapa* L. ssp. rapa), *Cucurbita maxima* Duchesne, *Cucumis melo* L., *Solanum lycopersicon* L., and *Solanum melongena* L. Apple (*Pyrus communis* L.), apricot (*Prunus armeniaca* L.), and sea-buckthorn (*Hippophae rhamnoides* L.) are the major horticulture crops in the Indian cold deserts (Kala, 2015; Oinam et al., 2007; Pant et al., 2018; Saxena et al., 2021).

3 Protected Cultivation and Impacts on Crop Diversity

Protected cultivation techniques, such as polyhouses, greenhouses, low tunnels, and mulching, have been widely adopted in the Himalayan cold deserts to overcome harsh climatic constraints and extend the growing season (Shukla et al., 2024). These technologies have enabled farmers to grow high-value vegetables and flowers throughout the year, significantly increased productivity and farm income. During summer season, the prominent vegetables as grown in the Ladakhi polyhouses are tomato, capsicum, brinjal, cucumber, bottle gourd, long melon, muskmelon, and watermelon, whereas during autumn and winter, cabbage, cauliflower, and broccoli are grown in the polyhouses (Angmo et al., 2017; Lamo et al., 2020). The adoption of protected cultivation often leads to a focus on a limited set of commercially lucrative crops, such as tomatoes, capsicum, cucumbers, and flowers, which thrive under controlled conditions and fetch higher market prices. This economic incentive can result in reduced cultivation of traditional, diverse crops that are typically grown in open fields or as part of mixed cropping and rotational farming systems (Oinam et al., 2007).

Protected cultivation has both positive and negative effects on crop diversity and production. It allows for the introduction of new crop varieties and hybrids that would not survive in open fields due to extreme cold, thus technically increasing the range of crops that can be grown in the region (Mishra et al., 2010). It enables off-season production, which can help bridge gaps in food availability and dietary diversity (Patniak & Mohanty, 2021). By growing high-value crops under protection small and marginal farmers, including women, gain access to new livelihood opportunities to improve household income. Whereas the economic focus on a few high-value crops under protected structures can reduce on-farm crop diversity, as farmers may abandon traditional grains, pulses, and indigenous vegetables in favor of more profitable options. Traditional practices like mixed cropping and crop rotation, which are important for maintaining agrobiodiversity and soil health, may decline as monoculture becomes more prevalent in protected environments. Over time, this can lead to genetic erosion of local crop varieties and a loss of traditional agricultural knowledge (Banoo et al., 2022).

Protected cultivation significantly enhances the overall productivity of crops in the Himalayan cold deserts by creating controlled microclimates that mitigate harsh environmental conditions. Polyhouses and low tunnels may increase crop productivity by 3–4 times compared to open-field cultivation (Shukla et al., 2024). This is due to improved temperature regulation, reduced exposure to frost and cold winds, and extended growing seasons. These techniques allow farmers to grow high-value vegetables and flowers throughout the year, including off-season periods when outdoor cultivation is not feasible due to extreme cold (Lamo et al., 2020). For example, plastic low tunnels in Ladakh enable harvesting 45–60 days earlier than normal seasons. Protected cultivation reduces evaporation losses and allows for efficient irrigation methods such as drip and sprinkler systems, conserving scarce water resources in these arid regions while optimizing nutrient uptake (Banoo et al., 2022; Patniak & Mohanty, 2021). Controlled environments help to minimize pest and disease infestation, leading to healthier plants, a better-quality produce and yield, which fetches higher market prices. The increased productivity and quality translate into higher profitability and better livelihoods for smallholder farmers. The initial investment in protected structures is often recovered within 3–5 years, with internal rates of return ranging from 31 to 73% (Banoo et al., 2022). This also generates employment opportunities and supports rural youth engagement in agriculture. In Spiti block of Himachal Pradesh, at present, the cash crops have replaced almost 80–90% of traditional crops, which has increased the farmer's income to Rs. 1,92,292 per capita by 2022 (Spehia et al., 2024).

4 Climate Change and Shift in Cropping Diversity

Over the past two decades, there has been a significant shift in the Himalayan cold deserts, such as Lahaul-Spiti and Ladakh, from traditional, water-resilient crops like barley to these more economically lucrative but water-intensive cash crops (Manocha, 2015). The cultivation area for crops like cauliflower has expanded dramatically, with cauliflower alone increasing from 33 ha in 2005–2006 to 658 ha in 2016–2017 in Lahaul-Spiti, making up nearly 15% of the total cultivated area. Cabbage and potato have followed similar trends. However, the cultivated area for barley has shrunk from 608 to 400 ha from 2005–2006 to 2016–2017 (Shashni & Sharma, 2023; Sherpa, 2024). Traditional crops are richer and more viable than the introduced cash crops in terms of providing energy and nutrition.

The adoption of cauliflower, cabbage, and potato has put additional pressure on already scarce water resources, as these crops require much more irrigation compared to traditional crops like barley. This shift is considered climatically unsustainable given the rapidly depleting snowmelt and glacial water that serve as the main irrigation sources in these cold desert regions (Nikolaou et al., 2020; Shaheen et al., 2013). Cauliflower, cabbage, and potato are currently the most water-intensive crops being grown in the Himalayan cold deserts, replacing traditional, less water-demanding crops such as barley and thereby intensifying water scarcity in these fragile ecosystems. This transition increases pressure on already limited water resources, making agriculture less sustainable under changing climate conditions (Sharma et al., 2023). The adoption of water-intensive crops without adequate irrigation infrastructure exacerbates the risk of crop failure and threatens food security (Gupta, 2022).

Agriculture in Himalayan cold deserts, such as Lahaul-Spiti and Ladakh, depends mostly on snowmelt for irrigation, as rainfall is scanty due to the rain shadow effect of the mountains. Declining snowfall and earlier snowmelt, driven by shifting precipitation patterns, have led to severe water shortages during key crop sowing and growing periods, causing frequent droughts and crop failures (Qin et al., 2020). In 2022, for example, 76 villages in Lahaul-Spiti faced a drought, with crop losses ranging from 50 to 100%, due to insufficient irrigation water (Gupta, 2022). Traditional irrigation systems (kuhls) and water management practices are under strain due to unpredictable water supply and receding glaciers, resulting in adverse impacts on sustaining subsistence system of livelihood (Saxena et al., 2021; Vishvakarma et al., 2005).

Erratic precipitation, including sudden intense rainfall and cloudbursts, leads to flash floods that erode agricultural land, destroy crops, and damage infrastructure such as bridges and irrigation channels (Kala, 2014). These events disrupt transportation, resulting in post-harvest losses when crops cannot be marketed in time. Reduced and erratic precipitation patterns contribute to soil erosion, declining soil fertility, and degradation of pastures, further impacting crop yields and livestock productivity (Tewari & Kapoor, 2013). Increased incidence of crop pests and livestock diseases has also been linked to climate-induced changes in precipitation and temperature (Kala & Silori, 2012).

5 Climate Change Mitigation and Adaptation in Indian Cold Deserts Farming

5.1 Mitigation Strategies

Mitigating the adverse effects of climate change in the cold deserts of the Indian Himalaya is an important aspect which can be accomplished by involving a multi-pronged approach such as transitioning to renewable energy, restoring ecosystems, innovation in water management, promoting sustainable climate-smart agriculture, reducing disaster risks, and empowering local communities (Mehta et al., 2023). These strategies, when implemented together, can reduce greenhouse gas emissions, enhance resilience, and secure the livelihoods of indigenous populations in one of India's most fragile and unique environments. Planting and protecting native woody species around villages and agricultural fields are required to help stabilizing soils, reducing wind erosion, and enhancing carbon sequestration (Kala, 2025; Rawat & Everson, 2013).

Indigenous communities have long managed land, water, and vegetation collectively, using customary rules to prevent overexploitation and ensure equitable resource distribution. This includes rotational grazing, regulated wood collection, and community forests, which help to maintain carbon sinks and prevent land degradation (MoEFCC, 2023). There is a need to accelerate construction and maintenance of traditional water retention structures—such as zings (small reservoirs), ponds, and diversion channels—to capture and store snowmelt and rainwater for use during dry periods (Norphel & Tashi, 2015). These systems minimize water loss, support irrigation, and reduce pressure on glacial sources, indirectly mitigating climate impacts by sustaining agriculture and vegetation cover. Crop-livestock integrated system has been practiced by the Himalayan cold desert communities (Dolker, 2018). Mixed farming, crop rotation, and the use of organic manure (from livestock) maintain soil fertility and reduce the need for chemical inputs. Cultivation of traditional, hardy crop varieties adapted to arid and cold conditions ensures food security while minimizing resource use.

5.2 *Adaptation Strategies*

Farmers in the Himalayan cold deserts face unique climate challenges, including water scarcity, erratic weather, and land degradation (Kumar et al., 2025; Tuladhar et al., 2023). The most effective adaptation strategies for farmers in the Himalayan cold deserts are those that integrate water resource management, crop and livelihood diversification, soil and land conservation, community knowledge, and climate-smart technologies (FAO, 2013; Kumar et al., 2025). These approaches, supported by training and extension services, help to build resilience and sustain livelihoods amid the challenges of climate change. Besides, combining traditional knowledge with innovative practices, tailored to the region's fragile environment and limited resources, may help in adaptation to climate change.

In Ladakh, the Indus and Shayok rivers only benefit 10–15% of agriculture whereas rest of the agriculture depends on traditional water management systems and snowmelt streams (Norphel & Tashi, 2015). Construction and maintenance of artificial glaciers (ice stupas), water reservoirs, and ponds to store water for irrigation during dry periods are being taken care, at present. Besides, there is a need of adopting diverse engineering solutions as integral part of watershed management such as restoration and extension of traditional diversion canals, gravity canal, lift irrigation scheme, and snow harvesting and irrigation systems to optimize water distribution and also adoption of efficient irrigation methods, such as drip and sprinkler systems, to maximize water use efficiency (Norphel & Tashi, 2015; Saxena et al., 2021). Crop diversification helps to adapt to climate change apart from rotating crops and adjusting planting times to suit changing climatic conditions and reduce pest and disease risks (Prasanna & Joshi, 2024). Farmers maintain and cultivate a wide variety of traditional crops and landraces, which are often more resilient to changing climatic conditions, pests, and diseases. For example, Himalayan communities like the Lahaulas (local people in Lahaul valley) grow dozens of indigenous crop varieties, ensuring food security even during adverse weather (Oinam et al., 2007). This diversity buffers against crop failure and supports adaptation to unpredictable rainfall and temperature extremes. Communities like the Gaddi in Himachal Pradesh traditionally combine small-scale subsistence agriculture with nomadic herding of sheep and goats (Mishra & Pandey, 2019). This dual-livelihood approach spreads risk and provides flexibility in the face of unpredictable weather, droughts, and resource scarcity. Many families diversify further by engaging in seasonal labor or trade, helping buffer against agricultural losses.

Traditional techniques such as terracing, agroforestry, and the use of organic fertilizers (made from animal excreta, plant residue, and compost) help to conserve soil and water, reduce erosion, and maintain soil fertility for farming on the steep erosion-prone Himalayan slopes. Trees are required to be planted on the bunds to combat soil erosion and improve microclimates (Kala, 2010; Rawat & Everson, 2013). Community-managed irrigation systems and water-harvesting structures are also widely used to cope with water scarcity and shifting precipitation patterns (Vishvakarma et al., 2005). Practices such as mixed cropping, crop rotation, and organic farming enhance ecosystem resilience and reduce dependence on external inputs. These methods help maintain soil health, suppress pests naturally, and adapt to shifting growing seasons (Gamage et al., 2023).

The crops and or crop varieties that require less water may be continued or reintroduced at large scale. Farmers are adapting to warmer temperatures and shorter winters by sowing crops earlier in the season and selecting varieties better suited to the changing climate, such as off-season vegetables and cash crops (Saxena et al., 2021; Sharma et al., 2023). There is a growing emphasis on farming traditional, hardy crops and medicinal plants (e.g., *Allium stracheyi*, *Carum carvi*, *Hippophae rhamnoides*), which are resilient to climatic extremes and have high market value. Traditional crops like millet and mustard play a significant role in helping farmers in the Indian cold deserts adapt to climate change through several key mechanisms. Millets are exceptionally resilient to drought and high temperatures, thriving in the arid, low rainfall, and poor-soil conditions typical of cold desert regions. Their ability to grow with minimal irrigation makes them a reliable crop as rainfall becomes more erratic and water scarcity intensifies (Pant et al., 2018; Kheya et al., 2023).

Millets require fewer chemical fertilizers and can grow well in low-nutrient soils, reducing costs and environmental impact for smallholder farmers. Cultivating a diverse range of millet varieties increases resilience to pests, diseases, and weather extremes, helping ensure stable yields and food security even when other crops fail (Kala, 2010; Raut et al., 2023). Seed conservation and post-harvest support further maintain this diversity, enabling farmers to adapt to changing conditions. Millets are highly nutritious and serve as both food and fodder, supporting household nutrition and livestock, which are crucial for mixed farming systems in the cold deserts (Pant et al., 2018). Mustard is well-suited to the short-growing seasons and cool temperatures of the Himalayan cold deserts. It can be sown and harvested quickly, fitting into the narrow agricultural window available in these regions. As a major oilseed, mustard provides a vital source of income and edible oil for local communities, helping diversify livelihoods and reduce dependence on a single crop. While mustard is somewhat sensitive to changing rainfall patterns, it can still be cultivated in regions with limited irrigation, making it a valuable option as water resources become less predictable (Sharma et al., 2024).

Despite their resilience, millet and mustard growers in the Indian cold deserts compromise with water scarcity, short-growing seasons, poor soils, labor and technology gaps, weak market access, insufficient policy support, and the emerging impacts of climate change on yields. There are reports on decline in farming of mustard in Lahual valley (Kuniyal et al., 2004). Addressing these challenges requires integrated efforts in water management, infrastructure, research, market development, and policy support. The number of livestock needs to be adjusted to match available resources and changing climate conditions (Pant et al., 2018; Saxena et al., 2021).

The farmers in cold desert do believe in some traditional indicators and observation-based forecasting (such as animal behavior or plant phenology) to anticipate weather changes, allowing timely adjustments in agricultural activities and disaster risk reduction. For example, changes in bird migration, unusual animal activity, flowering or fruiting patterns, and shifts in wind or cloud formations are interpreted as early warning signs of storms, droughts, or heavy snowfall (Meena et al., 2023). Such local knowledge and belief systems play a crucial role in helping farmers to take precautionary measures to combat climate change by taking necessary decisions for maintenance of their farms and crop productivity, especially during extreme weather events. Indigenous knowledge fosters strong community cohesion and collective management of resources, including forests, pastures, and water systems (Gaur et al., 2024). This collective approach enables effective disaster preparedness, equitable resource sharing, and rapid response to climate-induced hazards like floods and landslides. While indigenous knowledge provides a strong foundation for adaptation, many communities are now considering to combine traditional practices with modern scientific approaches to further strengthen resilience and address new climate challenges (Hussain et al., 2024).

Indigenous knowledge and traditional practices are central to climate adaptation for Himalayan farmers. They provide time-tested, locally tailored solutions for managing land, water, and biodiversity, and foster strong social networks for collective resilience (Singh, 2009; Witharana et al., 2025). These approaches, especially when integrated with modern innovations, are vital for building climate-resilient agricultural systems in the Himalayas. The cultivation of a wide range of traditional varieties supports agrobiodiversity, which buffers against crop failure due to pests, diseases, or extreme weather. Diverse cropping systems, including mixed and intercropping with perennial drought-tolerant species, help to maintain soil fertility and ecosystem health (Dias et al., 2014; Mohanty et al., 2024). Traditional crops often need fewer chemical inputs (fertilizers and pesticides), reducing costs and environmental impact while making farming more sustainable under changing climate conditions (Kala, 2010; Mohanty et al., 2024; Witharana et al., 2025).

6 Policy Interventions

Agriculture, water supply, irrigation, canals, and land rights are the state subjects as per the 7th schedule of the Indian constitution that empowers the state governments to make laws on these subjects. The cold deserts in India span over different states and agriculture being the state subject, the policy and legislations are not uniform across the cold desert states. However, the motive of all the state governments is to provide congenial environment to farmers so that they can increase the crop productivity, which subsequently contribute in the state economy. There are different ways and means adopted by state governments, including preparing a policy through which the agricultural productivity is expected to be enhanced. Such policies help in changing the agriculture structures, for instance change in cropping patterns from staples to income-elastic foods such as fruits and vegetables (Timmer, 1988).

In cold desert of India, the government policies are mainly focused on supporting sustainable agriculture, improving livelihoods, and integrating farmers into the mainstream economy. In Spiti valley of Himachal Pradesh, the policy of allotting land to the landless was introduced during 1990s, which supported the expansion of agriculture land areas (Chandrasekhar et al., 2011). The financial support for apple cultivation, value addition, and marketing by government has further promoted apple farming (Sharma & Chauhan, 2008). However, the climate change and increasing temperature has shifted the apple production belt in the Himachal cold desert areas (Basannagari & Kala, 2013).

Government subsidies play a pivotal role in shaping what crops that farmers choose to grow. In Ladakh, the Department of Agriculture provides 75% subsidy to the farmers for construction of greenhouse (Banoo et al., 2022). Subsidized seeds for crops like green pea, mustard, radish, and apple saplings along with subsidized chemical fertilizers and water management infrastructure (water tanks, canals), and subsidies for polyhouses (greenhouses), where the government covers a significant portion of the cost, are being provided to the farmers making protected cultivation possible even to the small farmers (Maikhuri et al., 2015). By reducing the cost of key agricultural inputs—such as seeds, fertilizers, irrigation equipment, and machinery—subsidies make it more financially viable for farmers to cultivate certain crops over others (Banoo et al., 2022; Sharma & Chauhan, 2008). When subsidies are targeted toward high-value or water-intensive crops, farmers are incentivized to shift away from traditional, less resource-intensive crops to those that promise higher returns, even if they require more water or inputs.

In the Indian cold deserts, government support has increasingly promoted protected cultivation (e.g., polyhouses) for high-value crops like vegetables and flowers. These subsidies help small landholders invest in infrastructure that enables the cultivation of water and nutrient intensive crops round the year, enhancing productivity and profitability even in harsh climates. The result is a shift toward crops such as cauliflower, cabbage, and potato, which are more lucrative but also more demanding on water resources.

Input subsidies—especially for fertilizers and irrigation—tend to favor crops that respond well to these inputs, often leading to monocultures of water-intensive or high-yield varieties and also overuse of chemical inputs, which may impose detrimental effects on ecosystems (Madhu, 2022). This pattern in farming has been observed across India, where subsidies have historically promoted the cultivation of staple crops like rice and wheat, but in the cold deserts, similar patterns are emerging with vegetables and cash crops (Maikhuri et al., 2015). Subsidies also indirectly influence crop choice by stabilizing farmer incomes, reducing risk, and supporting market access (e.g., through cold storage and transport subsidies). These factors make it more attractive for farmers to invest in perishable, high-value crops that require reliable infrastructure. Besides, crop insurance is being offered which is an important component of agricultural safety nets. However, if government subsidies are not aligned with sustainability goals, they can exacerbate resource depletion and environmental degradation.

7 Conclusion

Farmers in the cold deserts depend on rainfed and irrigated agriculture systems where they cultivate on the less fertile or fragmented lands. With climate change causing erratic rainfall, longer dry spells, and reduced water discharge in springs, the cold deserts farmers experience more severe declines in traditional crop yields and greater food insecurity compared to socially privileged groups who may have access to better land or irrigation. However, over the period, they have also advanced to protective cultivation that has changed traditional farming systems including cropping patterns from traditional water-resilient crops like barley to more economically lucrative but water-intensive cash crops. This transition increases pressure on already limited resources, especially when the cold deserts have been reeling under adverse effects of global climate change. Embracing traditional agricultural knowledge with innovative practices may help to mitigate the adverse impacts of climate change on local agriculture ecosystems. Restoration of climate-resistant indigenous crops, crop diversification, organic farming, agroforestry and improved water management may help in maintaining the agricultural biodiversity and ecosystem in the environmentally sensitive Indian cold deserts areas.

References

Angmo, S., Angmo, P., Dolkar, D., Norbu, T., Paljor, E., Kumar, B., & Stopdan, T. (2017). All year round vegetable cultivation in trenches in cold arid trans-Himalayan Ladakh. *Defence Life Science Journal, 2*, 54–58.

Banoo, A., Hussain, S., Khan, F. A. S., Dar, S. R., Dolkar, R., & Rasool, R. S. (2022). A review on economic aspect of protected cultivation in Ladakh. *The Pharma Innovation Journal, 11*(12), 4726–4731.

Banerji, G., & Basu, S. (2010). Adapting to climate change in Himalayan cold deserts. *International Journal of Climate Change Strategies and Management, 2*(4), 426–448.

Basannagari, B., & Kala, C. P. (2013). Climate change and apple farming in Indian Himalayas: A study of local perceptions and responses. *PLoS ONE, 8*(10), Article e77976. https://doi.org/10.1371/journal.pone.0077976

Chandrasekhar, K., Rao, K. S., & Saxena, K. G. (2011). Structure and function of agro-ecosystems in cold desert of Spiti catchment. In K. G. Saxena, L. Liang, & X. Xue (Eds.), *Global change, biodiversity and livelihoods in cold desert Region of Asia* (pp. 91–106). Bishen Singh Mahendra Pal Singh, Dehradun.

Dias, T., Dukes, A., & Antunes, P. M. (2014). Accounting for soil biotic effects on soil health and crop productivity in the design of crop rotations. *Journal of the Science of Food and Agriculture, 95*(3), 447–454.

Dolker, P. (2018). An overview of transition in traditional agriculture of Ladakh. *Journal of Himalayan Ecology and Sustainable Development, 13*, 26–48.

FAO. (2013). *Climate smart agriculture sourcebook* (557 pp). Food and Agriculture Organization of the United Nations. https://www.fao.org/climate-smart-agriculture-sourcebook/about/en/

Gamage, A., Gangahagedara, R., Gamage, J., Jayasinghe, N., Kodikara, N., Suraweera, P., & Merah, O. (2023). Role of organic farming for achieving sustainability in agriculture. *Farming System, 1*(1), Article 100005. https://doi.org/10.1016/j.farsys.2023.100005

Gaur, M. K., Goyala, R. K., Choudharyb, V., Kanwara, M. S., Singhc, N., Chichagharea, A. R., & Raghuvanshid, M. S. (2024). Traditional ecological knowledge and natural resource management in the cold desert of Leh-Ladakh, India. *Ecofeminism and Climate Change, 5*(2), 81–88.

Gupta, V. (2022). *Farming becomes tough in Himachal's cold desert with changing climate and crop pattern*. Mongabay.

Hussain, S., Sharma, S., Bhatti, R. C., & Singh, A. N. (2024). Ecosystem services of the trans-himalayan region with special reference to Ladakh: An overview. In S. Joseph, & A. Pradeep-kumar (Eds.), *Ecosystem services valuation for sustainable development*. Singapore: Springer. https://doi.org/10.1007/978-981-97-4688-0_5

Kala, C. P. (2003). *Medicinal plants of Indian trans-Himalaya: Focus on Tibetan use of medicinal resources*. Bishen Singh Mahendra Pal Singh, Dehradun, India.

Kala, C. P. (2010). Status of an indigenous agro-forestry system in changing climate: A case study of the middle Himalayan region of Tehri Garhwal, India. *Journal of Forest Science, 56*(8), 373–380.

Kala, C. P. (2014). Deluge, disaster and development in Uttarakhand Himalayan region of India: challenges and lessons for disaster management. *International Journal of Disaster Risk Reduction, 8*, 143–152.

Kala, C. P. (2015). *Nanda's Neelkhanth* (p. 340). Partridge Publishing.

Kala, C. P. (2025). Agroforestry in a changing climate: Challenges, opportunities and solutions. *Ecological Frontiers, 45*(2), 269–276. https://doi.org/10.1016/j.ecofro.2024.11.003

Kala, C. P., & Mathur, V. B. (2002). Patterns of plant species distribution in the Trans-Himalayan region of Ladakh, India. *Journal of Vegetation Science, 13*(6), 751–754.

Kala, C. P., & Silori, C. S. (2012). *Biodiversity, Communities and Climate Change*. TERI Publications, New Delhi.

Kheya, S. A., Talukder, S. K., Datta, P., Yeasmin, S., Rashid, M. H., Hasan, A. K., Anwar, M. P., Islam, A. K. M. A., & Islam, A. K. M. M. (2023). Millets: The future crops for the tropics—Status, challenges and future prospects. *Heliyon, 9*(11). https://doi.org/10.1016/j.heliyon.2023.e22123

Kumar, P., Sarda, R., Yadav, A., Ashwani, Gonencgil, B., & Rai, A. (2025). Farmer's perception of climate change and factors determining the adaptation strategies to ensure sustainable agriculture in the cold desert region of Himachal Himalayas, India. *Sustainability, 17*(6), 2548. https://doi.org/10.3390/su17062548

Kuniyal, J. C., Vishvakarma, S. C., & Singh, G. S. (2004). Changing crop biodiversity and resource use efficiency of traditional versus introduced crops in the cold desert of the northwestern Indian Himalaya: A case of the Lahaul valley. *Biodiversity and Conservation, 13*, 1271–1304. https://doi.org/10.1023/B:BIOC.0000019404.48445.27

Lamo, K., Kuma, P., Namgyal, D., Angchuk, S., & Kacho, N. F. (2020). Protected cultivation: Indispensable for cold Arid Ladakh. *International Journal of Advances in Agricultural Science and Technology, 7*, 75–78.

Madhu, P. (2022). Impact of agricultural subsidies on farm income and productivity: A case study. *International Journal of Research and Analytical Review, 9*(2), 282–289.

Maikhuri, R. K., Rao, K. S., Anarbaev, M., Safaralikhonov, A. B., Saxena, K. G., & Liang, L. (2015). Farming and environmental conservation in Asian cold deserts: A review. *International Journal of Ecology and Environmental Sciences, 41*(3–4), 217–241.

Manocha, V. (2015). Farmers in Ladakh are among biggest victims of climate change. *Down To Earth*.

Meena, H. M., Manjunatha, B. L., Raghuvanshi, M. S., Verma, A., Angchok, D., Dorjay, N., & Spalbar, E. (2023). Weather prediction using traditional knowledge in cold arid high altitude region of Ladakh in India. *Indian Journal of Traditional Knowledge, 22*(4), 864–873.

Mehta, D., Pandey, R., Gupta, A. K., & Juhola, S. (2023). Nature-based solutions in Hindu Kush Himalayas: IUCN global standard based synthesis. *Ecological Indicators, 154*, Article 110875. https://doi.org/10.1016/j.ecolind.2023.110875

Mishra, G. P., Singh, N., Kumar, H., & Singh, S. B. (2010). Protected cultivation for food and nutritional security at Ladakh. *Defence Science Journal, 61*, 219–230.

Mishra, H., & Pandey, B. W. (2019). Navigating the impacts of social and environmental changes to traditional lifestyle: A case study of Gaddi transhumance of Chamba district in Himachal Pradesh. *The Oriental Anthropologist, 19*(2), 326–337.

MoEFCC. (2023). *National action plan to combat desertification and land degradation through forestry interventions*. New Delhi: Ministry of Environment Forest and Climate Change, Government of India.

Mohanty, L. K., Singh, N. K., Raj, P., Prakash, A., Tiwari, A. K., Singh, V., & Sachan, P. (2024). Nurturing crops, enhancing soil health, and sustaining agricultural prosperity worldwide through agronomy. *Journal of Experimental Agriculture International, 46*(2), 46–67.

Nikolaou, G., Neocleous, D., Christou, A., Kitta, E., & Katsoulas, N. (2020). Implementing sustainable irrigation in water-scarce regions under the impact of climate change. *Agronomy, 10*(8), 1120. https://doi.org/10.3390/agronomy10081120

Norphel, C., & Tashi, P. (2015). Snow water harvesting in the cold desert in Ladakh: An introduction to artificial glacier. In H. Nibanupudi & R. Shaw (Eds.), *Mountain hazards and disaster risk reduction. Disaster risk reduction*. Tokyo: Springer. https://doi.org/10.1007/978-4-431-55242-0_11

Oinam, S. S., Kuniyal, J. C., Rawat, Y. S., Vishvakarma, S. C. R., & Pandey, D. C. (2007). Disrupting land-use changes cause land degradation in fragile cold desert ecosystems: A case of the Lahaul. *Journal of Water and Land Use Management, 7*(2), 197–217.

Pant, S., Rinchen, T., & Butola, J. S. (2018). Indigenous knowledge on bio-resources management for sustainable livelihood by the cold desert people, Trans-Himalaya, Ladakh, India. *Indian Journal of Natural Products and Resources, 9*(2), 168–173.

Patniak, R. K., & Mohanty, S. (2021). Protected cultivation: Importance, scope and status. *Food and Scientific Reports., 2*, 19–21.

Prasanna, G. S. S., & Joshi, J. L. (2024). Crop diversification for disease mitigation in changing climates. In N. L. Prasanna, U. S. Rajput, J. S. Padsala, R. Nigam & Alok (Eds.), *Plant pathology in the age of climate change* (pp. 264–283). Golden Leaf Publishers.

Qin, Y., Abatzoglou, J. T., Siebert, S., Huning, L. S., Kouchak, A. A., Mankin, J. S., Hong, C., Tong, D., Davis, S. J., & Muelleret, N. D. (2020). Agricultural risks from changing snowmelt. *Nature Climate Change, 10*, 459–465. https://doi.org/10.1038/s41558-020-0746-8

Raut, D., Sudeepthi, B., Gawande, K. N., Reddy, G., Vamsi, S., Padhan, S. R., & Panigrahi, C. K. (2023). Millet's role as a climate resilient staple for future food security: A review. *International Journal of Environment and Climate Change, 13*(11), 4542–4552.

Rawat, Y. S., & Everson, C. S. (2013). Availability and use of willow species in representative cold desert areas of northwestern Himalaya, India. *Journal of Mountain Science, 10*, 472–481.

Richards, P. (1985). *Indigenous agriculture revolution: Ecology and food production in West Africa.* Westview Press.

Saxena, A., Raghuvanshi, M. S., & Suna, T. (2021). Traditional water management of subsistence agriculture system in cold arid Ladakh: A review. *International Journal of Scientific and Engineering Research, 12*(4), 1019–1028.

Shaheen, F. A., Wani, M. H., Wani, S. A., & Norphel, C. (2013). Climate change impact in cold arid desert of North–Western Himalaya: Community based adaptations and mitigations. In *Knowledge systems of societies for adaptation and mitigation of impacts of climate change* (pp. 239–256). Berlin, Heidelberg: Springer Berlin Heidelberg.

Sharma, A., Shashni, S., & Rathore, S. (2023). Traditional knowledge system for sustainable agriculture practices of rural communities of North-Western Himalaya, India. In *Traditional ecological knowledge of resource management in Asia* (pp. 191–210). Springer International Publishing: Cham, Switzerland.

Sharma, A., Shashni, S., & Rathore, S. (2024). Traditional practices of agroforestry systems in the cold desert area of Himachal Himalayas. In A. Borthakur & P. Singh (Eds.), *Addressing the climate crisis in the Indian Himalayas.* Cham: Springer. https://doi.org/10.1007/978-3-031-50097-8_6

Sharma, D. P., & Chauhan, J. S. (2008). Potential of fruit growing in cold desert areas of North-Western Himalayas: Problems, prospects and future strategies. *Acta Horticulturae, 772*, 175–178.

Shashni, A., & Sharma, S. (2023). Sustainable agriculture in a cold desert: Case study of Lahaul and Spiti district of Himachal Pradesh. In R. K. Bhattacharjya, B. Talukdar, & K. L. Katsifarakis (Eds.), *Sustainable Water Resources Management: Proceedings from SWARM 2020. Book Series: Advances in Sustainability Science and Technology* (pp. 228–235). Singapore: Springer Nature. https://doi.org/10.1007/978-981-16-7535-5_25

Sherpa, D. (2024). *Environmental shifts and dwindling agricultural sustainability: A review of Lahaul-Spiti District.* Centre of Excellence for Himalayan Studies, Shiv Nadar Institution of Eminence, Delhi. Issue Brief. No.6. June 1–7.

Shrestha, U. B., Gautam, S., & Bawa, K. S. (2012). Widespread climate change in the Himalayas and associated changes in local ecosystems. *PLoS ONE, 7*(5), Article e36741. https://doi.org/10.1371/journal.pone.0036741

Shukla, R. M., Nazir, N. B., Pandey, Y., & Dadhich, S. M. (2024). Influence of protected cultivation structures on vegetable crops. *Journal of Agriculture and Ecology Research International, 25*(1), 116–122.

Singh, A. K. (2009). Probable agricultural biodiversity heritage sites in India: I. The cold-arid region of Ladakh and adjacent areas. *Asian Agri-History, 13*(2), 83–100. https://www.asianagrihistory.org/pdf/volume13/1.pdf

Spehia, R. S., Dev, I., Dhiman, A., Dhiman, B., & Chandel, R. S. (2024). Agriculture in the cold desert of Spiti Valley is difficult, but there are ways to make it easier. *Down To Earth.* Retrieved March 14, 2024, from https://www.downtoearth.org.in/agriculture/agriculture-in-the-cold-desert-of-spiti-valley-is-difficult-but-there-are-ways-to-make-it-easier-95024

Tewari, V. P., & Kapoor, K. S. (2013). Western Himalayan cold deserts: Biodiversity, eco-restoration, ecological concerns and securities. *Annals of Arid Zone, 52*, 225–232.

Timmer, C. P. (1988). The agricultural transformation. In *Handbook of development economics* (vol. 1, pp. 275–331).

Tuladhar, S., Hussain, A., Baig, S., Ali, A., Soheb, M., Angchuk, T., Dimri, A. P., & Shrestha, A. B. (2023). Climate change, water and agriculture linkages in the upper Indus basin: A field study from Gilgit-Baltistan and Leh-Ladakh. *Frontiers in Sustainable Food Systems, 6*, 1012363. https://doi.org/10.3389/fsufs.2022.1012363

Verma, S., Vashisth, S., Bhushan, L., & Acharya, C. L. (2025). Blending traditional knowledge of farmers in agriculture with modern scientific technologies in cold desert region of Himachal Pradesh. In C. L. Acharya, R. Chaudhary, P. Gurumurthy, & A. S. Rao (Eds.), *Blending Indian Farmers' Traditional Knowledge in Agriculture with Modern Scientific Technologies*. Singapore: Springer. https://doi.org/10.1007/978-981-96-1020-4_10

Vishvakarma, S. C. R., Rawat, Y. S., Oinam, S. S., & Kuniyal, J. C. (2005). Irrigation (*Kuhls*): A community management system in cold desert of the Lahaul valley, northwestern Himalaya, India. In P. Pradhan, & U. Gautam (Eds.), *Proceedings: Third International Seminar on Farmer Managed Irrigation Systems and Governance Alternatives* (pp. 425–444). Kathmandu, Nepal.

Witharana, L., Chen, D., Curio, J., & Burman, A. (2025). Traditional ecological knowledge in High Mountain Asia: A pathway to climate resilience in agriculture amidst changing climates. *Advances in Climate Change Research, 16*(1), 167–182. https://doi.org/10.1016/j.accre.2025.01.009

Climate Change Impacts on Indian Cold Deserts Vegetation and Conservation Implications

Chandra Prakash Kala

1 Introduction

One-third of the earth's surface is covered with deserts, of which the cold desert region shapes a unique biome by having the major precipitation in the form of snow (Kala, 2011; Walker, 1982). The harsh, arid, and high-altitude environment in the cold deserts of Indian sub-continent has evolved exceptionally unique vegetation that can withstand high temperature and limited rainfall (Kala, 2000). At present, these unique and sparsely distributed vegetation in Indian cold deserts faces several critical issues related to environmental degradation, climate change, and socio-economic challenges. Deglaciation caused by climate change and changes in soil moisture levels have subsequently impacted the vegetation and the cold desert ecosystems (Dimri et al., 2015; Mohanty et al., 2024). Many of the plant species are sensitive indicators of environmental change and shifts in such plant species and communities, loss of endemic species, and changes in their phenology provide early warnings of broader ecological disruptions caused by climate change in cold desert ecosystems (Snyder et al., 2019). Vegetation in cold deserts also plays a vital role in the local hydrological cycle, influencing soil moisture, runoff, and the availability of water from glacier and snowmelt. Further, changes in vegetation due to anthropogenic climate change can exacerbate water scarcity, increase the frequency and severity of droughts, and affect the timing and distribution of water resources critical for both ecosystems and human communities (Mohanty et al., 2024; Snyder et al., 2019).

Landscape level investigation of plant species is considered as an important component in determining the diversity of life forms and various ecological patterns with space variability (Farina, 1998; Kala, 2011). Understanding the interrelationships between vegetation and various environmental factors, including soil

C. P. Kala (✉)
Indian Institute of Forest Management, Nehru Nagar, Bhopal, Madhya Pradesh, India
e-mail: cpkala@yahoo.co.uk

Y. S. Rawat and C. P. Kala (eds.), *Cold Deserts*,
https://doi.org/10.1007/978-3-032-29119-6_4

and water, is a need of hour due to changes in climate. The observations on relationship between vegetation and altitude were first made in the early-nineteenth century (Bunzhuo et al., 1997). Following a century or more, studies on mountain vegetation along the altitudinal gradients were focused on themes such as the interrelationships with latitudinal and climatic variables, continuity of individual species, plant community types over elevation ranges, and the systematic variation of soil properties (Bunzhuo et al., 1997; Kala, 2000; Sharma & Kala, 2022). The distinct eco-physiological and biological factors are also studied and considered responsible for sparse plant density and reduced plant growth in Indian cold deserts (Chaurasia & Singh, 1996).

There are a few studies available on vegetation distribution in cold desert of India (e.g., Kachroo et al., 1977; Chaurasia & Singh, 1996; Kala & Mathur, 2002); however, the studies on impacts of climate change on cold desert plant species of Indian Himalaya are meagre. Many plant species in cold deserts are endemic and adapted to narrow ecological niches, making them especially susceptible to extinction as environmental conditions change (Kala, 2005). Degradation of high-altitude cold deserts and loss of vegetation directly threaten the livelihoods of indigenous and agro-pastoral communities who depend on these ecosystems for grazing, agriculture, and other resources (Kala, 2000, 2003). Studying climate change impacts on cold desert vegetation is vital for conservation of biodiversity, sustaining ecosystem services, protecting vulnerable human communities, and understanding broader environmental feedbacks and risks associated with global climate change. In this context, the present chapter aims to review the distribution of plant species and community composition in cold desert region of India. It also attempts to examine various impacts of climate change on vegetation and vulnerability of Indian cold desert areas.

2 Ecosystems and Floristic Diversity Within Indian Cold Deserts

The cold desert ecosystems in the Indian Himalaya are unique and diverse, shaped by extreme climatic conditions, high altitudes, and limited precipitation in the form of snow. The Indian cold deserts are categorized into alpine pastures, dry alpine scrub, dwarf juniper scrubs, sub-nival zone, steppe, and alpine grasslands (FSI, 2015; Leng et al., 2023). The alpine pastures are located in high-altitude regions and predominated by herbaceous plants dotted with dwarf woody shrubs such as Astragalus, and Juniperus. These ecosystems are critical for grazing and support endemic plant species. The dry alpine scrubs are characterized by sparse vegetation dominated by shrubs such as *Juniperus* and *Salix*. These areas are adapted to dry conditions with alkaline soils and limited water retention capacity. The dwarf juniper scrubs are composed of woody shrubs like *Juniperus*, which serve as an important source of, fuel and small timber for local communities (Rawat & Everson, 2012). The steppes and alpine grasslands are dominated on the slopes in the high-altitude areas mostly above 4,500 m (Kala, 2011).

Apart from the vegetation composition, the landforms are the major factors for the establishment of vegetation in the Indian cold deserts. Based on the diverse landscape units or landforms, Kala and Mathur (2002) have stratified cold desert of Ladakh into undulating land areas (rolling grounds adjacent to broader river valleys, connecting riverbeds to high-rising barren ridges), riverbeds (banks of river and valley bottom characterized by Salix spp. and dominance of *Hippophae rhamnoides* shrub), moraines (glacier-driven boulders of small to huge dimensions on lower slopes of mountains and also include loose rocks and stone debris on riverbed), scree slopes (barren mountain slopes covered with very small, loose, and sharply angular stones), narrow gorges (deep, narrow mountain gorges cut through by small, seasonal rivulets), crags and ridges (rough and barren, high-rising crags and ridges), and tablelands (high-altitude plateaus or flat land areas located at an elevation of c. 4,500–5,500 m).

The geomorphological diversity across the Indian cold deserts supports a high plant species diversity and so that the faunal diversity. A total of 647 species of vascular plants (e.g., Angiosperms, Gymnosperms, and Pteridophytes) are reported to inhabit the high-altitude cold deserts of Ladakh region in India (Kala, 2011). These species are distributed over 243 genera and 58 families. Asteraceae (89 species), Poaceae (66 species), Brassicaceae (54 species), Fabaceae (43 species), Caryophyllaceae (31 species), Polygonaceae (28 species), Ranunculaceae (25 species), Scrophulariaceae (23 species), Lamiaceae (23 species), Rosaceae (22 species), and Apiaceae (22 species) are the dominant families of plant species. Astragalus is the most dominant genus as it has 24 species, followed by Polygonum (18 species), Artemisia (15 species), Potentilla (14 species), Salix (12 species), Nepeta (11 species), Pedicularis (10 species), and Poa (10 species) in the cold desert of Ladakh (Kala, 2011). The cold deserts are also known for the high diversity of medicinal plants. In Indian cold desert, a total of 337 medicinal plant species have been reported by Kala (2003), of which many species are threatened. The most prominent rare and endangered plant species that also contain medicinal properties are *Ferula jaeschkeana, Saussurea bracteata, Saussurea gnaphaloides, Hyoscyamus niger, Dactylorhiza hatagirea, Bergenia stracheyi, Rheum moorcroftianum, Gentiana kurroo, Aconitum violaceum, Aconitum rotundifolium,* and *Lancea tibetica* (Kala, 2003, 2005). These plant species in the cold desert regions of the Indian Himalaya significantly impact the local ecosystems through various ecological functions and adaptations.

3 Environmental Factors and Plant Distribution Patterns

The vegetation in the cold desert regions of the Indian Himalaya though sparsely distributed, but it is highly specialized due to extreme climatic conditions that restrict the plant growth for a short-growing season depending on elevations (Kala & Mathur,

2002). In Ladakh region of India, the vegetation encompasses of a typical high-altitude desert characteristic, which is indicated by the presence of typical xerophytic species such as *Echinops cornigerus, Caragana brevifolia, Ephedra gerardiana, Capparis spinosa, Salsola collina,* etc. The distribution of vegetation in the cold desert is further controlled by distinct eco-physiological, environmental, and biological factors (Webber, 1978; Chaurasia & Singh, 1996; Marcela et al., 1998; Kala, 2000). Distinct landscape units (Downes et al., 1998) and elevations (Kala, 1998) are among the eco-physiological factors that primarily regulate the species diversity and richness on the mountains. Similarly, in Ladakh, plant species richness increases with altitude up-to 4,500 m, and above this elevation range, it decreases sharply due to increasing frequency of crags and ridges, scarcity of nutrients in soil, and heavy snowfall round the year. Thus, 4,500 m altitude forms a borderline for most species distribution in cold desert of Ladakh though the vegetation such as *Ladakiella klimesii, Poa attenuata,* and *Draba altaica* occurs up-to 6,150 m (Dolezal et al., 2016; Kala, 2011; Kala & Mathur, 2002).

The distribution of vegetation across landscape and altitudinal amplitudes is resulted into a hump-shaped frequency histogram that seems to be an artifact of historical biogeography (Kala & Mathur, 2002). This distribution pattern is also found remarkably similar in the avian communities of Ladakh (Kala & Jayapal, 1999). Apart from distinct landscape unit and altitude, the presence of water and availability of moisture in soil cause the variation in distribution of mountain vegetation (Kala, 1998, 2005). These variations are enhanced by microclimate, aspect, exposure, relief, and the nature of surface materials that further accentuate the differences in moisture and heat balances resulting in distinct vegetation communities (Bunzhuo et al., 1997; Rey Benayas et al., 1999). Relatively good growth of alpine vegetation is found in moist depressions and slopes moistened by melting snow and glacial streams. The hierarchical structure of Ladakh ecosystem is illustrated as bottom-up nested domain of landscape unit and altitude (Kala & Mathur, 2002).

Plant species in cold deserts are adapted to harsh conditions with features like waxy leaves to reduce water loss, cushion-like growth forms for protection against cold and dry wind, deep taproots for accessing underground water, and thick stems for water storage. These plants exhibit physiological adaptations, including stomatal closure and changes in leaf morphology, to withstand extreme temperatures and water scarcity (Billings, 1974; Bliss, 1962). Plants like junipers and Artemisia help to stabilize soil in arid conditions, reducing erosion caused by strong winds and limited vegetation cover. Their root systems bind the soil, preventing degradation and maintaining the integrity of the fragile desert landscape (Pierson et al., 2007). Dominant species such as *Astragalus* and *Ephedra gerardiana* provide habitats for smaller organisms, contributing to biodiversity conservation in an otherwise harsh environment. These plants support pollinators and other fauna, ensuring ecosystem functionality. Species such as *Hippophae rhamnoides* exhibit adaptations like deep root systems and osmotic adjustments, enabling efficient water use and survival during droughts. The cold desert plant species and the eco-physiological mechanisms help in maintaining soil health, supporting biodiversity, and ensuring ecosystem sustainability in the cold deserts of the Indian Himalaya. Their resilience further helps to maintain ecosystem stability under changing climatic conditions (Pierson et al., 2007; Quoreshi et al., 2022).

4 Impacts of Climate Change on Vegetation

The majority of plant species in the Indian cold deserts though are acclimatized to harsh climatic conditions and limited resources nonetheless, changing climate, rising temperature, and erratic rainfall pose great stress on these plant species and the structures and functions of their ecosystems. These plant species are subjected to vulnerability and extinction due to various anthropogenic pressures, including anthropogenic climate change. There are reports on reduced snowfall and increased ice melt rates due to warming apart from changes in snow melt timings and high variability in seasonal snow cover (Bhutiyani et al., 2008; Passang et al., 2022; Shafiq et al., 2016). All such climate-induced changes have profound effects on eco-physiological characteristics of plant species through alterations in growing season length (Sharma et al., 2024). In addition, the plant species in high altitudes experiences varied degrees of stress due to elevation-dependent warming, with higher elevations experiencing more stress than lower elevations (Mountain Research Initiative EDW Working Group, 2015).

Biological invasion is considered as one of the major drivers of global environmental change across the world (IPBES, 2019), and highland areas have been previously considered at a relatively lower risk of biological invasions (Pauchard et al., 2009). However, there are reports of biological invasion of 104 alien plant species in cold desert areas of Ladakh, of these 24 alien species are under cultivation, while 80 species are aliens growing in wild, with 55 naturalized and 25 invasive (Zargar et al., 2025). These non-native species often outcompete native plants for resources such as water, nutrients, and sunlight. This leads to the decline or local extinction of endemic species, reducing biodiversity. Non-native plants can disrupt natural ecological processes (Zargar et al., 2024). For instance, they may change soil chemistry or nutrient cycling, making it unsuitable for native vegetation (Pyšek & Richardson, 2006). An invasive species *Phragmites australis* has become a major ecological threat for agri-silvi-pastoral system and biodiversity in Ladakh due to its expansion not only in marshes and riverbeds but also in the agriculture land (Raghuvanshi et al., 2023). Invasive species can also alter hydrology by consuming excessive water or changing water infiltration rates, which is especially problematic in water-scarce cold deserts. The introduction of non-native species can lead to habitat degradation for native fauna by changing plant community composition and decline in food availability and shelter for wildlife.

Invasive species leads to a decline in species richness, ecosystem productivity, and accelerate soil erosion by destabilizing the ground cover. For instance, an invasive species, *Cirsium arvense* in Ladakh has adapted to roadside habitats (a resource-poor environment) by conserving resources and also to a resource-rich environment (agricultural and marshy land habitat) by acquiring them due to its ability to invade diverse habitats through trait adaptations and resource-use strategies (Hakim et al.,

2023). There are many plant species in the Indian cold desert areas which have become rare and endangered primarily due to a combination of overexploitation, habitat degradation, and the impacts of climate change (Kala, 2005). *Saussurea costus* (Falc.) Lipsch., a threatened species, valued for its medicinal properties, has experienced population reduction from overexploitation and environmental changes in the cold desert of Indian Himalaya (Kala, 2003). Alien species in cold desert of Ladakh is most prevalent in disturbed habitats, such as arable land and roadside habitats, which occupy the highest numbers of alien species, followed by wastelands. Fortunately, a comparatively a smaller number of alien species are found in other natural habitats, such as riverbanks, which are dynamic and frequently disturbed by water flow, and screes, which are characterized by loose stone debris (Zargar et al., 2024, 2025).

A large number of plant species in Indian cold deserts are also used for making medicines. Climate change impacts medicinal plants by causing habitat loss, shifting species ranges, altering life cycles and chemical composition, and threatening both biodiversity and traditional healthcare systems. Species with restricted distributions and high medicinal value are at greatest risk (Kala, 2005). Climate change may alter the chemical composition of many medicinal plant species, particularly the secondary metabolites, which are crucial for their therapeutic properties (Gairola et al., 2010). Changes in temperature, rainfall, and other environmental variables can influence the concentration and efficacy of these compounds (Qaderi et al., 2023). At higher altitudes, plants typically invest more in secondary metabolites like flavonoids to enhance their antioxidant capacity, which is crucial for survival in increased UV exposure and other abiotic stresses (Rodríguez-Hernández, 2019). Climate change, through factors such as elevated temperature, increased CO_2, altered precipitation, and enhanced UV radiation, directly affects the synthesis and accumulation of secondary metabolites compounds like phenolics, terpenoids, alkaloids, and flavonoids that are crucial for the medicinal value of high-altitude plants. This can enhance or reduce the medicinal value of these plants, depending on the specific compounds affected and the nature of the climatic changes (Gairola et al., 2010; Li et al., 2020; Rodríguez-Hernández, 2019).

Climate change driven by rising temperature, shifting precipitation patterns, and altered snow cover, with wide-ranging ecological consequences, alters plant phenology or the timing of key life cycle events in the cold desert regions of the Indian Himalaya (Kala, 2013). There are reports claiming that growing season has advanced in the high altitudes of the Himalaya, which is linked to warmer spring temperatures and earlier snowmelt (Khanduri et al., 2008; Telwala et al., 2013). Besides, the length of the growing season has increased with plants experiencing both earlier green-up and, in some areas, delayed senescence (the end of the growing season). Alterations in the timing of flowering, fruiting, and leaf-fall disrupt plant reproductive cycles and ecological interactions (Nitnaware, 2023). A study on *Ferula jaeschkeana* reports that there was a significant reduction in the number of umbels per flowering stem, umbellules per umbel, and flowers per umbellule with increasing elevations (Yaqoob & Nawchoo, 2017). Phenological shifts affect interactions between species (e.g., plants and their pollinators or seed dispersers), due to mismatch of timing of resource availability, increase in species competition, and subsequently decline in both plant and animal populations, which leads to instability of ecosystem. When the timings of plant growth no longer align with the migration schedules of herbivores, it can significantly impact the migratory patterns of herbivores in the Himalayas. The phenological shifts affect plant–pollinator relationships, as well (Chauhan & Uniyal, 2024). Altered phenology and climate conditions may favor invasive species that can outcompete native plants, further threatening local biodiversity and ecosystem balance.

5 Uniqueness of the Indian Cold Deserts and Global Contrasts

Deserts around the world exhibit a wide range of landscapes and geomorphological processes. One of the main factors influencing their character is the climate. The Thar, Kalahari, and Australian deserts have better moisture contents in comparison to hyper-arid Atacama, Libyan, and Namib deserts (Goudie, 2009). The cold deserts in Indian Himalaya are particularly distinctive due to their high levels of solar radiation and larger-temperature fluctuations, which are further influenced by elevation and location within the Himalayan rain shadow (Santra, 2015). Ladakh region of Indian cold desert receives only 50–70 mm of average annual rainfall. Apart from rainfall, the snow melt is the major source of water on which the all forms of life depend in the Indian cold desert areas. The trend analysis data of past one century from 1901 to 2000 has confirmed decrease in summer precipitation in Ladakh region. Such an erratic behavior of precipitation regime can have far reaching consequences on the vegetation distribution and subsequently the socio-economic conditions of the local people (Shafiq et al., 2016).

The low amount of rainfall in the Indian cold deserts has evolved unique vegetation composition, which is sparsely distributed, adapted to harsh, and arid environment. However, the changing climatic conditions have imposed pressures on the existing vegetation cover. The plants either have to withstand the prevailing climatic conditions or to develop a mechanism for adaption or shift their distribution range to suitable climatic conditions (Kala, 2013). While many plant species in the Indian cold deserts are vulnerable to anthropogenic climate change, there is an evidence of adaptive responses, including changes in functional traits, physiological acclimation, and upward range shifts (Gairola et al., 2010; Telwala et al., 2013). There is a report, which claims that about 87% of 124 endemic plant species in Sikkim has shifted their upper elevation limits upward (by 23–998 m), resulting in increased species richness in upper alpine zones compared to the nineteenth century (Telwala et al., 2013).

Surveys of alpine summits between 2014 and 2018 in Kashmir Himalaya found increased species richness and cover of dominant shrubs, graminoids, and forbs on lower summits, indicating resilience and community-level adaptation. However, species richness decreased at the highest elevations, highlighting limits to upward migration (Hamid et al., 2020). To thrive in harsh conditions, plants can respond to environmental variability through phenotypic plasticity (ability of an individual plant to display different phenotypes depending on its environment, even if the genotype remains the same) or genetic adaptation to the local conditions (Becker et al., 2006). For alpine plants, phenotypic plasticity occurs more frequently at higher altitudes as it enables fast adaptation to a highly variable environment (Pluess et al., 2011; Yaqoob & Nawchoo, 2017). These adaptations enhance survival and community resilience, but also underscore the need for long-term monitoring to assess the sustainability of such changes and the risks of local extinctions of plant species at the highest elevations.

Grasslands store approximately one-third of the global terrestrial carbon stocks and act as an important soil carbon sink (Bai & Cotrufo, 2022). Similarly, the cold desert vegetation composed of grasses, herbs, and dwarf woody species sequesters and stores a good amount of carbon, hence mitigating the adverse impacts of global warming and increasing CO_2 concentrations. Besides, the cold desert plants play a crucial role in maintaining ecosystem stability by preventing soil erosion, supporting unique food webs, and sustaining animal species, some of which are critically endangered (e.g., snow leopard, Tibetan antelope). The conservation of cold desert flora in India is crucial for maintaining unique biodiversity, fostering ecological balance, sustaining traditional knowledge, and enhancing the welfare of local communities.

6 Conservation and Management of Cold Desert Vegetation in Changing Climate

The cold desert vegetation in Indian Himalaya contributes to mitigate the impacts of climate change through carbon sequestration and maintaining soil and atmospheric temperature and humidity by evaporation. Being rhizomatous species, besides aboveground biomass accumulation, they also store carbon as below ground biomass, enhance soil carbon storage by contributing organic matter through litter accumulation, and fine root biomass (Kala, 1998). Deep-rooted woody species like junipers stabilize soil and improve its structure, enabling better carbon retention (Rawat & Everson, 2012). Shrubs such as *Hippophae rhamnoides* contribute significantly to soil aggregation and stability, promoting carbon sequestration in fragile desert soils (Tundup et al., 2017). Cold desert plants exhibit physiological adaptations, such as slow growth rates and water-efficient mechanisms, which reduce CO_2 outflow from soils and maximize carbon retention. In general, the cold desert plant species plays a vital role in enhancing soil carbon storage through their root systems, litter contribution, and adaptive strategies (Acharya et al., 2012). Their ability to sequester carbon is critical for ecological restoration and combating climate change in these fragile environments. Considering the significance of cold desert vegetation, attempts have been made to conserve this unique plant species diversity along with their ecosystems.

In view of protecting the biodiversity of cold desert areas in Indian Himalaya, a number of protected areas, including national parks and sanctuaries, have been established by the respective state governments. The Hemis National Park, Karakoram Wildlife Sanctuary, and Changthang Wildlife Sanctuary in Ladakh and Pin Valley National Park and Kibber Wildlife Sanctuary in Himachal Pradesh have been established (Kala, 2005). Besides, the United Nations Educational, Scientific and Cultural Organization (UNESCO) has recognized a part of the cold desert in Himachal Pradesh as a biosphere reserve under its Man and the Biosphere Programme, highlighting the importance of this region for biodiversity and its conservation. The climate change is a quite unique problem, which may not be resolved by fencing and protection nonetheless establishing protected areas can help in safeguarding the plant and animal species (Kala & Silori, 2013). Participation of local communities for mitigating the adverse impacts of climate change is equally important as these communities face complex and interlinked challenges in adapting to climate change.

7 Role of Monasteries in Plant Diversity Conservation

Buddhist monasteries in the cold deserts of the Indian Himalaya are vital actors in mitigating climate change impacts and conserving this sensitive ecosystem. Monasteries serve as centers of spiritual and cultural life, fostering a deep sense of environmental responsibility rooted in Buddhist teachings of interdependence and respect for all forms of life (Karmapa & Dorje, 2011). Through conservation, innovation (like ice stupas), cultural leadership, education, and collaboration with broader policy initiatives, they can help build resilience and inspire sustainable adaptation in cold deserts of Indian Himalaya. Monastic leaders, such as the Dalai Lama and the Karmapa Lama, actively promote environmental awareness and encourage both monks and local communities to adopt sustainable practices to combat adverse effects of climate change (Bhutia, 2022; Lama & Alt, 2020). Some monasteries have collaborated with scientists and NGOs to establish research and learning centers focusing climate adaptation, further bridging traditional knowledge with modern science. The sacred status of monasteries and their lands has contributed to the preservation of natural ecosystem and environments in the cold deserts areas of Indian Himalaya. This cultural reverence has been recognized in regional climate adaptation declarations and conservation frameworks, highlighting the synergy between traditional beliefs and government-backed climate action.

Monasteries promote tree planting, the preservation of natural landscapes, and the establishment of organic gardens. Woody plants near monasteries have been found to possess higher aboveground biomass and better quality compared to other areas, demonstrating the positive influence of monastic stewardship on local biodiversity and conservation of vegetation (Zhang et al., 2024). Buddhist monasteries designate forests and mountains as sacred sites, which enjoy higher protection due to their religious significance. Harvesting of plants and unsustainable resource extraction is either banned or discouraged in such sacred sites. Besides, monasteries encourage to actively plant trees and restore degraded lands, especially on monastery-owned or community lands (Tsering et al., 2023). They often select native species for reforestation, which helps to integrate new plantations into the local ecosystem and ensures to increase their survival rates. Monasteries also promote sustainable extraction of natural resources and discourage over-harvesting, ensuring that natural vegetation can recover and regenerate over time. Monasteries also help in regulating grazing on pasture lands, limiting the number of cattle and sometimes fencing off areas to allow natural regeneration. Rotational use of forest patches ensures some areas to rest and recover, which improves biodiversity and forest health. By setting an example and mobilizing local support, monasteries influence broader community attitudes and behaviors toward forest protection and regeneration (Shen et al., 2012; Zhang et al., 2024).

8 Conclusion

The cold deserts in India, particularly in areas like Ladakh, Lahaul-Spiti, and parts of Uttarakhand, harbor some unique diversity of plant species exclusively adapted to extreme conditions such as low precipitation, high altitude, and temperature fluctuations. Many of these plant species exhibit specialized adaptations, including deep root systems, waxy foliage, and tissues that store water, which contribute to the stability of the ecosystem amidst the severe climatic conditions prevalent in these cold deserts. However, at present, the cold desert vegetation faces threats from climate change, overgrazing, habitat loss, and unsustainable harvesting, which jeopardize both plant diversity and the essential services they provide. Many cold desert plants in Indian Himalaya are endemic and possess medicinal properties, hence their conservation is imperative for human welfare and global biodiversity. Conservation strategies, such as establishing protected areas and promoting sustainable land use practices, including use of traditional knowledge systems, are necessary to safeguard these unique plant resources for future generations. Besides, the buddhist monasteries in Indian cold deserts continue to be roped in and encouraged for safeguarding the cold deserts landscape by protecting sacred landscapes, leading reforestation efforts, managing grazing, and fostering collaboration with broader conservation initiatives.

References

Acharya, S., Charan, G., Singh, N., & Srivastava, R. B. (2012). Soil organic carbon sequestration of cold desert Ladakh. *Range Management and Agroforestry, 33*(1), 79–82.

Bai, Y., & Cotrufo, M. F. (2022). Grassland soil carbon sequestration: Current understanding, challenges, and solutions. *Science, 377*(6606), 603–608. https://doi.org/10.1126/science.abo238

Becker, U., Colling, G., Dostal, P., Jakobsson, A., & Matthies, D. (2006). Local adaptation in the monocarpic perennial *Carlina vulgaris* at different spatial scales across Europe. *Oecologia, 150*, 506–518.

Bhutia, K. D. (2022). Caring for the Land, Caring for the Dharma: The environmental history of Buddhism at Pemayangtse Monastery, Sikkim, as a resource for contemporary conservation initiatives. In *Religion and nature conservation* (pp. 15–28). Routledge.

Bhutiyani, M. R., Kale, V. S., & Pawar, N. J. (2008). Changing streamflow patterns in the rivers of northwestern Himalaya: Implications of global warming in the 20. *Current Science, 95*, 618–626.

Billings, W. D. (1974). Adaptations and origins of alpine plants. *Arctic and Alpine Research, 6*(2), 129–142.

Bliss, L. C. (1962). Adaptations of Arctic and alpine plants to environmental conditions. *Arctic, 15*(2), 117–144.

Bunzhuo, P., Lijie, P., Haosheng, B., & Higgitt, D. L. (1997). Vertical zonation of landscape characteristices in the Namjagbarwa massif of Tibet, China. *Mountain Research and Development, 17*, 43–48.

Chauhan, M., & Uniyal, V. P. (2024). Pollinators and pollination management through indigenous knowledge based adaptation in Indian trans-himalayan region: Climate change perspective. In A. Borthakur & P. Singh (Eds.), *Learning 'from' and 'with' the locals. Advances in Asian human-environmental research.* Cham: Springer. https://doi.org/10.1007/978-3-031-51696-2_4

Chaurasia, O. P., & Singh, B. (1996). *Cold desert plants* (vols. I–IV). Leh, India: Field Research Laboratory, DRDO.

Dimri, A. P., Niyogi, D., Barros, A. P., Ridley, J., Mohanty, U. C., Yasunari, T., & Sikka, D. R. (2015). Western disturbances: A review. *Reviews of Geophysics, 53*(2), 225–246.

Dolezal, J., Dvorsky, M., Kopecky, M., Liancourt, P., Hiiesalu, I., Macek, M., Altman, J., Chlumska, Z., Rehakova, K., Capkova, K., Borovec, J., Mudrak, O., Wild, J., & Schweingruberet, F. (2016). Vegetation dynamics at the upper elevational limit of vascular plants in Himalaya. *Scientific Reports, 6*, 24881 (2016). https://doi.org/10.1038/srep24881

Downes, B. J., Lake, P. S., Schreiber, E. S. G., & Glaister, A. (1998). Habitat structure and regulation of local species diversity in a stony upland stream. *Ecology, 68*, 237–257.

Farina, A. (1998). *Principles and methods in landscape ecology* (p. 235). Chapman and Hall.

FSI. (2015). *India State of forest report.* Forest Survey of India, Dehradun, Uttarakhand, India.

Gairola, S., Shariff, N. M., Bhatt, A., & Kala, C. P. (2010). Influence of climate change on production of secondary chemicals in high altitude medicinal plants: Issues needs immediate attention. *Journal of Medicinal Plants Research, 4*(18), 1825–1829.

Goudie, A. S. (2009). Global deserts and their geomorphological diversity. In A. J. Parsons & A. D. Abrahams (Eds.), *Geomorphology of desert environments.* Dordrecht: Springer. https://doi.org/10.1007/978-1-4020-5719-9_2

Hakim, N., Ahmad, M., Rathee, S., Sharma, P., Kaur, S., Batish, D. R., & Singh, H. P. (2023). Invasive *Cirsium arvense* displays different resource-use strategies along local habitat heterogeneity in the trans-Himalayan region of Ladakh. *Environmental Monitoring and Assessment, 195*(6), 730. https://doi.org/10.1007/s10661-023-11221-w

Hamid, M., Khuroo, A. A., Malik, A. H., Ahmad, R., & Singh, C. P. (2020). Assessment of alpine summit flora in Kashmir Himalaya and its implications for long-term monitoring of climate change impacts. *Journal of Mountain Science, 17*(8), 1974–1988.

IPBES, et al. (2019). Summary for policymakers of the global assessment report on biodiversity and ecosystem services of the Intergovernmental Science-Policy Platform on Biodiversity and Ecosystem Services. In S. Díaz, J. Settele, E. S. Brondizio, H. T. Ngo, M. Guèze, J. Agard, A. Arneth, P. Balvanera, K. A. Brauman, & S. H. M. Butchart (Eds.), *IPBES secretariat.* Bonn.

Kachroo, P., Sapru, B. L., & Dhar, U. (1977). *Flora of Ladakh* (172 pp.). Dehra Dun, India: Bishen Singh Mahendra Pal Singh.

Kala, C. P. (1998). *Ecology and conservation of alpine meadows in the valley of flowers National Park, Garhwal Himalaya.* Ph.D. thesis, Forest Research Institute (Deemed University), Dehradun, India.

Kala, C. P. (2000). Status and conservation of rare and endangered medicinal plants in the Indian trans-Himalaya. *Biological Conservation, 93*(3), 371–379.

Kala, C. P. (2003). *Medicinal plants of Indian trans-Himalaya: Focus on Tibetan use of medicinal resources*. Dehradun, India: Bishen Singh Mahendra Pal Singh.

Kala, C. P. (2005). Indigenous uses, population density, and conservation of threatened medicinal plants in protected areas of the Indian Himalayas. *Conservation Biology, 19*(2), 368–378.

Kala, C. P. (2011). Floral diversity and distribution in the high altitude cold Desert of Ladakh, India. *Journal of Sustainable Forestry, 30*, 360–369. https://doi.org/10.1080/10549811.2011.534036

Kala, C. P. (2013). Climate change and challenges of biodiversity conservation. In C. P. Kala & C. S. Silori (Eds.), *Biodiversity, communities and climate change* (pp. 259–269). Teri Publications.

Kala, C. P., & Jayapal, R. (1999). *Community characterization of vegetation and birds of Ladakh: An autumn survey* (91 pp). Project report of WII-USFWL. Dehradun, India: Wildlife Institute of India.

Kala, C. P., & Mathur, V. B. (2002). Patterns of plant species distribution in the trans-Himalayan region of Ladakh. *India. Journal of Vegetation Science, 13*(6), 751–754.

Kala, C. P., & Silori, C. S. (2013). *Biodiversity, Communities and Climate Change*. TERI Publications, New Delhi.

Karmapa, H., & Dorje, O. T. (2011). Walking the path of environmental Buddhism through compassion and emptiness. *Conservation Biology, 25*(6), 1094–1097.

Khanduri, V. K., Sharma, C. M., & Singh, S. P. (2008). The effects of climate change on plant phenology. *The Environmentalist, 28*, 143–147.

Lama, D., & Alt, F. (2020). *Our only home: A climate appeal to the world*. Toronto, Ontario, Canada: Harlequin Enterprises ULC.

Leng, R., Harrison, S., & Anderson, K. (2023). Himalayan alpine ecohydrology: An urgent scientific concern in a changing climate. *Ambio, 52*(2), 390–410. https://doi.org/10.1007/s13280-022-01792-2

Li, Y., Kong, D., Fu, Y., Sussman, M. R., & Wu, H. (2020). The effect of developmental and environmental factors on secondary metabolites in medicinal plants. *Plant Physiology and Biochemistry, 148*, 80–89. https://doi.org/10.1016/j.plaphy.2020.01.006

Marcela, F., Ana, C., Cecilia, E., & Donaldo, B. (1998). High-Andean vegetation and environmental gradients in northwestern Patagonia, Argentina. *Journal of Vegetation Science, 9*, 307–316.

Mohanty, R. B., Mishra, A. K., Mishra, K., Yadava, A. K., Quamar, M. F., Barua, I. C., & Kar, R. (2024). Early onset of aridity in the past millennium: Insights from vegetation dynamics and climate change in the alpine, cold-desert region of Trans Himalaya, India. *Plos One, 19*(1), Article e0295785. https://doi.org/10.1371/journal.pone.0295785

Mountain Research Initiative EDW Working Group. (2015). Elevation-dependent warming in mountain regions of the world. *Nature Climate Change, 5*, 424–430. https://doi.org/10.1038/nclimate2563

Nitnaware, H. (2023). Climate change severely affecting biodiversity in Hindu Kush Himalayas: ICIMOD. Down to Earth, 20 June 2023.

Passang, S., Schmidt, S., & Nüsser, M. (2022). Topographical impact on snow cover distribution in the trans-Himalayan region of Ladakh, India. *Geosciences, 12*(8), 311. https://doi.org/10.3390/geosciences12080311

Pauchard, A., Kueffer, C., Dietz, H., Daehler, C. C., Alexander, J., Edwards, P. J., Arévalo, J. R., Cavieres, L. A., Guisan, A., Haider, S., Jakobs, G., McDougall, K., Millar, C. I., Naylor, B. J., Parks, C. G., Rew, L. J., & Seipel, T. (2009). Ain't no mountain high enough: Plant invasions reaching new elevations. *Frontiers in Ecology and Environment, 7*(9), 479–486. https://doi.org/10.1890/080072

Pierson, F. B., Bates, J. D., Svejcar, T. J., & Hardegree, S. P. (2007). Runoff and erosion after cutting western juniper. *Rangeland Ecology & Management, 60*(3), 285–292.

Pluess, A. R., Frei, E., Kettle, C. J., Hahn, T., & Ghazoul, J. (2011). Plant growth and fitness of *Scabiosa columbaria* under climate warming conditions. *Plant Ecology & Diversity, 4*(4), 379–389.

Pyšek, P., & Richardson, D. M. (2006). The biogeography of naturalization in alien plants. *Journal of Biogeography, 33*(12), 2040–2050.

Qaderi, M. M., Martel, A. B., & Strugnell, C. A. (2023). Environmental factors regulate plant secondary metabolites. *Plants, 12*(3), 447. https://doi.org/10.3390/plants12030447

Quoreshi, A. M., Kumar, V., Adeleke, R., Qu, L., & Atangana, A. R. (2022). Soils and vegetation in desert and arid regions: Soil system processes, biodiversity and ecosystem functioning, and restoration. *Frontiers in Environmental Science, 10*, Article 962905. https://doi.org/10.3389/fenvs.2022.962905

Raghuvanshi, M. S., Sawal, R. K., Landol, S., Dorje, N., Pandey, L., Enoch, S., Raza, M., & Saxena, A. (2023). Invasive common reed (*Phragmites Australis* (Cav.) Trin. ex Steud.): A serious ecological threat to Ladakh Tethys Himalayan biodiversity in changing climate. *American Journal of Climate Change, 12*(3), 321–334. https://doi.org/10.4236/ajcc.2023.123015

Rawat, Y. S., & Everson, C. S. (2012). Ecological status and uses of juniper species in the cold desert environment of the Lahaul valley, North-western Himalaya, India. *Journal of Mountain Science, 9*, 676–686.

Rey-Benayas, J. M., Colomen, M. G. S., & Catherine, L. (1999). Effects of area, environmental status and environmental variation on species richness per unit area in Mediterranean wetlands. *Journal of Vegetation Sciences, 10*, 275–280.

Rodríguez-Hernández, D. (2019). Secondary Metabolites as a Survival Strategy in Plants of High Mountain Habitats. Boletín Latinoam. *Y Del Caribe Plantas Med Y Aromática, 18*, 444–458. https://doi.org/10.35588/blacpma.19.18.5.29

Santra, P. (2015). Scope of solar energy in cold arid region of India at Leh Ladakh. *Annals of Arid Zone, 54*(3&4), 109–117.

Shafiq, M. U., Bhat, M. S., Rasool, R., Ahmed, P., Singh, H., & Hassan, H. (2016). Variability of precipitation regime in Ladakh region of India from 1901–2000. *Journal of Climatology & Weather Forecasting, 4*(2), 165. https://doi.org/10.4172/2332-2594.1000165

Sharma, M. K., Hopak, N. E., & Chawla, A. (2024). Alpine plant species converge towards adopting elevation-specific resource-acquisition strategy in response to experimental early snow-melting. *Science of the Total Environment, 907*, Article 167906. https://doi.org/10.1016/j.scitotenv.2023.167906

Sharma, N., & Kala, C. P. (2022). Patterns in plant species diversity along the altitudinal gradient in Dhauladhar mountain range of the North-West Himalaya in India. *Trees, Forests and People, 7*, Article 100196. https://doi.org/10.1016/j.tfp.2022.100196

Shen, X., Lu, Z., Li, S., & Chen, N. (2012). Tibetan sacred sites: Understanding the traditional management system and its role in modern conservation. *Ecology and Society, 17*(2), 13. https://doi.org/10.5751/ES-04785-170213

Snyder, K. A., Evers, L., Chambers, J. C., Dunham, J., Bradford, J. B., & Loik, M. E. (2019). Effects of changing climate on the hydrological cycle in cold desert ecosystems of the Great Basin and Columbia Plateau. *Rangeland Ecology & Management, 72*(1), 1–12. https://doi.org/10.1016/j.rama.2018.07.007

Telwala, Y, Brook, B. W., Manish, K., & Pandit, M. K. (2013). Climate-induced elevational range shifts and increase in plant species richness in a Himalayan biodiversity epicentre. *PLoS One, 8*(2), e57103. https://doi.org/10.1371/journal.pone.0057103

Tsering, P. (2023). Tibetan buddhist monastery-based rangeland governance in Amdo Tibet, China. . *Land Use Policy, 131*, Article 106756. https://doi.org/10.1016/j.landusepol.2023.106756

Tundup, P., Wani, M. A., Dawa, S., & Dorjay, N. (2017). Soil conservation methods and revegetation of trans-Himalayan cold desert region-Ladakh. *International Journal of Current Microbiology and Applied Sciences, 6*(3), 1996–2001.

Walker, A. S. (1982). Deserts of China: Deserts now make up more than 13 percent of the land area in China, and various methods are being used to transform them into farmland. *American Scientist, 70*(4), 366–376.

Webber, P. J. (1978). Spatial and temporal variation of the vegetation and its production, Barrow, Alaska. In L. L. Tieszen (Eds.), *Vegetation and production ecology of an alaskan arctic tundra. Ecological studies* (vol. 29, pp. 37–112). New York: Springer-Verlag.

Yaqoob, U., & Nawchoo, I. A. (2017). Impact of habitat variability and altitude on growth dynamics and reproductive allocation in *Ferula jaeschkeana* Vatke. *Journal of King Saud University-Science, 29*(1), 19–27. https://doi.org/10.1016/j.jksus.2015.10.002
Zargar, S. A., Khuroo, A. A., Reshi, Z. A., & Ganie, A. (2024). Recent species' range expansion in the Himalayan Highlands: New distribution records of naturalized alien plants from Ladakh, India. *Biology Bulletin, 51*, 576–585. https://doi.org/10.1134/S1062359023606079
Zargar, S. A., Malik, R. A., Khuroo, A. A., Ganie, A. H., & Reshi, Z. A. (2025). Alien flora of the Himalayan Highlands: Naturalised and invasive plants in the Trans-Himalayan region of Ladakh, India. *Biological Invasions, 27*(4), 1–19. https://doi.org/10.1007/s10530-025-03562-z
Zhang, S., Wu, T., & Guo, L. (2024). The positive influence of Tibetan Buddhist monasteries on forest conservation. *Heritage Science, 12*, 413. https://doi.org/10.1186/s40494-024-01535-4

Plant Life and Adaptations in High Mountainous Regions

Suman Chandra and Hemant Lata

1 Introduction

High-altitude regions constitute reservoirs of alpine, subalpine-aromatic, ornamental, medicinal, and other economically valuable plant species. Among the Himalayan ranges, the Garhwal Himalaya exhibits the highest floristic diversity. Vegetation at these elevations is predominantly xerophytic, with increasing altitude driving dwarfism, multi-branched growth, and xeromorphic leaf adaptations. Alpine flora comprises evergreen and deciduous prostrate shrubs, short-stemmed herbaceous perennials (grasses, sedges, cushion, and rosette plants), lichens, and mosses. Most species are perennial, with perennation via roots, rhizomes, bulbs, bulbils, runners, and suckers. Underground structures serve as carbohydrate storage organs and often exceed shoot biomass in dry weight (Chen et al., 2021; Wang et al., 2024).

A substantial portion of alpine and arctic vegetation biomass is subterranean. In wet arctic tundra, majority of living plant material resides below the soil surface during summer (D'Imperio et al., 2018), with similar patterns observed in moist alpine meadows (Zhu et al., 2024). Drier sites exhibit lower belowground biomass (Scott & Billings, 1964). These structures are critical for carbohydrate storage, water relations, and mineral nutrition (Mooney & Billings, 1960; Wang et al., 2024). Despite low temperatures, high-altitude plants capture and store energy per growing season comparable to, or exceeding, that of lowland species (Scott & Billings, 1964;

S. Chandra · H. Lata
High Altitude Plant Physiology Research Center, HNB Garhwal University, Srinagar, Uttarakhand, India

S. Chandra (✉) · H. Lata
National Center for Natural Products Research, School of Pharmacy, University of Mississippi, University, MS 38677, USA
e-mail: suman@olemiss.edu

Department of BioMolecular Sciences, School of Pharmacy, University of Mississippi, University, MS 38677, USA

Y. S. Rawat and C. P. Kala (eds.), *Cold Deserts*,
https://doi.org/10.1007/978-3-032-29119-6_5

Zhang & Zhang, 2019). However, the brevity of the alpine growing season results in lower annual energy accumulation relative to lowland plants.

Alpine plants exhibit rapid growth following snowmelt. Perennial buds and dormant annual seeds resume development promptly, completing the life cycle within the brief summer season. Snowmelt functions as a key environmental trigger, terminating dormancy, inducing rapid germination, and initiating vegetative growth within days. In our laboratory at High Altitude Plant Physiology Research Center (HAPPRC), we have worked on various aspects of alpine plants, exploring their ecological adaptations, physiological responses, and survival strategies across different environmental gradients. Our research has focused on understanding how these unique species cope with extreme mountain conditions and how their distribution and functions may shift under changing climates. Through field studies, experimental analyses, and long-term monitoring, we aim to contribute valuable insights into the conservation and sustainable management of alpine plant communities (Chandra, 2004; Chandra & Lata, 2022, 2026; Chandra & Joshi, 2002; Chandra et al., 2011, 2012, 2013, 2026; Chandola et al., 2026; Purohit et al., 1998; Singh et al., 1999). A set of representative high-value Himalayan flora from high altitudes is illustrated in Fig. 1.

Fig. 1 Selected high-value alpine medicinal plants from the Himalayas: **a** *Aconitum balfourii*, **b** *Aconitum heterophyllum*, **c** *Nardostachys grandiflora*, **d** *Picrorhiza kurroa*, **e** *Rheum emodi*, and **f** *Podophyllum hexandrum*

2 Plant Growth and Development at High Altitudes

Alpine plants take advantage of favorable conditions to grow. Most of them grow quickly when the temperature and moisture are optimum, although the pattern of growth varies widely for different plant groups. The period of dormancy in alpine plants can end anywhere from mid-spring to late summer, with the exact timing determined by the surrounding environmental conditions. Plants usually start growing again when the snow melts or, in places without snow, when the ground and air warm up to about 0 °C. Having liquid water is crucial for early plant growth. In some cases, plants can grow a little even under leftover snow, as long as water is present near the soil surface or in the top layers of soil. Longer daylight hours also seem to help certain plants' buds come out of dormancy, but without temperatures above freezing, extra daylight alone cannot trigger growth (Billings et al., 1965; Maurya & Bhalerao, 2017). Once dormancy ends, shoots start growing quickly. Stored sugars in the roots and underground stems are rapidly transported to the new shoots and leaves to support their development (Mooney & Billings, 1960). These carbohydrates are used not only to build new tissues but also to produce anthocyanins—a type of pigment. As a result, young and actively growing shoots often appear dark red. These anthocyanins help protect plant tissues by absorbing ultraviolet (UV) radiation, reducing the risk of damage (Li et al., 2020).

A characteristic feature frequently observed in the stems of rapidly growing herbaceous species is the absence of pith. This structural adaptation results in the formation of a hollow stem, which promotes material efficiency and facilitates internal photosynthesis (Billings & Godfray, 1967; Tarvainen et al., 2018). Such stems are capable of elongating several centimeters per day, and the elevated respiration rate during these early growth phases supplies additional carbon dioxide that is utilized within the hollow stem. Notably, the internal stem temperature under sunlight can exceed the ambient air temperature by as much as 20 °C. Consequently, the initial stages of shoot development occur with limited dependence on ambient temperature. However, once leaf expansion and chlorophyll accumulation progress, subsequent growth becomes increasingly regulated by temperature. In the later part of the growing season, factors such as drought stress, photoperiod, and carbohydrate storage further influence growth dynamics (Billings & Mooney, 1968; Palacio et al., 2021).

By the end of the growing season, large quantities of carbohydrates accumulate in the roots and rhizomes. During winter, low temperatures and dormancy minimize the utilization of these underground reserves. When dormancy ends in early summer, the plant draws on its stored carbohydrates to fuel the rapid early growth and development of new shoots, which significantly reduces these energy reserves. At this stage, growth occurs so vigorously that respiration surpasses photosynthesis, causing carbohydrates to be consumed more rapidly than they are produced (Zhang et al., 2022). Once sufficient vegetative growth has been achieved, newly synthesized photosynthates are transported back to the underground organs. As a result, by the time the plant enters dormancy, carbohydrate reserves in these storage tissues are replenished to their annual maximum levels. The general pattern of the carbohydrate cycle in alpine plants was outlined by Mooney and Billings (1960). Alpine plants exhibit a notably efficient energy fixation process despite their short growing season and reduced stature relative to temperate herbaceous communities (Bliss, 1966). Their rapid flowering and fruiting represent key adaptations to the brief period of growth. This swift spring development is possible only when floral initiation takes place in the preceding year, followed by successful overwintering.

Although many alpine plant species are capable of producing fruits and seeds in adequate quantities, many others generate very few seeds, particularly in harsher environmental conditions. Bliss (1956) observed that species which yield abundant seeds in meadow habitats produce few, if any, seeds in areas exposed to strong winds or in regions where snow melts late. In numerous instances, alpine plants may bloom profusely yet fail to produce viable seeds (Klatt et al., 2018). Most alpine seeds do not fully mature until autumn and therefore remain ungerminated until the following spring (Wang et al., 2015). Seed production at high altitudes is often irregular and limited, with a low proportion of viable seeds. Studies in alpine regions (Bliss, 1956) have shown that many species flower but fail to set seed, and that reproductive success declines with increasing environmental severity. In some species, seed weight also decreases with altitude (Amici et al., 2021). Although certain alpine plants produce seeds regularly and in large quantities, these typically ripen between mid-July and September and are dispersed shortly thereafter. However, in most species, maturation occurs too late for germination within the same growing season.

Dormancy in alpine plant species has been known for a long time (Schwienbacher et al., 2011). Back in 1913, Braun observed that many alpine plants would not germinate under laboratory conditions, even following exposure to low temperatures. Years later, Amen (1966) proposed that most of these species were actually not dormant at all. However, the poor germination of some alpine plants has puzzled both scientists and gardeners for many years (Orsenigo et al., 2015). Dormancy in alpine plants is induced by photoperiod, reduced thermoperiods and lower soil water potential in alpine regions (Bernareggi et al., 2016; Rosbakh et al., 2022). In a review by Baskin et al. (1998), it was shown that over 70% of arctic-alpine species have dormant seeds, mostly with physiological dormancy, and to a lesser extent, physical dormancy. Despite several thorough early investigations into seed germination in alpine habitats (Körner, 2003) and an early summary by Amen (1966), we still do not fully understand what causes dormancy in many alpine species (Baskin et al., 1998).

3 Germination of Seeds

Seed germination is a critical event in the plant life cycle, determining the probability of successful seedling establishment and eventual survival to maturity. The germination ecology of alpine plant species has been extensively investigated (Mondoni et al., 2016; Rosbakh et al., 2022; Vázquez-Ramírez & Venn, 2021). In many temperate species, there is often a delay between seed maturation or dispersal and the onset of germination, primarily due to environmental conditions unfavorable for germination. During such intervals, seeds enter a state of conditional dormancy (Song et al., 2023), allowing them to survive until conditions become suitable for growth.

3.1 Influence of Light and Temperature on Germination

Among environmental factors, light and temperature are of paramount importance in regulating germination and determining a species' ecological distribution. Populations of the same species may exhibit variations in their responses to these factors (Fletcher et al., 2020; Maleki et al., 2022). Investigations by Haberlandt (1888) established the minimum, optimum, and maximum temperature thresholds for seed germination. Harrington (1963) demonstrated that lower temperatures prolong the germination period, while Kuleshov (1963) reported that seedling emergence requires higher temperatures than initial germination. Seeds of many alpine plants display optimal germination under low temperatures, underscoring the adaptive significance of this trait in cold environments. Likewise, the role of light in promoting germination has been well documented since the late nineteenth century.

3.2 Genetic and Environmental Interactions in Germination Behavior

Germination requirements are shaped by the interaction between a species' genetic constitution and environmental factors (Imaizumi et al., 2025; Wittington, 1973), as well as by dormancy mechanisms that promote survival during unfavorable seasons (Rosbakh et al., 2020). Investigations on alpine flora of the Garhwal Himalaya (Semwal & Purohit, 1980; Semwal et al., 1983; Zhang et al., 2024) revealed that extreme climatic conditions limit flowering and fruiting, often resulting in low seed viability. Despite these constraints, most alpine species exhibit minimal intrinsic dormancy, and their germination occurs more effectively under light than in darkness (Mooney & Billings, 1961; Semwal & Purohit, 1980). Although extensive research has addressed environmental influences on seed longevity, inconsistencies in experimental control and data recording have limited understanding of the precise mechanisms leading to viability loss (Roberts, 1972; Visscher et al., 2016).

3.3 Seed Longevity and Dormancy Relationships

Seed longevity is profoundly affected by storage conditions, particularly temperature and moisture content. Studies indicate that low moisture and temperature levels extend seed viability (Villiers, 1974). However, excessive-drying or high-drying temperatures can be detrimental, depending on the initial seed moisture content. Extremely low moisture contents can also delay germination, although this effect can be reversed by gradual rehydration under high humidity (Nautiyal et al., 2023). While a direct correlation between dormancy and longevity remains uncertain, dormancy is widely regarded as an adaptive mechanism enabling prolonged seed survival in soil environments (Odum, 1974; Rehmani et al., 2022). Furthermore, seed survival patterns observed in natural soils differ markedly from those found under controlled storage conditions (Roberts, 1972).

4 Establishment of Seedling

Once the requirements for germination have been fulfilled, the subsequent and equally crucial developmental phase is seedling establishment. Compared to the relatively well-documented germination phase, the processes governing seedling establishment in alpine plant species remain insufficiently understood. In alpine environments, seed germination typically occurs in early summer, leaving seedlings only a brief period in which to develop an adequate root system and accumulate sufficient carbohydrate reserves to endure the forthcoming winter. Consequently, seedling mortality in alpine taxa tends to be exceptionally high (Bonde, 1968; Rosbakh et al., 2022). Bonde (1968) suggested that within cushion plant communities, seedling mortality may arise from competition for essential resources such as water, nutrients, or light, although in open microsites, soil drought appeared to be the principal cause of mortality. In contrast, Wager (1938), working in East Greenland, identified low carbohydrate production rates and persistently low temperatures as the main constraints on seedling growth and survival, noting that neither soil water availability nor nutrient supply appeared to be limiting factors. Wager further reported that annual mortality among seedlings younger than 5 years was approximately 50%, and that these plants typically produced only one new set of leaves per year, as carbohydrate assimilation was insufficient to support additional leaf formation.

Experimental investigations by Billings and Mooney (1968) using *Oxyria* seedlings demonstrated that during the initial growing season, resource allocation is primarily directed toward root system development, a strategy that enhances survival under conditions of late-summer surface soil desiccation common in alpine habitats. Bonde (1968) observed that although the number of leaves remained relatively constant between the first and third growing seasons, the dry biomass of seedlings increased approximately threefold over this period. Furthermore, when seedlings were transplanted to greenhouse conditions for 2 months, they exhibited prolific shoot production and a substantial increase in leaf number from an average of 3.8–15 leaves per individual, indicating a pronounced positive response to elevated temperature.

5 Underground Storage of Food-Formation of Tubers

The development of tubers and other subterranean storage structures exhibits several notable physiological and morphological analogies to floral and fruiting processes in plants. In alpine perennials, the accumulation of reserve materials within underground organs serves a dual purpose: first, as an adaptive mechanism for surviving adverse environmental conditions, and second, as a means of vegetative propagation, compensating for the typically low efficiency of viable seed production in these taxa. These storage organs commonly occur in the form of tubers, bulbs, or rhizomes. In alpine perennial dicotyledons, the belowground biomass is substantially greater than the aerial portion and functions predominantly as a carbohydrate reservoir (Mooney & Billings, 1960; Scott & Billings, 1964).

The formation of storage organs is generally associated with the suppression of elongation growth. Moser and Hess (1968) reported that reduced shoot extension coincides with the initiation of tuberous root development. Furthermore, experimental evidence indicates that the process of tuberization can be stimulated through the application of growth regulators or retardants that inhibit vegetative elongation (Moser & Hess, 1968; Sharma et al., 2021). As with many plant developmental phenomena, tuber and storage organ formation is regulated by environmental cues such as photoperiod, light intensity, and temperature. Numerous studies have demonstrated the influence of photoperiod on tuberization (Chen et al., 2025; Hamaoka et al., 2022; Plantenga et al., 2016).

There is broad consensus that tuber initiation is hormonally regulated (Chen et al., 2022; Saidi & Hajibarat, 2021). Tsuda et al. (2019) demonstrated that exogenous gibberellic acid (GA) application can affect the yield and tuber initiation. Auxins were first implicated in potato tuberization and subsequently in onion bulb formation observed that auxin application inhibits tuberization in *Begonia* and that endogenous auxin concentrations in leaves decline during short-day induction. Conversely, cytokinins have been shown to strongly promote tuberization in *Begonia* and potato. Cytokinin-like substances have also been detected in developing tubers and tuberous roots of sweet potato (*Ipomoea batatas*). Later it was identified that *cis*-zeatin riboside and a novel cytokinin in the tubers of this species. Although cytokinins can be readily extracted from the tubers of potato and *Begonia* (Esashi & Leopold, 1968b), their primary effect appears to be stimulation of cell enlargement rather than cell division, suggesting that normal tuber growth involves a more complex hormonal interplay. Additionally, a threefold increase in ethylene production during short-day induction of tuberization in *Dahlia*, preceding the onset of visible tuber development was reported.

In most *Solanum* species, tuber formation is promoted by long nights and low temperatures. In *Begonia evansiana*, tuber formation occurs even under noninductive photoperiods if low night temperatures are provided (Esashi et al., 1964b). Similarly, in Japanese artichoke (*Stachys sieboldii*), high temperatures favor stolon development, while subsequent exposure to lower temperatures induces tuber formation (Li et al., 2024). Burt (1964) found that low-temperature treatment leads to earlier tuber initiation, coinciding with increased sugar concentration in stolon tissues.

The role of leaves as the primary organs perceiving photoperiodic signals has also been reported (Kumar et al., 2016). The majority of tuber filling occurs toward the end of the growing season (Gómez et al., 2018), and subsequent utilization of stored reserves involves the enzymatic hydrolysis of tuber polysaccharides (Zhang et al., 2020).

6 Adaptive Features of High-Altitude Plants

Bradshaw (1965) suggested that an organism's capacity to adjust to its environment depends on achieving an optimal balance of traits that enhance beneficial effects and minimize harmful ones. Adaptations, therefore, are inherited modifications in an organism's structure and function that improve its survival and reproductive success within a particular environment (Carmona et al., 2017). In plants, these adaptive traits vary according to climatic conditions and can be broadly classified into four categories: (1) morphological, (2) physiological, (3) biochemical, and (4) ecological.

6.1 Morphological Adaptations

At high altitudes, plants exhibit several morphological modifications that hold significant adaptive value (Bai et al., 2019; Billings & Mooney, 1968; Harrison & Fitter, 2019; Luxmoore & Millington, 1971). One of the most distinctive adaptations to the harsh alpine environment is the tendency for plants to exhibit reduced stature and to adopt predominantly herbaceous growth forms (Billings & Mooney, 1968). Although the dwarf stature of alpine plants is primarily a genetic trait (Landolt, 1975), phenotypic plasticity enables notable morphological adjustments in response to environmental stress. Most trees at high altitudes are shrubby, twisted, and densely branched, which helps them withstand strong winds. In some species, particularly those from wet meadow communities, the pith is absent in the stem, creating a hollow cavity that maintains a localized phyto-microclimate where temperatures remain slightly above the freezing point of the surrounding air. Additionally, stomatal frequency tends to increase with altitude (Grace & Russell, 1977; Xu et al., 2017), accompanied by the development of epidermal appendages and marginal sclerenchyma. In many alpine plants, both the foliage and reproductive shoots are covered in a layer of dense, wool-like hairs, which protect the non-dormant buds from extreme cold. At night, large wooly leaves curl inward to shield the growing tip from frost damage.

6.2 Physiological and Biochemical Adaptations

Alpine plants exhibit remarkable physiological and biochemical adaptations that enable them to survive and reproduce under extreme environmental conditions. One of the distinctive features of these species is the absence of seed dormancy in many alpine plants (Amen, 1966; Bliss, 1958), allowing germination whenever conditions become favorable. These seeds often maintain viability for several years; for instance, *Poa annua* seeds remain viable for long periods and show seasonal germination periodicity (Daws & Turner, 2017). Germination typically occurs between 20 and 30 °C, while it is inhibited below 10 °C (Sayers & Ward, 1966; Zhou et al., 2020). Consequently, most germination takes place soon after snowmelt in early summer (Billings & Mooney, 1968).

Following snowmelt, the rapid development of shoots and leaves is supported by the mobilization of carbohydrates and lipids from storage organs such as roots, rhizomes, and old evergreen leaves. During this phase, respiration rates are high, accelerating metabolic processes. However, photosynthesis initially lags behind respiration due to the time required for chlorophyll synthesis (Bliss, 1966). Once sufficient chlorophyll is formed, net photosynthesis becomes positive during daylight, marking a critical transition for plant growth.

Annual alpine species generally exhibit higher photosynthetic rates than perennials, a trait believed to have adaptive significance by promoting rapid shoot growth and completion of life cycles within a short growing season (Billings & Mooney, 1968). Moreover, alpine plants typically attain their maximum photosynthetic rates at lower temperatures compared to lowland species grown under similar conditions. Some alpine taxa, such as *Oxyria digyna* (Mooney & Billings, 1961) and *Thalictrum alpinum* (Mooney & Johnson, 1965), show photosynthetic optima at relatively higher temperatures, indicating ecological variation within alpine populations.

Interestingly, several alpine species are capable of photosynthesizing under very low light intensities (Bliss, 1963; Körner & Paulsen, 2015). As light intensity increases, the optimum temperature for photosynthesis also rises (Scott & Billings, 1964). The lower light saturation point observed in alpine plants may result from the accumulation of assimilates (Warren-Wilson, 1966). These plants are also able to maintain photosynthesis at low CO_2 concentrations, a feature attributed to their greater inherent capacity to absorb CO_2 even under the reduced partial pressure of this gas at high elevations (Gao et al., 2020; Körner & Asshoff, 2018).

Three major photosynthetic pathways—the Calvin cycle (C_3), the C_4 dicarboxylic acid cycle, and Crassulacean Acid Metabolism (CAM)—confer adaptive advantages under specific environmental conditions (Flexas & Medrano, 2016). Among alpine plants, C_3 photosynthesis predominates (Ostendorf & Körner, 2019). Based on physiological adaptability, it is hypothesized that with increasing aridity and high light intensity, C_3 species could be gradually replaced by C_4 and CAM plants at higher elevations. The nocturnal CO_2 fixation in CAM plants offers an advantage where large diurnal temperature fluctuations occur under low O_2 and CO_2 concentrations. Nonetheless, C_4 plants may not benefit under such cold alpine conditions, and possible modifications within the C_3 mechanism in high-altitude species cannot be ruled out.

In alpine perennials, the photosynthate produced during the growing season is utilized not only for immediate growth and reproduction but also stored in roots, rhizomes, and evergreen leaves. A substantial portion of these assimilates accumulates toward the end of the growing season, serving as energy reserves for rapid spring growth and early flowering (Mooney & Billings, 1960). Typically, carbohydrates are stored in the roots and shoots, whereas lipids accumulate in the leaves (Billings & Mooney, 1968). During cold nights, starch is degraded to sucrose or possibly transformed into osmiophilic globules (likely lipids) within the chloroplasts of *Oxyria digyna* (Chabot & Billings, 1972). This ability to convert starch to sucrose at low temperatures and efficiently translocate it is considered a key evolutionary adaptation of alpine plants to high-mountain environments.

6.3 Ecological Adaptation

Plants growing at high altitudes have evolved a range of adaptations that enable them to survive under extreme environmental conditions. Heat absorbed within the canopy is partly transferred to the adjacent air, allowing shoot tissues to maintain temperatures 10–20 °C higher than the ambient air (González & Sánchez, 2016). The alpine vegetation of various mountain ranges, though shaped by distinct topography and climate, exhibits both similarities and differences in community composition (Körner & Ohsawa, 2016). In general, herbaceous perennials dominate these ecosystems, with annual species occurring only in small patches. The physiognomy of these perennials transitions gradually from hydrophilous forms in moist zones to arid steppe-like species in drier areas. Across temperate alpine regions, plant communities display a consistent pattern of distribution along environmental gradients (Bliss, 1966).

Alpine vegetation forms a range of ecological communities according to habitat conditions. On barren rock surfaces and among large boulders, petrophilous communities occur, typically composed of xerophytes, mesophytes, and hygrophytes—including liverworts, mosses, and eurotypic species. In the glacial zones, communities are dominated by cryophytic small herbs, while grassy meadows just above the timberline consist mainly of mountain frigid steppe species—a mixture of mesophytes and hygrophytes. At still higher elevations (above 5000 m), steppes and cryophytes become increasingly dominant, and mesophytic species become rare (Raman & Singh, 2019).

The local distribution of life forms within alpine vegetation is strongly influenced by soil moisture availability. Where moisture is adequate, graminoids such as grasses, sedges, and rushes predominate. In contrast, stemless rosette plants and cushion species occupy drier, rocky ridges and disturbed sites, often acting as pioneer species in ecological succession. Evergreen-leaved plants are more abundant in areas protected by winter snow and nourished by summer meltwater (Mora & Wang, 2016).

Based on vegetational characteristics, alpine plant communities are commonly classified into three main types: the Sedge–Willow type, the Fell-field Sedge–Forb type, and the Late Snowbank type (Bliss, 1956). The Sedge–Willow community typically develops in level, marshy areas near lakes and ponds, the Fell-field type occurs on rocky slopes where soil formation has begun, and the Late Snowbank community is found in sites with extremely short growing seasons.

According to Billings and Mooney (1968), the most severe alpine habitats occur in late-lying snowbanks and on windswept, arid ridges—areas where plant growth is minimal. Under these conditions, the flora is dominated by cryptogamic communities, with just a small community of vascular plants, including graminoids, diminutive herbaceous species, cushion growth forms, and trailing shrubs, persisting. On windy ridge lines, the flora becomes restricted to tight, moss-like cushions punctuated by occasional grass tufts, illustrating the highest elevations at which vascular plants can withstand the environment. In contrast, softer, delicate herbs are often found within rock crevices, where they benefit from shelter and adequate moisture supply.

7 Changes in Growth Parameters Along Altitudinal Gradients

The survival of plants in any environment depends on their capacity for modification, which is determined by maintaining an appropriate balance between reversible and irreversible changes in both structure and function. The genotype of an individual plant may either remain constant or undergo alterations in response to environmental variation. The interaction between the environment and the plant's genetic characteristics is commonly referred to as its "response."

In natural ecosystems, increasing altitude represents a complex combination of multiple environmental stresses. Consequently, those plants least influenced by altitudinal changes are considered to possess the greatest potential resistance to such stresses. In this context, several researchers undertook pioneering studies on plant behavior under varying climatic conditions to explore their range of adaptability (Chaves & Oliveira, 2017; Morin & Thuiller, 2018; Valladares & Pearcy, 2016).

7.1 Morphological Changes

Recent studies examining the morpho-physiological characteristics of plant species distributed along altitudinal gradients have shown that increasing elevation leads to a progressive decline in several growth parameters. These include cumulative height, extension growth, leaf production, leaf area, dry matter accumulation, nodulation, and relative growth rate—regardless of the plants' photosynthetic pathway or ecological diversity (Bhatt & Lee, 1981; Nautiyal & Purohit, 1980; Saxena, 1980;

Singh & Soni, 2021; Thapliyal, 1982). The degree of inhibition caused by environmental factors associated with altitude varies among different species. Among the species studied, CAM plants exhibited the least fluctuation in phenological behavior compared with C_4 plants (López & Medina, 2016; Nautiyal, 1980). Overall, most morpho-physiological traits demonstrated a notable flexibility in response to altitudinal variation.

Variations in soil and air temperatures, particularly low night temperatures, are largely responsible for the gradual reduction in plant height observed at higher altitudes (Saxena, 1980; Sharma, 1980). This supports the view that dwarfism in alpine species is predominantly an environmentally induced trait. The decline in cumulative height and extension growth rate is generally more pronounced in lowland species than in high-altitude species (Todaria, 1980). Studies have further shown that the decrease in dry matter production with increasing elevation is greatest in C_4 species, moderate in C_3 species, and least in CAM species. Conversely, dry matter per unit fresh weight tends to increase with altitude, being highest in C_3, lower in C_4, and lowest in CAM plants. Despite this, C_4 species exhibit superior overall dry matter production and net assimilation rates compared with the other two groups (Nautiyal, 1980; Saxena, 1980). According to Todaria (1980), high-altitude plants are generally more succulent than their lowland counterparts.

In alpine ecosystems, the ratio of belowground to aboveground biomass is typically very high, indicating more efficient downward translocation of assimilates (Saxena, 1980; Sharma, 1980). This difference is most pronounced in CAM species. Furthermore, while lowland species tend to store more dry matter in stems and roots, high-altitude species accumulate greater quantities in leaves and roots (Todaria, 1980). With increasing altitude, flowering, fruiting, and seed maturation in several crop plants become progressively delayed (Saxena, 1980; Sharma, 1980). Among the various growth stages, the reproductive phase is more adversely affected by altitudinal environmental changes than the vegetative phase.

At higher elevations, the interdependence among morphological and physiological traits becomes more pronounced. There exists a direct correlation between plant height and dry matter accumulation, and this relationship shows increasing dependence on leaf characteristics as altitude rises. Moreover, at greater heights, leaf differentiation and leaf area production display a stronger dependence on total plant dry matter (Sharma, 1980). According to Bhatt and Lee (1981), leaf reflectance decreases with altitude, suggesting that high-altitude plants possess a greater capacity for solar energy absorption than lowland species.

Investigations on Artemisia species by Nautiyal and Purohit (1980) revealed that stomatal number and pore width decline with increasing altitude. However, the stomatal frequency per unit leaf area was found to be higher during morning and evening hours at lower altitudes than at higher ones. Their study also indicated that leaf thickness decreases at high elevations, though the size and structure of epidermal, mesophyll, and spongy parenchyma cells vary among species. Similarly, Saxena (1980) observed in *Eleusine coracana* and *Paspalum miliaceum* that the epidermal layer becomes thinner at high altitudes, accompanied by the development of an additional mesophyll layer.

7.2 *Biochemical Studies*

A notable feature of high-altitude plants is the high concentration of anthocyanin pigments in their leaves (Billings & Mooney, 1969; Sundqvist & McDonald, 2018). The interaction of anthocyanins with chlorophyll imparts a dark purple or nearly black coloration to the foliage. This dark pigmentation reduces light reflection and transmission, enabling the leaves to absorb a greater proportion of incoming solar radiation, which in turn raises leaf temperature above the surrounding air. Beyond their role in temperature regulation, anthocyanins help plants manage surplus sugars that accumulate in cool, bright conditions or in nutrient-poor environments with limited nitrogen or phosphorus, and they provide protection by absorbing damaging ultraviolet radiation (Winkel-Shirley & Han, 2019; Zoratti & Muthusamy, 2017).

In a study on ten wild and six cultivated species, Todaria et al. (1980) observed that total chlorophyll and carotenoid concentrations decreased with increasing altitude. Additionally, they reported an increase in the chlorophyll a/b ratio at higher elevations, primarily due to a substantial decline in chlorophyll "b" content. Bhadula et al. (1981) investigated seed proteins and isozymes of esterase and acid phosphatase in several *Rhododendron* species collected from different altitudes. Their findings indicated that, generally, protein profile diversity decreases with increasing altitude, whereas the diversity of esterase isozymes increases in species from higher elevations.

Enzyme activities in plants of different photosynthetic groups have also been found to vary with altitude. Studies on PEP-carboxylase, glycolate oxidase, and NADH hydroxypyruvate reductase in C_4 leguminous, C_3 leguminous, and non-leguminous C_3 plants revealed that even species within the same photosynthetic category display distinct enzyme activity patterns when grown at low versus high altitudes (Pandey & Purohit, 1980; Rao & Sharma, 2017). For instance, the activity of PEP-carboxylase decreases in *Eleusine coracana* at high altitude but increases in *Amaranthus paniculatus* under similar conditions. Similarly, NADH hydroxypyruvate reductase exhibits species-specific responses to altitude, whereas glycolate oxidase activity shows a consistent pattern within the same photosynthetic group. Specifically, C_4 plants experience a threefold reduction in glycolate oxidase at high altitude, while leguminous C_3 species display increased activity under the same conditions.

In a related study, Bhadula et al. (1986) examined *Selinum vaginatum*, an alpine perennial herb that can also thrive at low altitudes. The survival of this species at lower elevation was associated with a twofold increase in glycolate oxidase, a threefold increase in PEP-carboxylase, and approximately 33% reduction in RuBP carboxylase activity. Since photorespiration is enhanced at low altitude, the elevated PEP-carboxylase system likely compensates for the greater carbon loss due to photorespiration, enabling the species to adapt successfully across altitudinal gradients.

8 Conclusion

In conclusion, plant life in high mountains is shaped by steep environmental gradients that demand exceptional adaptability, from enduring freezing temperatures to thriving in nutrient-poor soils and intense solar radiation. These resilient species form unique, finely balanced ecosystems that are highly sensitive to disturbance. Looking ahead, the future of alpine plants is increasingly uncertain as climate change alters temperature regimes, precipitation patterns, and the boundaries of suitable habitats. While some species may migrate upward or adjust their physiology, many face the risk of habitat loss as mountaintops offer limited space for further retreat. Protecting alpine biodiversity will therefore require sustained conservation efforts, climate-responsive management strategies, and continued scientific study to ensure these remarkable plant communities endure for generations to come.

References

Amen, R. D. (1966). The extent and role of seed dormancy in alpine plants. *The Quarterly Review of Biology, 41*(3), 271–281.

Amici, A., Puglielli, G., De Angelis, A., & Strohmeyer, D. (2021). Altitude and latitude have different effects on population characteristics of the widespread plant *Anthyllis vulneraria. Oecologia, 197*, 709–722.

Bai, X., Li, Y., & Wang, X. (2019). Morphological and physiological adaptations of alpine plants to high-altitude environments: Implications for future climate change. *Ecological Indicators, 103*, 194–202.

Baskin, J. M., Nan, X., & Baskin, C. C. (1998). A comparative study of seed dormancy and germination in an annual and a perennial species of Senna (Fabaceae). *Seed Science Research, 8*(4), 501–512.

Bernareggi, G., Carbognani, M., Mondoni, A., & Petraglia, A. (2016). Seed dormancy and germination changes of snowbed species under climate warming: The role of pre- and post-dispersal temperatures. *Annals of Botany, 118*(3), 529–539.

Bhadula, S. K., Thapliyal, A. P., & Purohit, A. N. (1981). Seed proteins, esterases and acid phosphatases in rhododendron species from different altitudes in Garhwal Himalaya. *Indian Journal of Experimental Biology, 19*, 199–200.

Bhadula, S. K., Pandey, O. P., & Purohit, A. N. (1986). Thermostability of acid phosphatase in Selinum vaginatum Clarke and Acer caesium Wall. grown at low and high altitudes. *Biologia Plantarum, 28*(6), 465–469.

Bhatt, R. N., & Lee, P. A. (1981). A scaling method for low temperature behavior of random antiferromagnetic systems. *Journal of Applied Physics, 52*(3), 1703–1707.

Billings, W. D., & Godfrey, P. J. (1967). Photosynthetic utilization of internal carbon dioxide by hollow-stemmed plants. *Science, 158*(3797), 121–123.

Billings, W. D., & Mooney, H. A. (1968). The ecology of arctic and alpine plants. *Biological Reviews, 43*(4), 481–529.

Bliss, L. C. (1956). A comparison of plant development in microenvironments of arctic and alpine tundras. *Ecological Monographs, 26*(4), 303–337.

Bliss, L. C. (1958). Seed germination in arctic and alpine species. *Arctic, 11*(3), 180–188.

Bliss, L. C. (1963). Alpine plant communities of the presidential range, New Hampshire. *Ecology, 44*(4), 678–697.

Bliss, L. C. (1966). Plant productivity in alpine microenvironments on Mt. Washington, New Hampshire. *Ecological Monographs, 36*(2), 125–155.
Bonde, E. K. (1968). Survival of seedlings of an alpine clover (*Trifolium nanum* Torr.). *Ecology, 49*(6), 1193–1195.
Bradshaw, A. D. (1965). Evolutionary significance of phenotypic plasticity in plants. *Advances in Genetics, 13*, 115–155.
Burt, R. L. (1964). Influence of short periods of low temperature on tuber initiation in the potato. *European Potato Journal, 7*(4), 197–208.
Carmona, C. P., Hild, A. L., & Gage, S. H. (2017). Trait plasticity and the adaptive response of plants to environmental stressors: A balance between benefits and costs. *Frontiers in Plant Science, 8*, 547.
Chabot, B. F., & Billings, W. D. (1972). Origins and ecology of the Sierran alpine flora and vegetation. *Ecological Monographs, 42*(2), 163–199.
Chandola, V., Sati, P., Chandra, S., Purohit, V. K., Nautiyal, A. R., & Chandra, S. (2026). Changes in growth and metabolism of Alpine plants along altitudinal and microhabitat gradients. In S. Chandra & H. Lata (Eds.), *High altitude medicinal plants* (pp. 37–58). Springer.
Chandra, S. (2004). Effect of altitude on energy exchange characteristics of some alpine medicinal crops from Central Himalayas. *Journal of Agronomy and Crop Science, 190*(1), 13–20.
Chandra, S., & Joshi, S. C. (2002). Diurnal and seasonal variation in CO_2 levels in the surface air of Garhwal Himalaya. *Indian Journal of Forestry, 25*(2), 205–208.
Chandra, S., & Lata, H. (2022). Effect of altitude on biochemical and photosynthetic characteristics of *Aconitum balfourii* and *Podophyllum hexandrum*: High value endangered medicinal herbs from Himalayas. *International Journal of Tropical Agriculture, 40*(3–4), 195–203.
Chandra, S. L., & Lata, H. (2026). *High altitude medicinal plants; botany, conservation and cultivation* (p. 561). Springer.
Chandra, S., Lata, H., & Joshi, S. C. (2011). Photosynthetic characteristics of Indian rhubarb (*Rheum emodi* wall.): An endangered medicinal plant. *Planta Medica, 77*(05), 8.
Chandra, S., Lata, H., & Joshi, S. C. (2012). Altitudinal variation in biochemical and photosynthetic characteristics of *Aconitum balfourii*: An Alpine endangered medicinal herb from Himalayas. *Planta Medica, 78*(05), 4.
Chandra, S., Lata, H., & Joshi, S. C. (2013). Gas and water vapor exchange characteristics of *Picrorhiza kurooa* Royle ex Benth, an endangered Alpine medicinal plant, at lower altitude. *Planta Medica, 79*(05), 14.
Chandra, S., Trivedi, V., Purohit, V. K., Nautiyal, M. C., & Chandra, S. (2026). Climatic change and physiological adaptations. Alpine medicinal plants. In S. Chandra & H. Lata (Eds.), *High altitude medicinal, plants* (pp. 11–35).
Chaves, M. M., & Oliveira, M. M. (2017). Mechanisms of plant stress tolerance: Insights into ecological range and resilience. *Ecology and Evolution, 7*(5), 1687–1697.
Chen, P., Yang, R., Bartels, D., Dong, T., & Duan, H. (2022). Roles of abscisic acid and gibberellins in stem/root tuber development. *International Journal of Molecular Sciences, 23*(9), 4955.
Chen, S., Wang, D., Lin, J., & Xu, Z. (2025). The comprehensive regulation of light intensity and photoperiod on growth and yield of virus-free potato under the same daily light integral. *Agronomy, 15*(4), 898.
Chen, T., Yun, H., Chen, C., & Liu, Y. (2021). Below-ground growth of Alpine plants, not above-ground growth, is linked to the extent of its carbon storage. *Plants, 10*(12), 2680.
D'Imperio, L., Arndal, M. F., Nielsen, C. S., Elberling, B., & Schmidt, I. K. (2018). Fast responses of root dynamics to increased snow deposition and summer air temperature in an Arctic wetland. *Frontiers in Plant Science, 9*, 1258.
Daws, M. I., & Turner, J. R. (2017). Seed longevity and germination periodicity in Poa species: Adaptive strategies to seasonal environments. *Seed Science Research, 27*(3), 133–142.
Fletcher, R. A., Varnon, K. M., & Barney, J. N. (2020). Climate drives differences in the germination niche of a globally distributed invasive grass. *Journal of Plant Ecology, 13*(2), 195–203.

Flexas, J., & Medrano, H. (2016). The functional significance of the three major photosynthetic pathways: Adaptations to water stress, temperature, and light. *Journal of Experimental Botany, 67*(16), 4813–4825.

Gao, Q., Zhang, X., & Li, X. (2020). Photosynthetic capacity and CO_2 fixation in alpine plants: Implications for survival in low CO_2 environments. *Functional Plant Biology, 47*(2), 128–136.

Gómez, A. R., González, C. E., & González, J. A. (2018). Seasonal dynamics of tuber growth and carbohydrate partitioning in potato (*Solanum tuberosum* L.) during the growing season. *Field Crops Research, 218*, 66–74.

González, M., & Sánchez, J. (2016). Heat dissipation and thermal acclimation in alpine plants: Maintaining shoot temperatures above ambient levels in high-altitude environments. *Environmental and Experimental Botany, 124*, 22–30.

Grace, J., & Russell, G. (1977). The effect of wind on grasses: III. Influence of continuous drought or wind on anatomy and water relations in *Festuca arundinacea* schreb. *Journal of Experimental Botany, 28*(2), 268–278.

Haberlandt, G. (1888). *Die Chlorophyllkörper der Selaginellen* (Vol. 71, p. 291). Flora.

Hamaoka, N., Nabeshima, M., Moriyama, T., Kozawa, Y., & Ishibashi, Y. (2022). Photoperiodic regulation of tuber enlargement in water yam (*Dioscorea alata L.*). *Agronomy, 12*(12), 2939.

Harrington, J. F. (1963). The value of moisture-resistant containers in vegetable seed packaging. *Bulletin of the California Agricultural Experiment Station, 792*(23), 6.

Harrison, S. P., & Fitter, A. H. (2019). Phenotypic plasticity in alpine plants: Adaptations to environmental stress in high-altitude habitats. *Functional Ecology, 33*(9), 1631–1641.

Imaizumi, T., Ohigashi, K., & Koarai, A. (2025). Maternal temperature imposes a longer-term effect on seedling emergence than does genetic variation in seed dormancy. *Plant, Cell & Environment, 48*(7), 5304–5316.

Klatt, S., Schinkel, C. C. F., Kirchheimer, B., Dullinger, S., & Hörandl, E. (2018). Effects of cold treatments on fitness and mode of reproduction in the diploid and polyploid alpine plant *Ranunculus kuepferi* (Ranunculaceae). *Annals of Botany, 121*(7), 1287–1298.

Körner, C. (2003). *Alpine plant life: Functional plant ecology of high mountain ecosystems* (2nd ed.). Springer.

Körner, C., & Asshoff, R. (2018). High-altitude plants: Adaptations to low CO_2 concentrations and low atmospheric pressure. *Plant Ecology & Diversity, 11*(3), 321–330.

Körner, C., & Ohsawa, M. (2016). Alpine plant communities: A comparison of vegetation types across high-altitude mountain ranges. *Journal of Mountain Science, 13*(2), 335–348.

Körner, C., & Paulsen, J. (2015). Alpine plant photosynthesis under low light: Adaptations to high-altitude environments. *Ecological Research, 30*(5), 825–834.

Kuleshov, N. N. (1963). *Agronomic seed* (p. 304). Selhozizdat.

Kumar, S., Jain, S., & Ghosh, A. (2016). Role of leaves in photoperiodic regulation of tuberization in potato (*Solanum tuberosum* L.). *Physiologia Plantarum, 157*(3), 390–401.

Landolt, E. (1975). Morphological differentiation and geographical distribution of the Lemna gibba-Lemna minor group. *Aquatic Botany, 1*, 345–363.

Li, G., Ma, J., Qu, L., Du, H., Guo, S., Wang, Z., Liu, D., Wang, X., Tan, Y., Meng, F., Ma, Z., Zhang, C., & Li, C. (2024). Transcriptomics and metabolomics analyses of Chinese artichoke tubers (*Stachys sieboldii* Miq.) during the developmental period. *Journal of the American Society for Horticultural Science, 149*(5), 282–295.

Li, W., Tan, L., Zou, Y., Tan, X., Huang, J., Chen, W., & Tang, Q. (2020). The effects of ultraviolet A/B treatments on anthocyanin accumulation and gene expression in dark-purple tea cultivar 'Ziyan' (*Camellia sinensis*). *Molecules, 25*(2), 354.

López, R. S., & Medina, E. (2016). Phenological patterns and environmental adaptability of CAM plants in arid ecosystems: Comparisons with C_4 species. *Environmental and Experimental Botany, 125*, 1–9.

Luxmoore, R. J., & Millington, R. J. (1971). Growth of perennial ryegrass (*Lolium perenne* L.) in relation to water, nitrogen, and light intensity: II. Effects on dry weight production, transpiration and nitrogen uptake. *Plant and Soil, 34*(1), 561–574.

Maleki, K., Baskin, C. C., Baskin, J. M., Kiani, M., Alahdadi, I., & Soltani, E. (2022). Seed germination thermal niche differs among nine populations of an annual plant: A modeling approach. *Ecology and Evolution, 12*(8), e9240.
Maurya, J. P., & Bhalerao, R. P. (2017). Photoperiod- and temperature-mediated control of growth cessation and dormancy in trees: A molecular perspective. *Annals of Botany, 120*(3), 351–360.
Mondoni, A., Probert, R., Rossi, G., Pauli, H., Rosbakh, S., Winkler, M., & Abeli, T. (2016). Seed dormancy and germination changes of snowbed species under climate warming: The role of pre- and post-dispersal temperatures. *Annals of Botany, 118*(3), 529–539.
Mooney, H. A., & Billings, W. D. (1960). The annual carbohydrate cycle of alpine plants as related to growth. *American Journal of Botany, 47*(7), 594–598.
Mooney, H. A., & Billings, W. D. (1961). Comparative physiological ecology of arctic and alpine populations of *Oxyria digyna*. *Ecological Monographs, 31*(1), 1–29.
Mooney, H. A., & Johnson, A. W. (1965). Comparative physiological ecology of an arctic and alpine population of *Thalictrum alpinum* L. *Ecology, 46*(5), 721–727.
Mora, C. V., & Wang, X. (2016). Snow cover and meltwater: Influences on plant distribution and growth patterns in alpine ecosystems. *Alpine Botany, 126*(1), 45–55.
Morin, X., & Thuiller, W. (2018). Ecological range shifts in plants: Environmental and physiological constraints on adaptation to climate change. *Trends in Ecology & Evolution, 33*(12), 929–939.
Nautiyal, A. C. (1980). Cyanophycean algal remains and paleoecology of the Precambrian Gangolihat dolomites formation of the Kumaun Himalaya. *Indian Journal of Earth Sciences, 7*, 1–11.
Nautiyal, P. C., Sivasubramaniam, K., & Dadlani, M. (2023). Seed dormancy and regulation of germination. *Seed Science and Technology, 52*, 39–66.
Nautiyal, S., & Purohit, A. N. (1980). High altitude acclimatization in plants: Stomatal frequency and anatomical changes in leaves of Artemisia species. *Biologia Plantarum, 22*(4), 282–286.
Odum, E. P. (1974). The strategy of ecosystem development. *Readings in Environmental Impact, 164*, 224.
Orsenigo, S., Abeli, T., Rossi, G., Bonasoni, P., Pasquaretta, C., Gandini, M., & Mondoni, A. (2015). Effects of autumn and spring heat waves on seed germination of high mountain plants. *PLoS One, 10*(7), e0133626.
Ostendorf, B., & Körner, C. (2019). Photosynthetic pathways in alpine plants: Predominance of C_3 photosynthesis and the occasional presence of C_4 species in subalpine and alpine habitats. *Alpine Botany, 129*(1), 15–23.
Palacio, S., Ferrio, J. P., Araus, J. L., & den Held, M. (2021). Carbon allocation to growth and storage depends on elevation provenance in an herbaceous alpine plant of Mediterranean climate. *Plants, 10*(6), 1240.
Pandey, O. P., & Purohit, A. N. (1980). Activity of pep-carboxylase and two glycolate pathway enzymes in C3 and C4 plants grown at two altitudes. *Current Science*, 263–265.
Plantenga, F. D. M., Siakou, M., Bergonzi, S., Heuvelink, E., Bachem, C. W. B., Visser, R. G. F., & Marcelis, L. F. M. (2016). Regulating flower and tuber formation in potato with light spectrum and day length. *Acta Horticulturae, 1134*, 267–275.
Purohit, A. N., Hemant Lata, H. L., Chandra, S., & Purohit, M. C. (1998). Some characteristics of four morphological variants of *Podophyllum hexandrum* Royle. *Plant Genetic Resources Newsletter, 114*, 51–52.
Raman, K., & Singh, N. (2019). Vegetation composition and ecological transitions along an elevational gradient in the Himalayas: The shift from mesophytic to cryophytic communities above 5000 m. *Journal of Vegetation Science, 30*(6), 1175–1186.
Rao, S. R., & Sharma, P. (2017). Altitudinal variation in enzyme activity and photosynthetic regulation in C_3, C_4, and CAM plants: A comparative study on PEP-carboxylase, glycolate oxidase, and NADH hydroxypyruvate reductase. *Environmental and Experimental Botany, 134*, 10–18.
Rehmani, M. S., et al. (2022). Seed dormancy and longevity: A mutual dependence or a trade-off? *Plant and Cell Physiology, 63*(8), 1029.

Roberts, E. H. (1972). Cytological, genetical and metabolic changes associated with loss of viability. In *Viability of seeds* (pp. 253–306). Springer Netherlands.
Rosbakh, S., Baskin, C. C., & Baskin, J. M. (2020). Nikolaeva et al.'s reference book on seed dormancy and germination. *Ecology, 101*(7), e03049.
Rosbakh, S., Fernández-Pascual, E., Mondoni, A., & Onipchenko, V. (2022). Alpine plant communities differ in their seed germination requirements along a snowmelt gradient in the Caucasus. *Alpine Botany, 132*(2), 223–232.
Saidi, A., & Hajibarat, Z. (2021). Phytohormones: Plant switchers in developmental and growth stages in potato. *Journal of Genetic Engineering and Biotechnology, 19*(1), 89.
Saxena, R. K. (1980). Palynology of Matanomadh formation in type area, north-western Kutch, India (part-3). *Discussion. Palaeobotanist, 26*(3), 279–296.
Sayers, R. L., & Ward, R. T. (1966). Germination responses in alpine species. *Botanical Gazette, 127*(1), 11–16.
Schwienbacher, E., Navarro-Cano, J. A., Neuner, G., & Erschbamer, B. (2011). Seed dormancy in alpine species. *Flora-Morphology, Distribution, Functional Ecology of Plants, 206*(10), 845–856.
Scott, D., & Billings, W. D. (1964). Effects of environmental factors on standing crop and productivity of an alpine tundra. *Ecological Monographs, 34*(3), 243–270.
Semwal, J. K., & Purohit, A. N. (1980). Germination of Himalayan Alpine and temperate *Potentilla*. In *Proceedings/Indian Academy of Sciences* (Vol. 89(1), pp. 61–65). Springer India.
Semwal, J. K., Purohit, A. N., & Gaur, R. D. (1983). Seed germination in some Himalayan alpine plants. *Seed Research, 11*(1), 42–46.
Sharma, H. K. (1980). Genetic analysis of plant height in pigeonpea. In *International Workshop on Pigeonpeas* (p. 55).
Singh, A., Purohit, A. N., Bhadula, S. K., Lata, H., Kuniyal, C. P., & Chandra, S. (1999). Seed production potential and germination behaviour in populations of *Podophyllum hexandrum* Royle. *Journal of Plant Biology, 26*, 51–58.
Singh, J. S., & Soni, P. (2021). Ecophysiological growth responses in high-altitude plants: Adaptation strategies for growth and productivity. *Frontiers in Plant Science, 12*, 652–663.
Song, Y., Zhang, M., Guo, Y., & Gao, X. (2023). Change in seed dormancy status controls seasonal timing of seed germination in Korean pine (*Pinus koraiensis*). *Journal of Plant Ecology, 16*(2), rtac067.
Sundqvist, M. K., & McDonald, K. L. (2018). Anthocyanin accumulation and its ecological significance in high-altitude plants: A protective adaptation against UV radiation and freezing stress. *Environmental and Experimental Botany, 151*, 70–79.
Tarvainen, L., Wallin, G., Lim, H., Linder, S., Oren, R., Ottosson Löfvenius, M., Räntfors, M., Tor-Ngern, P., & Marshall, J. (2018). Photosynthetic refixation varies along the stem and reduces CO_2 efflux in mature boreal *Pinus sylvestris* trees. *Tree Physiology, 38*(4), 558–569.
Thapliyal, V. (1982). Stochastic dynamic model for long range prediction of monsoon rainfall in peninsular India. *Mausam, 33*(4), 399–404.
Todaria, N. P. (1980). Growth behavior and functional dynamics of plant species from different altitudes in Garhwal Himalaya. D. Phil. Thesis, Garhwal University, Srinagar.
Todaria, N. P., Thapaliyal, A. P., & Purohit, A. N. (1980). Altitudinal effects on chlorophyll and carotenoid contents in plants species. *Photosynthetica, 14*, 236.
Tsuda, S., Tsuji, H., Tamiya, S., & Mori, M. (2019). Application of gibberellin to potato seed tubers enhances the yield of small tubers (*Solanum tuberosum L.*). *Japanese Journal of Crop Science, 88*(1), 1–8.
Valladares, F., & Pearcy, R. W. (2016). Plant responses to environmental variation: Adaptations and strategies for coping with climate change. *Global Change Biology, 22*(9), 3170–3183.
Vázquez-Ramírez, J., & Venn, S. E. (2021). Seeds and seedlings in a changing world: A systematic review and meta-analysis from high altitude and high latitude ecosystems. *Plants, 10*(4), 768.
Villiers, T. A. (1974). Seed aging: Chromosome stability and extended viability of seeds stored fully imbided. *Plant Physiology, 53*(6), 875–878.

Visscher, A. M., Seal, C. E., Newton, R. J., Latorre Frances, A., & Pritchard, H. W. (2016). Dry seeds and environmental extremes: Consequences for seed lifespan and germination. *Functional Plant Biology, 43*(7), 656–668.

Wager, H. G. (1938). Growth and survival of plants in the Arctic. *The Journal of Ecology,* 390–410.

Wang, G., Baskin, C. C., Baskin, J. M., Yang, X., Liu, G., Zhang, X., Ye, X., & Huang, Z. (2015). Timing of seed germination in two alpine herbs on the southeastern Tibetan plateau: The role of seed dormancy and annual dormancy cycling in soil. *Plant and Soil, 389*(3), 465–476.

Wang, J., Zhang, X., Wang, R., Yu, M., Chen, X., Zhu, C., Shang, J., & Gao, J. (2024). Climate factors influence above- and belowground biomass allocations in Alpine meadows and desert steppes through alterations in soil nutrient availability. *Plants, 13*(5), 727.

Wilson, J. W. (1966). An analysis of plant growth and its control in arctic environments. *Annals of Botany, 30*(3), 383–402.

Winkel-Shirley, B., & Han, J. (2019). Anthocyanin biosynthesis and its role in plant adaptation to stress. *Journal of Experimental Botany, 70*(6), 1825–1838.

Wittington, W. J. (1973). Genetic regulation of germination. In W. Heydecker (Ed.), *Seed ecology. University of Nottingham 19th Easter School in Agricultural Science Butterworths London* (pp. 5–30).

Xu, L., Chen, Q., & Wang, X. (2017). Altitudinal variation in leaf morphology and stomatal characteristics of Alpine plants on the Tibetan plateau. *Environmental and Experimental Botany, 141,* 142–151.

Zhang, H., Zhang, X., & Xu, Z. (2020). Enzymatic degradation of starch and soluble sugars in potato tubers during storage and sprouting: Role of amylases and invertases. *Postharvest Biology and Technology, 164,* 111158.

Zhang, T., Chen, Y., Yang, X., Zhang, H., Guo, Z., Hu, G., Bai, H., Sun, Y., Huang, L., & Ma, M. (2024). Warming reduces mid-summer flowering plant reproductive success through advancing fruiting phenology in an alpine meadow. *Proceedings of the Royal Society B, 291*(2033), 20241110.

Zhang, Y., Gao, Q., Wang, R., Zhu, C., & Gao, J. (2022). Seasonal and inter-annual variations of carbon dioxide fluxes and their determinants in an alpine meadow. *Frontiers in Plant Science, 13,* 894398.

Zhou, G., & Zhang, H. (2019). Temperature and light interactions in the photosynthetic performance of alpine plants: Implications for growth under varying environmental conditions. *Frontiers in Plant Science, 10,* 391.

Zhou, M., Zhang, X., & Li, Z. (2020). Temperature sensitivity of seed germination in temperate grasses: Implications for climate change adaptation. *Plant Species Biology, 35*(1), 16–23.

Zhu, X., Chen, D., Kogure, M., Hoy, E., Breen, A. L., Chatterjee, A., Davidson, S. J., Frost, G. V., Hollingsworth, T. N., Loboda, R. R., Macander, M. J., Miller, C. E., Miller, E. A., Natali, S. M., Raynolds, M. K., Rocha, A. V., Tsuyuzaki, S., Walker, D. A., Williams, M., Xu, X., Zhang, Y., French, N., & Goetz, S. (2024). A synthesized field survey database of vegetation and active-layer properties for the Alaskan tundra (1972–2020). *Earth System Science Data, 16,* 3687.

Zoratti, L., & Muthusamy, S. K. (2017). The multifunctional role of anthocyanins in plant stress responses: Implications for ecological adaptation. *Frontiers in Plant Science, 8,* 277.

The Use of Remote Sensing and Geographic Information Systems (GIS) for the Mapping of Cold Deserts' Bio-resources

Xiaoxiong Li

1 Introduction

Cold deserts stand as terrestrial sentinels of global change. These vast, extreme biomes, encompassing high-altitude and high-latitude landscapes like the Tibetan Plateau and the Patagonian Steppe, are defined by frigid temperatures, profound aridity, and intense solar radiation. Though seemingly stark, they support uniquely adapted biodiversity and maintain a delicate ecological balance. They are not merely remote wildernesses; they function as critical headwaters for major river systems—the "Water Towers of Asia" being a prime example (Immerzeel et al., 2010; Piao et al., 2010)—and support the livelihoods of indigenous and local communities, often through pastoralism (Chen et al., 2014). However, these sentinel landscapes are disproportionately affected by global climate change, experiencing accelerated warming rates that trigger a cascade of environmental responses, including permafrost degradation (Cheng & Wu, 2007), altered hydrological regimes (Immerzeel et al., 2010), and shifts in vegetation phenology (Piao et al., 2010; Shen et al., 2015a, 2015b). Compounded by anthropogenic pressures such as overgrazing and infrastructure development, the inherent vulnerability of these ecosystems is critically heightened (Chen et al., 2014).

In this context, accurate mapping and systematic monitoring of bio-resources are fundamental for science-informed conservation and management (Turner et al., 2003a, 2003b). Yet, the very characteristics that define cold deserts—their immense scale and inaccessibility—render traditional field-based survey methods logistically prohibitive and inherently limited in scope (Turner et al., 2003a, 2003b). A paradigm shift toward more efficient, scalable technologies is therefore necessary. To overcome these limitations, remote sensing (RS) and Geographic Information Systems (GIS)

X. Li (✉)
Center of Geographical Studies (CEG), Institute of Geography and Spatial Planning (IGOT), University of Lisbon, Lisboa, Portugal
e-mail: xiaoxiongli@edu.ulisboa.pt

Y. S. Rawat and C. P. Kala (eds.), *Cold Deserts*,
https://doi.org/10.1007/978-3-032-29119-6_6

offer an indispensable suite of tools (Lillesand et al., 2015). RS provides a synoptic, multi-temporal perspective that transcends the limitations of ground-based observation, while GIS provides the analytical framework to transform raw geospatial data into actionable intelligence (Campbell & Wynne, 2011; Longley et al., 2015). By leveraging the synergistic power of this integrated approach, researchers can move from sparse, point-based knowledge to a holistic, spatially explicit understanding of ecosystem function with unparalleled cost-effectiveness (USGS, 2023).

This chapter provides a comprehensive overview of the application of RS and GIS for monitoring bio-resources within cold desert ecosystems. It is structured to guide the reader from foundational concepts to advanced applications, exploring diverse remote sensing data sources, detailing core GIS analytical techniques, and showcasing major application areas. The chapter will also critically examine the unique challenges inherent in these environments and discuss future directions, illustrating these concepts with two case studies to bridge theory and practice.

2 Remote Sensing Data for Cold Deserts: A Multi-sensor Perspective

The efficacy of remote sensing in characterizing the bio-resources of cold deserts hinges critically on the selection of appropriate data, which is governed by the specific biophysical parameter of interest, the required spatial and temporal scales of analysis, and the inherent environmental constraints of the study area. Modern remote sensing offers a sophisticated suite of sensors operating across the electromagnetic spectrum, each providing a unique lens through which to view these fragile ecosystems. A comprehensive monitoring strategy rarely relies on a single data source; rather, it leverages the synergistic potential of multiple sensors. This section delineates the primary types of remote sensing data and the platforms that deliver them, emphasizing their specific utility and limitations in the context of cold desert environments. These data are acquired by sensors aboard various platforms, ranging from global-scale satellites to local-scale drones, which serve complementary roles in ecosystem monitoring.

2.1 Optical Multispectral and Hyperspectral Data

Optical remote sensing, which measures reflected solar radiation in the visible (VIS), near-infrared (NIR), and shortwave-infrared (SWIR) wavelengths, remains the workhorse for vegetation and land cover mapping.

Principle and Application: The methodology is predicated on the distinct spectral signatures of surface features. Healthy vegetation, for instance, exhibits strong absorption in the red band due to chlorophyll and high reflectance in the NIR band due to leaf cellular structure. This contrast is the foundation for numerous vegetation indices (e.g., the Normalized Difference Vegetation Index, NDVI) (Rouse et al., 1974), which serve as proxies for vegetation vigor, density, and biomass (Xue & Su, 2017). In cold deserts, optical data are fundamental for mapping the distribution of sparse grasslands, shrublands, and alpine meadows; delineating the extent of water bodies and wetlands; and monitoring snow and ice cover due to their high albedo.

Key Data Sources: A wealth of optical data is available from long-standing public archives, which are invaluable for climate change studies.

Moderate Resolution: The Landsat series (USGS, 2025) (30 m resolution, 16-day repeat) provides a multi-decadal record essential for long-term trend analysis. The Sentinel-2 mission from the European Space Agency (ESA) (ESA, 2025) offers enhanced capabilities with 10–20 m resolution, a 5-day revisit time (with two satellites), and additional spectral bands in the red-edge portion of the spectrum, which is highly sensitive to vegetation health. MODIS (Moderate Resolution Imaging Spectroradiometer) (NASA, 2025a) provides daily global coverage at a coarser resolution (250–1000 m), the Sentinel-2 mission (ESA, 2025) delivers high spatial resolution data (up to 10 m) and includes specific spectral bands in the red-edge range (Table 1). This capability allows for more precise detection of chlorophyll content and vegetation health in fragmented landscapes compared to broad-scale sensors, making it ideal for long-term trends analysis over extensive regions. High and Very High Resolution: Commercial satellites such as WorldView (Maxar Technologies, 2025) and Pléiades (Airbus, 2025) provide very high-resolution (VHR) data (<1 m), enabling detailed mapping of individual shrub clumps or small, critical resource patches, though at a higher cost and with smaller geographic footprints. And usually only available in the visible light band (VIS).

Limitations: The utility of optical data in cold deserts is frequently constrained by persistent cloud cover, especially in mountainous regions, and the significant influence of atmospheric scattering at high altitudes. Furthermore, the low percentage of vegetation cover means that spectral signals are often dominated by the background soil and rock, complicating the interpretation of vegetation indices (Karnieli et al., 2010). Topographic shadows in complex terrain also present a major analytical challenge (Riaño et al., 2003).

2.2 *Hydro-Thermal Monitoring: The Synergy of Thermal (LST) and Precipitation Data*

The diagnostic power of thermal remote sensing is profoundly enhanced when synergistically combined with satellite-derived precipitation products. This integrated hydro-thermal approach provides a far more complete picture of ecosystem water and energy dynamics than either dataset could alone (Kogan, 1997).

Table 1 Overview of MODIS key data product types (Moderate resolution imaging spectroradiometer NASA, 2025a) (This data is completely free and publicly available: https://modis.gsfc.nasa.gov/)

Domain	Key data products and indices	Key applications and description
Land	Surface reflectance vegetation indices (NDVI, EVI) Land cover type fire and thermal anomalies Land surface temperature (LST) Vegetation productivity (GPP, NPP)	Monitoring vegetation health, land use change, forest fires, agricultural productivity, drought, urban heat island effects, and the global carbon cycle
Atmosphere	Aerosol optical depth (AOD) Cloud properties atmospheric profiles (temp/moisture) Water vapor	Meteorological monitoring: Retrieving atmospheric temperature and humidity profiles to understand local weather patterns affecting bio-resources Precipitation drivers: Monitoring cloud dynamics and water vapor to estimate moisture availability for desert vegetation Climate indicators: Assessing radiative forcing and its impact on the fragile thermal balance of cold deserts
Cryosphere	Freeze/Thaw state Land surface temperature (LST)	Water availability Analysis: Monitoring snow cover extent and melt timing to estimate water supply for cold desert vegetation and bio-resources Habitat suitability: Using freeze/thaw states and temperature data to determine the growing season length and assess thermal stress on local ecosystems

Principle and Application: While thermal sensors measure land surface temperature (LST)—a proxy for the surface energy state—precipitation products (e.g., GPM IMERG, CHIRPS) quantify the primary water input (Beck et al., 2019; Funk et al., 2015). This coupling is fundamental for assessing ecosystem water stress; for instance, the co-occurrence of anomalously high LST and low precipitation is a robust indicator of developing drought (Karnieli et al., 2010). Furthermore, this synergy is critical for accurately modeling evapotranspiration (Bastiaanssen et al., 1998) and for understanding the complex dynamics of permafrost, where the ground thermal regime is heavily modulated by the insulating effects of snow cover (a key form of precipitation) (Zhang, 2005).

Key Data Sources: LST data are acquired by thermal bands on satellites like Landsat and MODIS (Wan et al., 2015), with specialized missions like ECOSTRESS offering high spatiotemporal detail. Their value is maximized when used in concert with gridded precipitation datasets to form a comprehensive picture of hydro-climatic conditions affecting cold desert bio-resources.

2.3 Microwave (Radar) Data

Radar (Radio Detection and Ranging) is an active sensing system that transmits its own microwave energy and records the backscattered signal (Lillesand et al., 2015).

Principle and Application: The primary advantage of radar is its ability to operate independently of solar illumination and to penetrate clouds, haze, and smoke, making it exceptionally valuable for systematic monitoring in cold desert regions where weather conditions are often adverse (Ulaby et al., 1981). The backscattered signal is sensitive to the geometric and dielectric properties of the surface, including surface roughness, topography, and most importantly, moisture content. This makes radar highly effective for mapping soil moisture (Wagner et al., 2009), detecting inundated areas, and delineating wet versus dry snow (Nagler & Rott, 2002). Furthermore, interferometric synthetic aperture radar (InSAR) techniques can measure subtle surface deformation with millimeter-level precision, enabling the monitoring of glacier velocity, permafrost thaw-related subsidence, and landslide risks (Bürgmann et al., 2000).

Key Data Sources: The Sentinel-1 mission provides free, open-access C-band radar data with a high temporal resolution (6–12 days), which has revolutionized operational monitoring of the cryosphere and land surface (Torres et al., 2012). Other relevant missions include the Japanese ALOS-2/PALSAR-2 (L-band), which offers greater penetration depth (Rosenqvist et al., 2014).

2.4 Gridded Climate and Environmental Datasets

While remote sensing provides a snapshot of the surface, understanding ecosystem dynamics requires contextual environmental data, which are often available as global or regional gridded products. These datasets are not directly sensed in the same manner but are derived from a combination of station data, interpolation techniques, and model reanalysis.

Principle and Application: These datasets provide spatially explicit information on key environmental drivers. They are indispensable for climate change impact assessments and are frequently used as predictor variables in species distribution and habitat suitability models. By correlating bio-resource patterns (e.g., vegetation greenness from NDVI) with climate variables, researchers can uncover fundamental ecological relationships.

Key Data Sources:

Climate Data: WorldClim (Fick & Hijmans, 2017) is a widely used resource providing historical and future climate data (e.g., temperature, precipitation) derived from interpolations and downscaled General Circulation Models (GCMs) (Fig. 1). TerraClimate (Abatzoglou et al., 2018) offers high-resolution monthly climate and

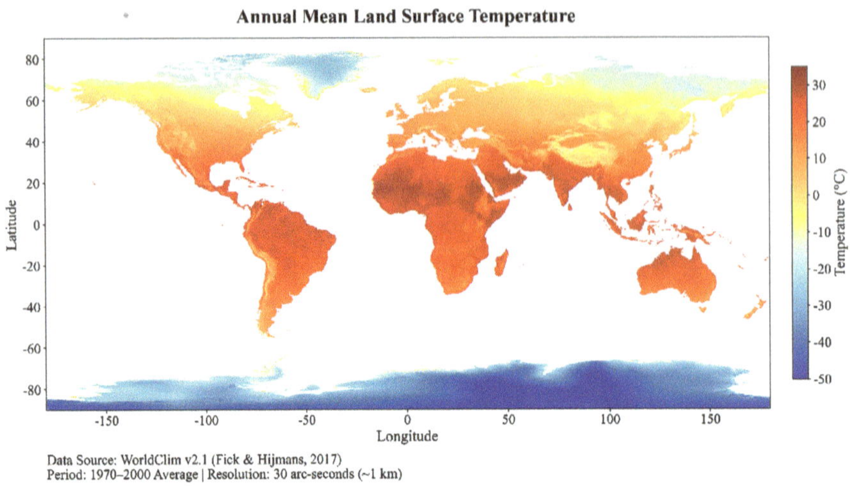

Fig. 1 Mean land surface temperature of global

climatic water balance data, blending station data with satellite-derived variables. CHIRPS (Funk et al., 2015) provide a quasi-global precipitation dataset, crucial for water-limited cold deserts.

Environmental Data: Datasets like **SoilGrids** (Poggio et al., 2021) offer global predictions for soil properties (e.g., organic carbon, pH), which are critical for vegetation and land degradation studies.

2.5 *Digital Elevation Models (DEMs) and Topographic Derivatives*

Topography is arguably the single most important controlling factor of ecological processes in the mountainous terrain characteristic of most cold deserts. Digital Elevation Models (DEMs), which provide a 3D representation of the Earth's surface, are therefore a foundational dataset for nearly all environmental analyses.

Principle and Application: Topography directly governs the distribution of energy (i.e., solar radiation) and matter (i.e., water, soil), thereby creating a mosaic of micro-climates and habitats. DEMs are used not only as a direct input for analysis but also as a crucial tool for the topographic correction of optical remote sensing imagery, mitigating the severe effects of illumination and shadow in complex terrain.

Key Data Sources: A variety of global DEMs are freely available, including the Shuttle Radar Topography Mission (SRTM) DEM (Farr et al., 2007) (~30 m), the ALOS PALSAR Radiometrically Terrain Corrected DEM (Tadono et al., 2014) (~12.5 m), and the Copernicus DEM (European Space Agency, 2022) (30 m).

For local-scale studies, ultra-high-resolution DEMs can be generated from airborne LiDAR or drone-based photogrammetry (Structure from Motion, SfM).

Topographic Derivatives: The true analytical power of DEMs is unlocked through the calculation of derivatives, including the following:

Slope: The gradient of the terrain, a primary driver of erosion potential, surface runoff velocity, and habitat suitability.

Aspect: The direction a slope faces, which critically controls solar insolation, temperature regimes, and moisture availability, often creating stark contrasts in vegetation cover on opposing slopes.

Curvature: The rate of the change of the slope, which helps identify zones of erosion (convex surfaces) and deposition (concave surfaces).

Topographic Wetness Index (TWI): A steady-state wetness model that combines slope and upslope contributing area to predict locations of potential soil moisture accumulation. It is highly effective for mapping potential wetlands and modeling the spatial distribution of soil moisture.

2.6 *Global Land Use/Land Cover (LULC) Products*

Ready-to-use LULC products, derived from satellite imagery, provide a foundational layer for assessing land use change, quantifying habitat loss, and parameterizing ecological models. Recent high-resolution products like ESA WorldCover (10 m) (Zanaga et al., 2022) offer a state-of-the-art information. However, caution is advised, as their broad thematic classes may not capture the nuances of cold desert vegetation communities and can be prone to inaccuracies in sparsely vegetated areas (Xu et al., 2024).

2.7 *Platforms: From Global Satellites to Local Drones*

The sensors described above are carried on two main classes of platforms, which serve complementary roles.

Satellites: Orbiting platforms are the backbone of regional to global-scale monitoring. Their systematic and repetitive data acquisition provides the synoptic view and long-term archives necessary to understand large-scale ecosystem processes and to place contemporary changes within a historical context (Campbell & Wynne, 2011).

Uncrewed Aerial Vehicles (UAVs. or Drones): UAVs represent a paradigm shift for local-scale, high-precision mapping (Anderson & Gaston, 2013). Equipped with lightweight optical, multispectral, or LiDAR sensors, they can acquire data at unprecedented spatial resolutions (cm-level) on demand. In cold deserts, UAVs are invaluable for (i) fine-scale mapping of vegetation composition and health in research plots, (ii) monitoring the success of restoration projects, and (iii) crucially, for providing high-quality ground-truth data to calibrate and validate products derived from satellite imagery, thereby bridging the critical gap between field measurements and satellite observations (Manfreda et al., 2018).

In summary, no single sensor or platform can capture the full complexity of cold desert bio-resources. The future of effective monitoring lies in an integrated approach, where data from multiple sources are fused to create composite products that leverage the unique strengths of each technology (Garcia-del-Real & Alcaráz, 2024), thereby providing a more holistic and robust understanding of these vital and vulnerable ecosystems Kumar et al. (2024).

3 Core GIS Analytical Techniques: From Data to Ecological Insight

The true value of remote sensing data is unlocked through the analytical power of Geographic Information Systems (GIS), particularly through the synergistic synthesis of optical, thermal, and radar (SAR) datasets. In the context of cold desert research, relying on a single data source is often insufficient due to the complex interplay between scarce precipitation and vegetation dynamics. GIS provides the framework to integrate these distinct modalities: optical data (e.g., MODIS, Landsat) captures surface reflectance and vegetation phenology; thermal infrared data reveals land surface temperature (LST) and evapotranspiration stress; and radar data characterizes surface roughness and soil moisture content independent of cloud cover. By fusing these datasets, researchers can move beyond simple mapping to sophisticated spatial modeling—such as combining LST and vegetation indices to assess drought severity or integrating SAR texture with optical spectral signatures to accurately classify sparse cold desert bio-resources.

3.1 Data Preprocessing and Foundational Image Analysis

Before any meaningful analysis can occur, raw remote sensing data must undergo a series of rigorous preprocessing steps to correct for sensor, atmospheric, and geometric distortions (Liang et al., 2002). In the context of cold deserts, certain corrections are particularly critical.

Preprocessing: Atmospheric correction is crucial to convert sensor-level radiance values to surface reflectance, accounting for the unique scattering and absorption properties of the high-altitude atmosphere (Chavez, 1996).

Vegetation Indices (VIs): VIs are mathematical combinations of spectral bands designed to enhance the spectral contribution of vegetation. The Normalized Difference Vegetation Index (NDVI) is the most widely used, serving as a robust proxy for vegetation photosynthetic activity and green biomass (Rouse et al., 1973). However, in sparsely vegetated cold deserts, the NDVI signal is often contaminated by the spectral properties of the underlying soil and rock. To address this, soil-adjusted indices like the Soil-Adjusted Vegetation Index (SAVI) (Huete, 1988) or the Enhanced Vegetation Index (EVI) (Huete et al., 2002) are often employed as they are designed to minimize soil background effects. The calculation formulas for the above three indices are as follows:

$$\mathrm{NDVI} = \frac{\mathrm{NIR} - \mathrm{Red}}{\mathrm{NIR} + \mathrm{Red}}$$

$$\mathrm{SAVI} = \frac{\mathrm{NIR} + \mathrm{Red} + L}{\mathrm{NIR} - \mathrm{Red}} \times (1 + L)$$

$$\mathrm{EVI} = G \times \frac{\mathrm{NIR} - \mathrm{Red}}{L + \mathrm{NIR} + C_1 \times \mathrm{Red} - C_2 \times \mathrm{Blue}}$$

Image Classification: This is the process of assigning pixels to thematic classes (e.g., grassland, bare rock, water, snow). Common methods include unsupervised classification (e.g., ISODATA), which identifies statistical clusters in the data, and supervised classification (e.g., Maximum Likelihood, Random Forest), which uses user-defined training samples (Fig. 2) (Belgiu & Drăguţ, 2016; Campbell & Wynne, 2011). In the heterogeneous landscapes of cold deserts, Object-Based Image Analysis (OBIA) is increasingly preferred. OBIA first segments an image into meaningful objects (e.g., clumps of shrubs) and then classifies these objects using their spectral, spatial, and textural properties, often yielding more accurate and cartographically pleasing results than traditional pixel-based methods (Blaschke, 2010).

Validation and Accuracy Assessment: No map or model should be used for decision-making without a quantitative assessment of its accuracy. This is achieved through validation, typically using an independent set of reference data. For a thematic map, this involves collecting ground-truth points in the field or interpreting very high-resolution imagery (e.g., from drones). A confusion matrix is then constructed to compare the map's classifications with the reference data, from which key metrics like overall accuracy and the Kappa coefficient are calculated (Congalton, 1991). This process, while challenging and resource-intensive in remote cold deserts, is a non-negotiable component of robust geospatial science, ensuring that the final products are a reliable representation of reality.

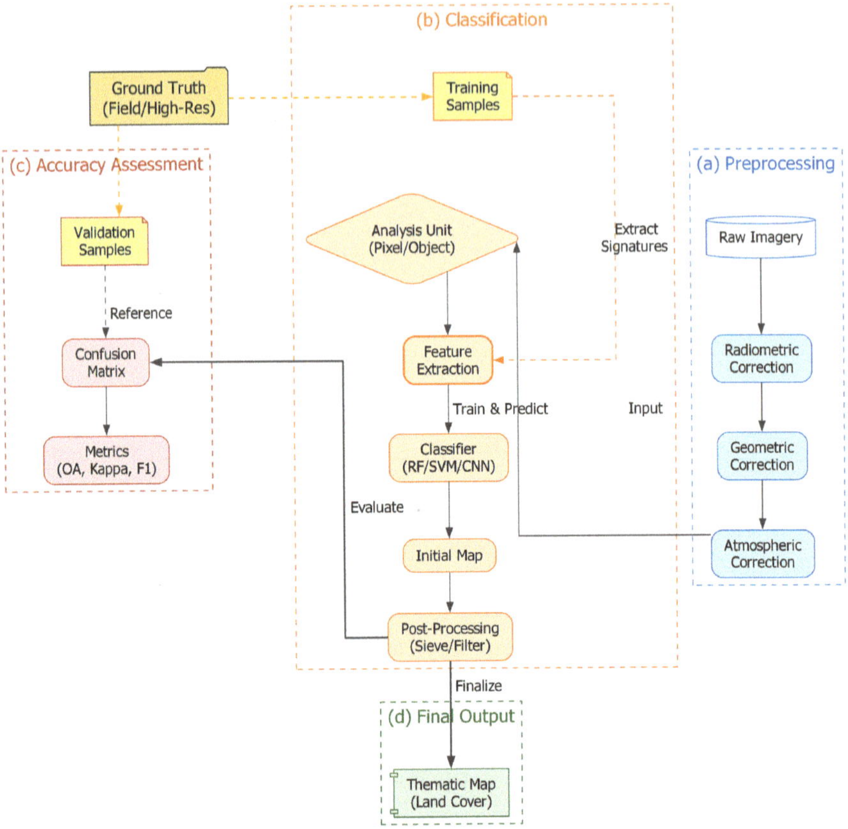

Fig. 2 Workflow of remote sensing image classification

3.2 *Thematic Mapping and Dynamic Monitoring*

Once data are processed and classified, GIS is used to generate thematic maps and to quantify changes over time, which are cornerstones of resource management and climate change assessment.

Thematic Mapping: This involves the creation of maps depicting the spatial distribution of a specific theme. Examples in cold desert research include producing baseline vegetation type maps, inventories of wetlands and glacial lakes, and maps of permafrost extent. These maps are not static products; they form the foundational data layers for virtually all subsequent spatial analysis and modeling.

Dynamic Monitoring (Change Detection): The multi-temporal nature of satellite archives enables powerful change detection analysis. This technique quantifies how landscapes have changed between two or more time points. Methodologies range from simple image differencing of VIs to identify areas of vegetation gain or loss, to the more robust post-classification comparison, which compares thematic maps from different years to identify specific class-to-class transitions (e.g., grassland converting to bare land) (Lu et al., 2004). Such analyses are critical for quantifying key climate change impacts, such as glacier retreat, the expansion or shrinkage of alpine lakes, and long-term trends in vegetation productivity (i.e., "greening" or "browning") (Zhu et al., 2016).

Cold Desert Bio-resource Mapping: To address the challenge of low vegetation cover (LAI < 0.5) in cold deserts, we integrated Soil-Adjusted Vegetation Indices (SAVI) to mitigate the overestimation of biomass common with traditional NDVI. In arid environments with less than 30% cover, the spectral dominance of soil backgrounds and biological soil crusts—which can exhibit elevated NDVI values (0.18–0.30)—often mimics vascular vegetation. By incorporating a soil-adjustment factor, SAVI effectively decouples the shrub signal from background brightness and microphytic activity, significantly improving the Producer's Accuracy for sparse shrubland classes (Alavipanah et al., 2016).

$$\mathrm{SAVI} = \frac{(\mathrm{NIR} - \mathrm{Red}) * (1 + L)}{\mathrm{NIR} + \mathrm{Red} + L}$$

3.3 *Spatial Modeling*

The highest level of GIS analysis involves using thematic data to model ecological processes and predict spatial patterns, with rigorous validation being essential for scientific credibility. This is a prime example of predictive analysis. Habitat suitability modeling (HSM) integrates species presence data (e.g., field sightings or GPS collar locations of fauna like the snow leopard) with a suite of environmental predictor layers (e.g., elevation, slope, vegetation cover, distance to water—all derived within the GIS) (Guisan & Zimmermann, 2000). Statistical algorithms, such as Maximum Entropy (MaxEnt) (Phillips et al., 2006), are then used to define the species' ecological niche and predict a continuous surface of habitat suitability across the landscape. The resulting maps are invaluable for identifying critical habitats, designing conservation corridors, and assessing the potential impact of climate change on species' ranges.

4 Major Application Areas: Geospatial Science in Action

Building on the thematic mapping and change detection techniques discussed in Sect. 3, one of the most fundamental applications of these geospatial methods is the creation of comprehensive resource inventories and baseline maps. The true measure of geospatial technologies lies in their ability to address pressing environmental and societal challenges. In cold deserts, where data is scarce and logistical constraints are severe, remote sensing and GIS have become indispensable tools for moving beyond theoretical understanding to informed action. This section highlights four major application domains, illustrating how these technologies underpin resource management, climate change assessment, and conservation efforts, directly supporting the core themes of vulnerability, adaptation, and mitigation.

4.1 Resource Inventory and Baseline Mapping

The foundational step in managing any ecosystem, particularly sensitive ones like cold deserts, is to create a comprehensive inventory of its resources and to continuously monitor their dynamics. Geospatial technologies provide the essential tools for both of these critical tasks, enabling the creation of accurate baseline maps and the tracking of environmental changes over time.

Resource Inventory and Baseline Mapping: Accurate, spatially explicit baseline maps are the essential reference against which all future change is measured.

Vegetation and Land Cover: Using multispectral satellite imagery (e.g., Sentinel-2) and advanced classification algorithms (e.g., Random Forest), researchers can produce detailed maps delineating key vegetation types like alpine meadows and steppe grasslands (Belgiu & Drăguţ, 2016). This is fundamental for estimating forage availability, carbon stocks, and habitat structure.

Water Resources: Techniques using spectral indices like the Normalized Difference Water Index (NDWI) allow for the precise delineation of glacial lakes, rivers, and wetlands (McFeeters, 1996). These inventories are critical for water management, especially in "Water Tower" regions where they support downstream communities.

Cryosphere and Permafrost: Mapping the extent of glaciers, snow cover, and permafrost probability zones provides a critical baseline for understanding the state of the cryosphere, a key indicator of climate change (Kääb et al., 2012).

Monitoring Climate Change and Land Degradation: Cold deserts are sentinels of climate change. Geospatial time-series analysis provides clear evidence of its impacts by comparing multi-temporal datasets to quantify environmental transformations.

Vegetation Dynamics: Long-term NDVI archives (e.g., from Landsat and MODIS) are used to detect trends in vegetation productivity, revealing patterns of landscape "greening" from warming or "browning" from drought and degradation (Zhu et al., 2016).

Cryospheric Decline: Comparing historical and recent satellite imagery provides a quantifiable record of glacier retreat. Advanced techniques like InSAR can detect land surface subsidence associated with permafrost thaw, a process that impacts infrastructure and releases stored carbon.

Land Degradation and Desertification: Recent advances in high-resolution remote sensing have significantly improved the detection of active degradation features, particularly soil erosion and sand dune expansion (Almalki et al., 2022). However, mapping bio-resources in cold deserts remains challenging due to the high spectral similarity between sparse vegetation and background soil, requiring specialized indices to accurately distinguish degradation from natural barrenness (Du et al., 2019).

4.2 Supporting Conservation and Sustainable Management

Once a comprehensive inventory and baseline understanding of bio-resources are established (as detailed in Sect. 4.1), geospatial intelligence directly informs on-the-ground conservation and management strategies, enabling a shift from reactive problem-solving to proactive, spatially targeted planning.

Rangeland Management: For pastoralist communities, rangelands are a vital resource. Remote sensing can estimate forage biomass over vast areas, allowing for the development of dynamic models of grazing capacity. This information supports sustainable practices like rotational grazing by identifying when and where to move herds, preventing overgrazing and promoting rangeland health.

Water Resource Management: By monitoring changes in the depth and volume of groundwater aquifers, managers can assess risks associated with over-extraction and land subsidence. Tracking key metrics like rainfall, soil moisture, and evaporation rates provides crucial input for groundwater models that predict sustainable yields and recharge rates, informing water allocation strategies for agriculture and urban consumption in arid regions (Chen et al., 2016).

Biodiversity Conservation: GIS-based Habitat Suitability Modeling is a cornerstone of modern conservation (Guisan & Zimmermann, 2000). By integrating species occurrence data with environmental layers (topography, vegetation, human disturbance), it is possible to map critical habitats and connectivity corridors for iconic and threatened species like the snow leopard or wild yak. These maps are invaluable for prioritizing conservation investments and designing effective protected area networks (Jun, 2022; Penjor, 2022).

4.3 Landscape Restoration Planning and Assessment

While sustainable management aims to prevent ecosystem decline, degradation can still occur. In these instances, as degradation pressures mount, ecological restoration has become a key mitigation strategy. Geospatial tools are vital in both the planning and monitoring phases of restoration projects.

Prioritization and Planning: Effective restoration targeting relies on GIS-based Multi-Criteria Decision Analysis (MCDA) that integrates multi-source datasets. To ensure accuracy, degradation assessment now couples vegetation trends with climatic data (e.g., Rain-Use Efficiency) to distinguish human impact from drought (Abolhasani et al., 2022), while erosion risk is determined using comprehensive models (e.g., RUSLE) that combine topography with rainfall erosivity and soil erodibility layers (Li et al., 2023).

Monitoring and Assessment: Remote sensing provides an objective and cost-effective means to assess the effectiveness of restoration interventions (Willis, 2015). By comparing high-resolution imagery (often from drones) or satellite time-series data from before and after a project, managers can quantify increases in vegetation cover, improvements in soil stability, and the overall trajectory of ecosystem recovery, enabling adaptive management and demonstrating project success (Padró et al., 2019).

5 Unique Challenges and Limitations in Cold Desert Environments

While remote sensing and GIS offer unprecedented opportunities for studying cold deserts, the application of these technologies is not without significant challenges. The inherent environmental characteristics of these regions, coupled with technical and logistical constraints, demand careful consideration and specialized methodologies. Acknowledging these limitations is essential for conducting robust scientific research and avoiding the misinterpretation of geospatial products.

5.1 Environmental and Biophysical Constraints

The cold desert landscapes create a suite of challenges that can confound standard remote sensing analyses (Fig. 3).

Low Vegetation Cover and Soil Background Effects: Low Vegetation Cover and Soil Background Effects: The most pervasive challenge is the sparse nature of the vegetation in cold deserts. In many areas, the remotely sensed signal is dominated by the spectral reflectance of the underlying soil, rock, and biological soil crusts rather than the vegetation itself. This "soil background noise" significantly reduces the sensitivity of traditional vegetation indices like NDVI.

Specifically, in these arid regions, NDVI values typically range between 0.05 and 0.20 (Kushwaha et al., 2024), while Leaf Area Index (LAI) often remains below 0.5 (Myneni et al., 2002). At these low ranges, the spectral signal of vegetation is easily overwhelmed by the background soil brightness, leading to misinterpretation of plant

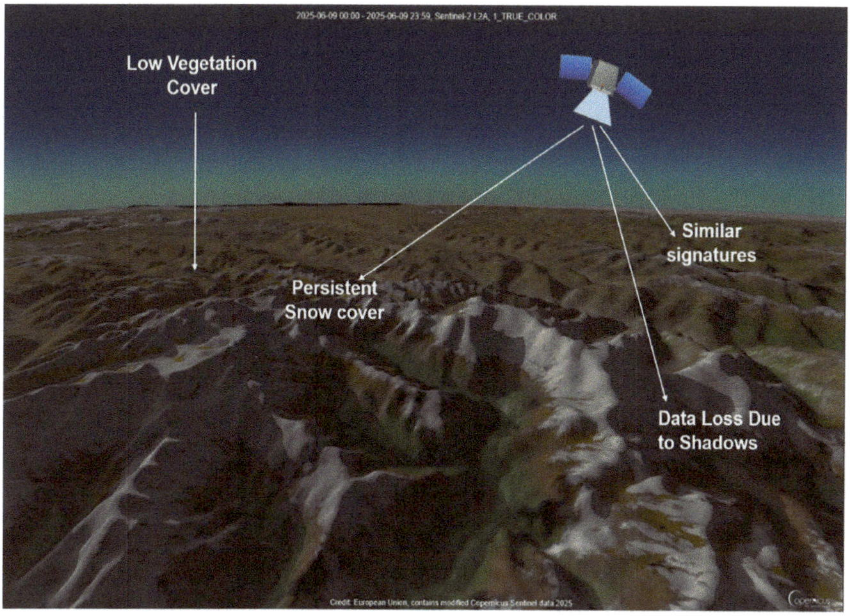

Fig. 3 Features identification from satellite data (Sentinel-2 L2A satellite data acquired on 09 June 2025, showing a Himalayas mountainous region)

health and biomass (Myneni et al., 2002).To overcome these challenges, specialized methods are required. Soil-Adjusted Vegetation Indices (e.g., SAVI, MSAVI) incorporate a soil-adjustment factor to minimize background influence (Ren & Zhou, 2014; Wu et al., 2019). Additionally, spectral mixture analysis (SMA) has proven effective in separating the fractional cover of vegetation from bare soil, providing a more accurate assessment than pixel-based indices alone (Khiry, 2007; Salih et al., 2017).

Spectral Confusion: The spectral signatures of different surface features in cold deserts can be remarkably similar, leading to high classification uncertainty. For example, senescent (dry) vegetation, certain types of rock, and bare soil can be spectrally indistinguishable in broad multispectral data. This makes the accurate mapping of specific plant communities or the delineation of degradation stages particularly difficult without the aid of advanced sensors or analytical techniques like spectral mixture analysis (Asner & Heidebrecht, 2002).

To resolve this in cold arid environments, multi-dimensional feature extraction is essential. First, exploiting the unique phenological metrics (time-series) of cold desert vegetation—capturing the short "green-up" pulse—allows for separation from static rock substrates and dormant dry grass (Snyder et al., 2016; Wang et al., 2015). Second, texture analysis (e.g., using GLCM or SAR fusion) effectively distinguishes the surface roughness of shrub communities from smooth saline-alkali soils, which often share identical spectral brightness (Shupe & Marsh, 2004; Zribi et al., 2003).

Finally, hyperspectral indices are increasingly used to identify biological soil crusts (BSCs), a critical cold desert component often confused with bare soil in broadband imagery (Panigada et al., 2019; Wang et al., 2023).

Persistent Snow and Ice Cover: Seasonal or permanent cryospheric components, while a key subject of study, also act as a significant data contaminant. Snow and ice cover obscure the underlying land surface for large portions of the year, drastically reducing the temporal window for effective land cover and vegetation analysis. The margins of melting snow patches can also be easily misclassified as other features, introducing error into change detection studies (Hall & Riggs, 2007). To mitigate these issues, robust atmospheric correction techniques (e.g., FLASH, ATCOR) are essential during preprocessing. These algorithms correct for the "adjacency effect"—where bright snow reflection contaminates neighboring dark pixels—and improve the spectral discrimination between snow and cloud cover (Kokhanovsky et al., 2020; Niraj et al., 2022).

Complex Topography and Shadows: The mountainous terrain typical of cold deserts creates extreme topographic variation. This results in extensive and dynamic shadowing, which drastically reduces the signal received by the sensor from shaded slopes. To mitigate these artifacts, specific preprocessing chains are essential: Haze correction (e.g., dark object subtraction) reduces atmospheric scattering, while topographic correction (e.g., C-correction, Minnaert) normalizes illumination in shadowed areas (Chinh et al., 2014; Hantson & Chuvieco, 2011).

5.2 *Technical, Data, and Logistical Constraints*

Beyond the environmental hurdles, researchers face a number of technical and practical limitations.

The Spatial-Temporal Resolution Trade-Off: A classic challenge in remote sensing is the trade-off between spatial and temporal resolution. Satellites that provide frequent observations (e.g., MODIS, with daily revisits, Sentinel-2, 3–5 daily revisits) do so at a coarse spatial resolution (250–500 m resolution, Sentinel-2 provides significantly higher spatial detail with its 10 m and 20 m bands), which may be insufficient to capture the fine-scale heterogeneity of cold desert landscapes. Conversely, very high-resolution commercial satellites provide exquisite spatial detail but have low revisit times and high data acquisition costs, making them unsuitable for continuous, large-area monitoring. Data fusion techniques are often required to overcome this limitation (Zhu et al., 2010).

Data Availability and Cloud Contamination: While the free and open data policies of programs like Landsat and Sentinel have been transformative, their optical sensors are frequently hampered by cloud cover, particularly during monsoonal periods in regions like the Tibetan Plateau. This can create significant gaps in time-series data, complicating phenological analysis and the detection of short-term events (Shen et al., 2015a, 2015b).

Scarcity of Validation Data: The credibility of any remote sensing product hinges on rigorous validation with independent, high-quality reference data. However, the extreme remoteness, logistical inaccessibility, and often harsh weather conditions of cold deserts make extensive fieldwork for ground-truthing prohibitively expensive and, at times, dangerous. This scarcity of validation data is a major bottleneck that can limit the certifiable accuracy of maps and models, forcing a reliance on less direct or opportunistic validation methods (Foody, 2002).

Requirement for Specialized Expertise: Effectively navigating these challenges requires a high level of technical expertise. The proper selection of imagery, execution of advanced preprocessing steps, and choice of appropriate analytical algorithms demand a sophisticated understanding of both remote sensing physics and the specific ecological context of cold deserts. This can be a barrier for researchers or management agencies without dedicated geospatial specialists (Pettorelli et al., 2018).

6 Selected Case Studies

To bridge the gap between theory and practice, this section presents two representative case studies from different cold desert regions. These examples illustrate how the integration of diverse geospatial datasets and analytical techniques provides critical insights for understanding ecosystem vulnerability and supporting adaptive management in the face of environmental change. The first case study demonstrates Habitat Suitability Modeling for a keystone species, while the second showcases the power of time-series analysis using vegetation indices to assess continental-scale trends.

Case Study 6.1: Predicting Climate-Induced Range Shifts of a Keystone Plant Species in the Tibetan Plateau using MaxEnt

The Tibetan Plateau, often termed the "Third Pole," is experiencing climate warming at a rate exceeding the global average. This rapid change threatens the stability of its unique alpine ecosystems. *Sabina tibetica* (Tibetan Juniper) is a foundational treeline species, forming extensive juniper woodlands at high elevations. These woodlands are critical for soil and water conservation, providing habitat, and serving as a key carbon sink in these high-altitude landscapes. Understanding how climate change will impact the distribution of this crucial species is paramount for assessing future ecosystem stability. This case study demonstrates the use of Maximum Entropy (MaxEnt) modeling to predict the potential current and future suitable habitat for *Sabina tibetica* (Li et al., 2025).

Methodology

Species Occurrence Data: High-quality occurrence records for *Sabina tibetica* were compiled from the Global Biodiversity Information Facility (GBIF) and field survey data. Regarding field sampling data, this is species distribution data I collected in the field on the Qinghai-Tibet Plateau in 2018–2019 using a handheld GPS receiver with a positional accuracy of approximately 3–5 m. The data were carefully cleaned to remove spatial duplicates and records with obvious geo-referencing errors.

Environmental Predictors (Current and Future): A suite of predictor variables was prepared. WorldClim Version 2.1, a widely used high-resolution global climate dataset, provides climate surfaces representing the 1970–2000 period. This baseline was selected due to the high quality and spatial coverage of weather station data available for this timeframe, making it a robust reference for modeling and assessing climate patterns (Fick & Hijmans, 2017). Nineteen bioclimatic variables from the WorldClim dataset and topographic variables (elevation, slope) from a DEM were used. To project into the future (e.g., 2070), equivalent bioclimatic variables were obtained from WorldClim, downscaled from a General Circulation Model (GCM) under two Shared Socioeconomic Pathways (SSP-4.5 and SSP-8.5). However, relying solely on climatic averages is often insufficient for complex cold desert ecosystems. To address this, recent studies have increasingly integrated multi-source satellite products to complement bioclimatic variables. For instance, optical data (e.g., Sentinel-2) is used to capture fine-scale vegetation structure, while thermal infrared sensors (e.g., Landsat TIRS) monitor surface moisture stress, and RADAR (e.g., Sentinel-1) provides structural information independent of cloud cover (Park et al., 2017; Preston et al., 2023; Wang et al., 2025). Recent case studies demonstrate that combining these dynamic satellite observations with static WorldClim predictors significantly improves species distribution modeling accuracy in heterogeneous landscapes (Wang et al., 2025).

Model Execution and Validation: The MaxEnt software was used to model the relationship between species occurrences and the environmental predictors. Seventy-five percent of the occurrence data were used for model training and 25% for testing. The model's performance was evaluated using the Area Under the Curve (AUC), and it was projected onto the future climate scenarios (Fig. 4).

Results and Discussion

The model for the current period performed well (AUC > 0.9) (Fig. 5), accurately predicting the known distribution of *Sabina tibetica* along the high-elevation slopes forming the alpine treeline (Li et al., 2025; Lissovsky & Dudov, 2021). The future projections revealed a significant potential range shift (Fig. 6). Under both scenarios, the model predicted a contraction of suitable habitat at the species' lower-elevation, warmer range limits. Concurrently, it predicted a potential upward elevational shift, with new suitable habitats emerging in higher-altitude tundra zones that are currently too cold to support tree growth. The high emissions scenario (SSP5-8.5) showed a far more drastic and fragmented future distribution (Li et al., 2025).

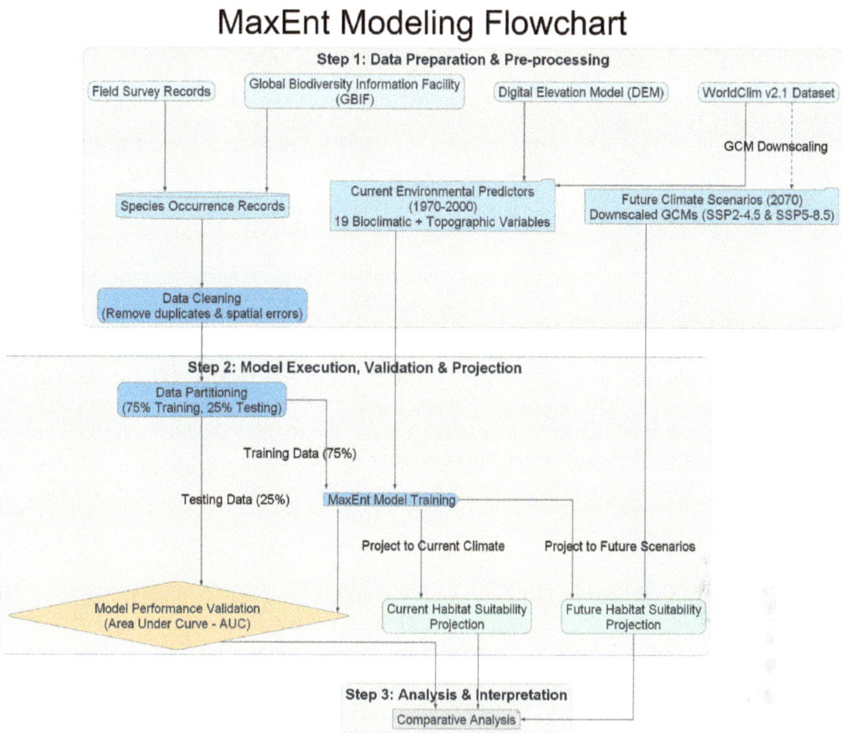

Fig. 4 Flowchart summarizing the processing methodology used in this study (Li et al., 2025)

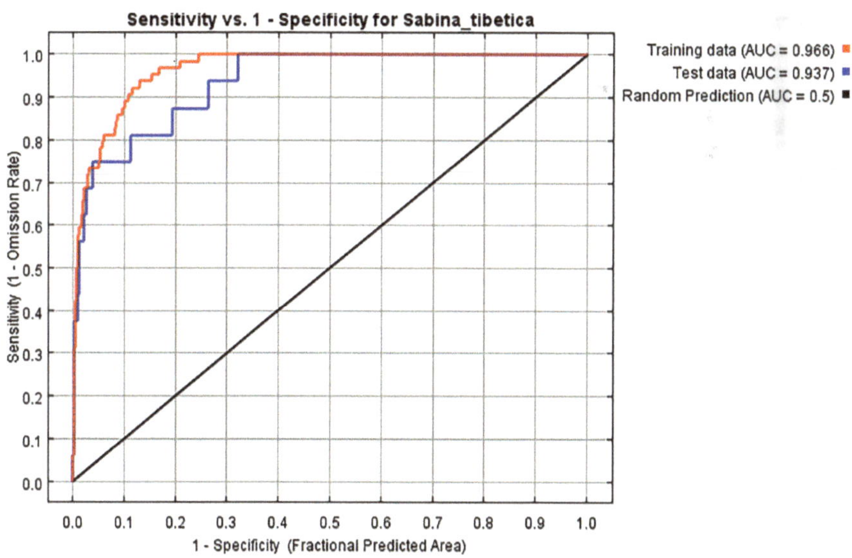

Fig. 5 ROC curve for the *Sabina tibetica* species distribution model (This figure was created using data from my own publication (Li et al., 2025))

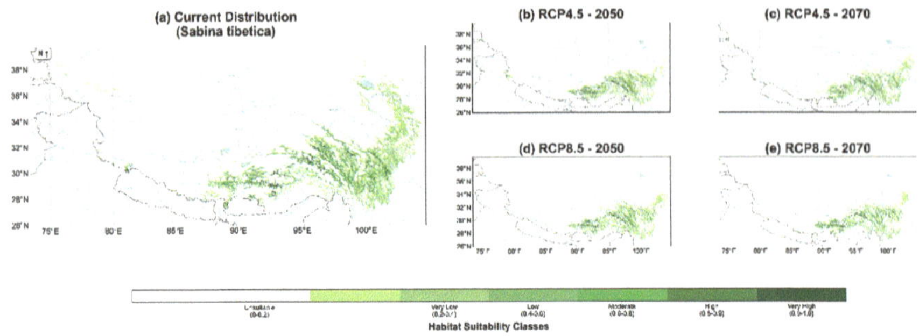

Fig. 6 A series of maps showing the predicted suitable habitat for *Sabina tibetica* under current and future climate scenarios (Li et al., 2025). **a** Predicted current distribution. **b, c** The predicted future distribution for 2050 and 2070, respectively, under the RCP4.5 (intermediate emissions) scenario. **d, e** The predicted distribution for the same years under the RCP8.5 (high emissions) scenario, which projects a greater reduction and fragmentation of the habitat

Management Implications

The maps identify "hotspots of decline," increasing the vulnerability of ecosystems and pastoral communities. They also highlight potential climate "refugia" for conservation prioritization and help guide long-term grazing management and restoration policies to avoid maladaptation.

Case Study 6.2: Assessing Vegetation Greening Trends in Southern Andes using MODIS NDVI Time-Series

Southern Andes encompasses diverse ecosystems ranging from the Amazon rainforest to the Patagonian Steppe, each responding differently to climate variability and change. The region includes water-limited cold desert ecosystems in Patagonia, temperate grasslands, and tropical forests that are highly sensitive to precipitation and temperature shifts. Understanding vegetation productivity trends across this continental scale is crucial for assessing ecosystem health, carbon dynamics, and climate change impacts. This study demonstrates the use of long-term satellite data to quantify vegetation trends across Southern Andes.

Methodology

Data Acquisition and Processing: MODIS NDVI time-series data from 2000 to 2023 were processed to generate annual maximum NDVI values for each pixel across Southern Andes, providing a 24-year record of vegetation productivity patterns.

Time-Series Analysis: Annual mean NDVI was calculated across the entire study region to assess continental-scale vegetation trends. Both linear regression and Mann-Kendall statistical tests were applied to detect significant trends while accounting for temporal autocorrelation and non-normal data distributions.

Trend Detection and Statistical Analysis: Linear regression analysis revealed the overall direction and magnitude of change, while the Mann-Kendall test provided robust, non-parametric trend detection less sensitive to outliers and data distribution assumptions.

Results and Discussion

The analysis revealed a significant greening trend across Southern Andes from 2000 to 2023 (Fig. 7). Key findings include the following:

Overall Trend: Annual mean NDVI increased at a rate of approximately 0.002563/year, representing a statistically significant greening pattern ($p < 0.05$).

Spatial Patterns: The continental-scale analysis shows distinct vegetation productivity zones, with highest NDVI values (>0.8) in the Amazon basin and Atlantic coastal forests (shown in green), moderate values (0.4–0.6) in the Cerrado and Pampas regions (yellow-orange), and lowest values (<0.4) in the Patagonian Steppe and Andean highlands (red-brown areas).

Temporal Variability: Analysis of the MODIS NDVI time-series reveals significant inter-annual variability despite the overall positive trend. Notable increases in vegetation productivity occurred in 2008, 2011, and 2020–2023, while distinct declines were observed during 2005, 2012, and 2015–2016. These fluctuations align closely with ENSO cycles, where La Niña phases (often associated with increased precipitation in this region) likely drove greening pulses, whereas El Niño-related moisture deficits led to productivity declines.

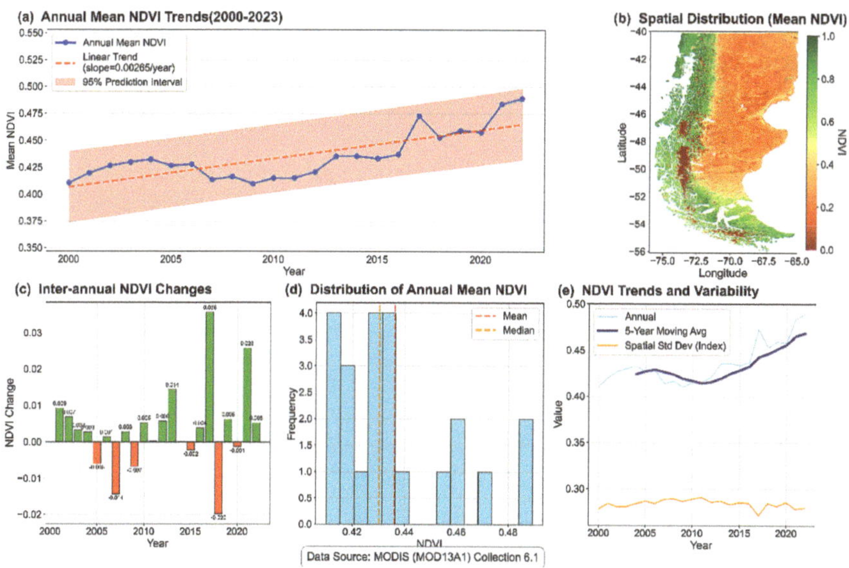

Fig. 7 Comprehensive NDVI trend analysis for southern andes (2000–2023): temporal dynamics, spatial patterns, and statistical assessment

Regional Implications: The sustained greening trend suggests enhanced vegetation productivity potentially driven by CO_2 fertilization effects, changing precipitation patterns, or recovery from historical land use changes. However, the high spatial heterogeneity visible in the NDVI distribution map indicates that local and regional factors significantly modulate these continental trends.

This analysis demonstrates the value of long-term satellite monitoring for detecting vegetation changes at continental scales, providing critical information for understanding ecosystem responses to global change drivers.

Management Implications

Vulnerability Assessment: The map provides a direct visualization of landscape-scale vulnerability. The large, contiguous areas of "browning" are hotspots of land degradation, signaling a decline in ecosystem services and increased risk for desertification.

Sustainable Land Management: This information is invaluable for regional planning agencies and livestock organizations. It can be used to adjust stocking rates in browning areas to reduce grazing pressure or to promote land conservation policies.

Climate Adaptation: By correlating these trends with climate data, researchers can better understand the specific drivers of change. This knowledge is crucial for developing targeted adaptation strategies, such as implementing water-harvesting techniques in vulnerable browning zones.

7 Future Directions: The Next Frontier in Geospatial Analysis of Cold Deserts

The field of geospatial science is in a state of rapid evolution, driven by parallel advancements in sensor engineering, computational power, and analytical paradigms. For cold desert research, these developments promise to overcome existing limitations and unlock unprecedented insights into ecosystem dynamics, vulnerability, and resilience. This section outlines the key future directions that will define the next generation of monitoring and modeling in these critical environments.

7.1 Advances in Sensor Technology and Data Acquisition

The quality and type of data fundamentally constrain what can be analyzed. The near future will see a proliferation of more capable sensors, providing data with enhanced spectral, spatial, and temporal resolutions.

Hyperspectral Imaging: The transition from multispectral to hyperspectral remote sensing represents a significant leap forward in Earth observation (Goetz et al., 1985).

Modern and forthcoming missions, such as EnMAP and the planned Surface Biology and Geology (SBG) mission (Cawse-Nicholson et al., 2021), are designed to acquire data in hundreds of narrow, contiguous spectral bands. This rich spectral information enables a shift beyond generic vegetation indices toward more advanced applications, including the direct identification of plant species, the detection of subtle physiological stress through changes in leaf biochemistry (Ustin & Gamon, 2010), and the precise mineralogical mapping of exposed soils and rocks—a crucial capability in sparsely vegetated environments like cold deserts.

Active Sensing Evolution: Advances in spaceborne active sensing, exemplified by LiDAR missions such as ICESat-2 and GEDI, are enabling increasingly dense and accurate measurements of key Earth system variables like vegetation canopy structure and snow depth (Urbazaev et al., 2022). Concurrently, new radar missions operating at different frequencies, such as the L-band NISAR mission, will provide enhanced capabilities for monitoring dynamic processes including permafrost thaw, soil moisture changes, and glacier flow (Kellndorfer et al., 2025).

The Drone Revolution and Sensor Miniaturization: The growing affordability and sophistication of Uncrewed Aerial Vehicles (UAVs) equipped with miniaturized sensors are fundamentally reshaping environmental monitoring (Anderson & Gaston, 2013). These platforms effectively bridge the critical scale gap between detailed, ground-based field plots and coarser-resolution satellite pixels by providing on-demand, ultra-high-resolution data from LiDAR, thermal, and hyperspectral sensors (Manfreda et al., 2018). This capability enables novel and precise applications, including quantitative ecosystem assessment and the essential calibration and validation of global satellite products (Tmušić et al., 2020).

7.2 *The Paradigm Shift: AI, Cloud Computing, and Data Fusion*

The impending deluge of data from new sensors necessitates a paradigm shift in how data is processed and analyzed. Artificial Intelligence (AI) and cloud computing are the enabling technologies at the heart of this transformation.

Artificial Intelligence and Machine Learning

Deep learning algorithms, particularly Convolutional Neural Networks (CNNs), are already outperforming traditional methods in image classification, object detection, and feature extraction from remotely sensed data (Ma et al., 2019; Zhu et al., 2017). In the future, AI will be routinely applied to automatically identify subtle patterns of land degradation, map individual animal herds from VHR imagery (Weinstein, 2018), and forecast ecosystem responses by learning complex, non-linear relationships within vast datasets that are beyond human interpretation (Zhu et al., 2017).

Big Data and Cloud Computing Platforms

The era of downloading and processing massive datasets on local machines is ending, giving way to a new paradigm of cloud-native geospatial analysis (Gorelick et al., 2017). Cloud-based platforms like Google Earth Engine (GEE), Microsoft's Planetary Computer, and the NASA Earthdata Cloud provide access to petabyte-scale archives of analysis-ready data and the distributed computational power to process it in place. This democratization of high-performance computing empowers any researcher, anywhere, to conduct continental-scale, multi-decadal analyses that were once the exclusive domain of major space agencies (Gorelick et al., 2017).

Intelligent Data Fusion

The true analytical frontier lies in the intelligent fusion of multi-source data (Schmitt & Zhu, 2016). Future analytical frameworks will seamlessly integrate disparate data streams—such as optical, thermal, radar, and LiDAR data—with climate model outputs and in situ sensor networks (Ghamisi et al., 2018). AI-driven fusion algorithms, in particular, will learn to weigh the relative strengths and weaknesses of each data source to produce synergistic data products that are more accurate and information-rich than the sum of their parts (Ghamisi et al., 2018; Lahat et al., 2015), providing a holistic, multi-dimensional view of ecosystem processes.

7.3 *Toward Integrated Methodologies and Digital Twins*

The ultimate goal is to move beyond monitoring to predictive, holistic understanding. This requires integrating geospatial data into more comprehensive modeling frameworks that couple ecological, hydrological, and socioeconomic systems.

Integrated Earth System Modeling: A new generation of coupled Earth System Models (ESMs) will be central to this effort. Within these models, geospatial data products will serve not only as inputs but as the primary drivers and validation layers. For instance, the assimilation of near-real-time remote sensing observations—such as snow cover, soil moisture, and vegetation productivity—into land surface models has been shown to significantly improve the accuracy of forecasts for river discharge, carbon fluxes, and overall ecosystem responses to climate change scenarios (Reichle, 2008).

Socio-Ecological System Analysis: A critical research frontier is the formal integration of socioeconomic data (e.g., grazing intensity, population density, infrastructure development) with biophysical data within a socio-ecological systems (SES) framework (Ostrom, 2009). The use of GIS is essential for this synthesis, enabling a more nuanced understanding of land use change drivers and a more robust assessment of the vulnerability of human communities that depend on ecosystem services and bio-resources (Turner et al., 2003a, 2003b).

The Rise of "Digital Twins": The culmination of these advancements points toward the creation of "Digital Twins" for entire ecosystems (Bauer et al., 2021). A Digital Twin is a dynamic, data-driven virtual replica of a real-world system, constantly updated with new remote sensing and field data. Such models will not only track the current state of the ecosystem but will also allow managers and policymakers to run "what-if" scenarios. By exploring the potential consequences of different management interventions or climate futures, Digital Twins can provide an unprecedented tool for adaptive, science-based decision-making and governance (Bauer et al., 2021; Holling, 1978).

8 Conclusions

This chapter has demonstrated that RS and GIS are not merely useful tools but are fundamentally indispensable for the scientific study and sustainable management of cold desert ecosystems. Faced with extreme climates, vast and inaccessible terrain, and increasing threats from climate change and human pressures, traditional field-based research methods are insufficient. Geospatial technologies provide the only viable means to overcome these limitations, offering a paradigm shift from sparse, point-based knowledge to a holistic, spatially explicit understanding of these fragile environments.

Summary of Capabilities and Contributions to Cold Desert Research: The core contribution of RS and GIS lies in their ability to provide objective, repeatable, and scalable data across vast landscapes. This chapter has shown how multi-sensor remote sensing—spanning optical, thermal, radar, and LiDAR—enables foundational scientific advancements. Crucially, it demonstrates the capacity to monitor these unique environments characterized by extremely sparse vegetation, where NDVI values typically fluctuate between 0.05 and 0.20 and LAI remains below 0.5. These include creating accurate baseline inventories of critical bio-resources like sparse vegetation, water bodies, and cryospheric components (glaciers and snowpack). The multi-temporal nature of satellite archives allows for the robust monitoring of environmental dynamics, quantifying long-term trends such as vegetation "greening" or "browning," glacier retreat, and permafrost thaw. Furthermore, by integrating these observations with topographic data from DEMs within a GIS framework, researchers can analyze the complex interplay of factors that govern ecosystem processes, transforming raw data into actionable ecological insight.

Support for Vulnerability Assessment, Adaptation, and Mitigation: Beyond fundamental science, these geospatial technologies directly empower the critical tasks of vulnerability assessment, adaptation, and mitigation. By detecting trends in vegetation productivity and land degradation, RS-GIS workflows identify "hotspots" of vulnerability, guiding targeted conservation efforts. The case studies illustrated how Habitat Suitability Modeling can predict species' range shifts under future climate scenarios, informing proactive conservation planning and the identification of potential climate "refugia." This information is vital for developing adaptation strategies, such as adjusting grazing patterns for pastoral communities or planning water management in "Water Tower" regions. For mitigation, these tools are essential for planning and prioritizing landscape restoration projects by integrating data on degradation severity and erosion risk. Subsequently, they provide a cost-effective means to monitor the success of these interventions, facilitating adaptive management and ensuring resources are used effectively.

Future Outlook and the Need for Integration: The future of cold desert research is being shaped by a rapid evolution in geospatial technology. As detailed, a new generation of advanced sensors—including hyperspectral, spaceborne LiDAR, and high-resolution radar—will provide unprecedented detail on ecosystem composition and function. The true power of this data deluge, however, will be unlocked by the paradigm shift toward Artificial Intelligence, cloud computing, and intelligent data fusion. These technologies will enable researchers to overcome persistent challenges like spectral confusion and data gaps while extracting complex, non-linear relationships from massive, multi-source datasets.

The ultimate frontier lies in integration: fusing data from different sensors, bridging scales from drones to satellites, and combining biophysical data with socioeconomic drivers in comprehensive models. This trajectory points toward the development of "Digital Twins" of cold desert ecosystems—dynamic, data-driven virtual replicas that can be used to test management scenarios and forecast future changes. By embracing this integrated, data-driven approach, the scientific and management communities can move from simply observing change to actively forecasting, shaping, and securing a more resilient and sustainable future for these vital and fragile ecosystems.

References

Abatzoglou, J. T., Dobrowski, S. Z., Parks, S. A., & Hegewisch, K. C. (2018). TerraClimate, a high-resolution global dataset of monthly climate and climatic water balance from 1958–2015. *Scientific Data, 5*(1), 1–12. https://doi.org/10.1038/sdata.2017.191

Abolhasani, A., Zehtabian, G., Khosravi, H., Rahmati, O., Alamdarloo, E. H., & D'Odorico, P. (2022). A new conceptual framework for spatial predictive modelling of land degradation in a semiarid area. *Land Degradation & Development, 33*(17), 3358–3374. https://doi.org/10.1002/ldr.4391

Airbus. (2025). Pléiades imagery and services. Airbus Defence and Intelligence. Retrieved September 21, 2025, from https://www.intelligence-airbusds.com/imagery/constellation/pleiades/

Alavipanah, S. K., Damavandi, A. A., Mirzaei, S., Rezaei, A., Hamzeh, S., Matinfar, H. R. et al. (2016). Remote sensing application in evaluation of soil characteristics in desert areas. *Natural Environment Change, 2*(1), 1–24. https://jnec.ut.ac.ir/article_59485_0.html

Almalki, R., Khaki, M., Saco, P. M., & Rodriguez, J. F. (2022). Monitoring and mapping vegetation cover changes in arid and semi-arid areas using remote sensing technology: A review. *Remote Sensing, 14*(20), 5143. https://doi.org/10.3390/rs14205143

Anderson, K., & Gaston, K. J. (2013). Lightweight unmanned aerial vehicles will revolutionize spatial ecology. *Frontiers in Ecology and the Environment, 11*(3), 138–146. https://doi.org/10.2307/23470549

Asner, G. P., & Heidebrecht, K. B. (2002). Spectral unmixing of vegetation, soil and dry carbon cover in arid regions: Comparing multispectral and hyperspectral observations. *International Journal of Remote Sensing, 23*(19), 3939–3958. https://doi.org/10.1080/01431160110115960

Bastiaanssen, W. G., Menenti, M., Feddes, R. A., & Holtslag, A. A. M. (1998). A remote sensing surface energy balance algorithm for land (SEBAL). 1 Formulation. *Journal of Hydrology, 212*, 198–212. https://doi.org/10.1016/s0022-1694(98)00253-4

Bauer, P., Stevens, B., & Hazeleger, W. (2021). A digital twin of earth for the green transition. *Nature Climate Change, 11*(2), 80–83. https://doi.org/10.1038/s41558-021-00986-y

Beck, H. E., Pan, M., Roy, T., Weedon, G. P., Pappenberger, F., Van Dijk, A. I., & Wood, E. F. (2019). Daily evaluation of 26 precipitation datasets using stage-IV gauge-radar data for the CONUS. *Hydrology and Earth System Sciences, 23*(1), 207–224. https://doi.org/10.5194/hess-23-207-2019

Belgiu, M., & Drăguţ, L. (2016). Random forest in remote sensing: A review of applications and future directions. *ISPRS Journal of Photogrammetry and Remote Sensing, 114*, 24–31. https://doi.org/10.1016/j.isprsjprs.2016.01.011

Blaschke, T. (2010). Object based image analysis for remote sensing. *ISPRS Journal of Photogrammetry and Remote Sensing, 65*(1), 2–16. https://doi.org/10.1016/j.isprsjprs.2009.06.004

Bürgmann, R., Rosen, P. A., & Fielding, E. J. (2000). Synthetic aperture radar interferometry to measure Earth's surface topography and its deformation. *Annual Review of Earth and Planetary Sciences, 28*(1), 169–209. https://doi.org/10.1146/annurev.earth.28.1.169

Campbell, J. B., & Wynne, R. H. (2011). *Introduction to remote sensing*. Guilford press.

Cawse-Nicholson, K., Townsend, P. A., Schimel, D., Assiri, A. M., Blake, P. L., Buongiorno, M. F., et al. (2021). NASA's surface biology and geology designated observable: A perspective on surface imaging algorithms. *Remote Sensing of Environment, 257*, Article 112349. https://doi.org/10.1016/j.rse.2023.113534

Chavez, P. S. (1996). Image-based atmospheric corrections-revisited and improved. *Photogrammetric Engineering and Remote Sensing, 62*(9), 1025–1035.

Chen, B., Zhang, X., Tao, J., Wu, J., Wang, J., Shi, P., et al. (2014). The impact of climate change and anthropogenic activities on alpine grassland over the Qinghai-Tibet Plateau. *Agricultural and Forest Meteorology, 189*, 11–18. https://doi.org/10.1016/j.agrformet.2014.01.002

Chen, J., Famigliett, J. S., Scanlon, B. R., & Rodell, M. (2016). Groundwater storage changes: Present status from GRACE observations. In *Remote sensing and water resources* (pp. 207–227). Springer International Publishing. https://doi.org/10.1007/978-3-319-32449-4_9

Cheng, G., & Wu, T. (2007). Responses of permafrost to climate change and their environmental significance, Qinghai-Tibet Plateau. *Journal of Geophysical Research: Earth Surface, 112*(F2). https://doi.org/10.1029/2006jf000631

Chinh Ke, L., Ngoc Tuong, T., & Van Hung, N. (2014). Correction of spectral radiance of optical satellite image for mountainous terrain for studying land surface cover changes. *Geodesy and Cartography, 63*(1). https://doi.org/10.2478/geocart-2014-0003

Congalton, R. G. (1991). A review of assessing the accuracy of classifications of remotely sensed data. *Remote Sensing of Environment, 37*(1), 35–46. https://doi.org/10.1016/0034-4257(91)90048-b

Du, J., Watts, J. D., Jiang, L., Lu, H., Cheng, X., Duguay, C., Farina, M., Qiu, Y., Kim, Y., Kimball, J. S., & Tarolli, P. (2019). Remote sensing of environmental changes in cold regions: Methods, achievements and challenges. *Remote Sensing, 11*(16), 1952. https://doi.org/10.3390/rs11161952

European Space Agency. (2022). *Copernicus DEM product handbook*. Document No. GEO.2018-1988-2.

European Space Agency (ESA). (2025). Sentinel-2 Mission. Copernicus Programme. Retrieved September 21, 2025, from https://sentinels.copernicus.eu/web/sentinel/missions/

Farr, T. G., Rosen, P. A., Caro, E., Crippen, R., Duren, R., Hensley, S. et al. (2007). The shuttle radar topography mission. *Reviews of Geophysics, 45*(2). https://doi.org/10.1029/2005rg000183

Fick, S. E., & Hijmans, R. J. (2017). WorldClim 2: New 1-km spatial resolution climate surfaces for global land areas. *International Journal of Climatology, 37*(12), 4302–4315. https://doi.org/10.1002/joc.5086

Foody, G. M. (2002). Status of land cover classification accuracy assessment. *Remote Sensing of Environment, 80*(1), 185–201. https://doi.org/10.1016/s0034-4257(01)00295-4

Funk, C., Peterson, P., Landsfeld, M., Pedreros, D., Verdin, J., Shukla, S., et al. (2015). The climate hazards infrared precipitation with stations—A new environmental record for monitoring extremes. *Scientific Data, 2*(1), 1–21. https://doi.org/10.1038/sdata.2015.66

Garcia-del-Real, J., & Alcaráz, M. (2024). Unlocking the future of space resource management through satellite remote sensing and AI integration. *Resources Policy, 91*, Article 104947. https://doi.org/10.1016/j.resourpol.2024.104947

Ghamisi, P., Rasti, B., Yokoya, N., Wang, Q., Hofle, B., Bruzzone, L. et al. (2018). Multisource and multitemporal data fusion in remote sensing. https://doi.org/10.48550/arXiv.1812.08287

Goetz, A. F., Vane, G., Solomon, J. E., & Rock, B. N. (1985). Imaging spectrometry for earth remote sensing. *Science, 228*(4704), 1147–1153. https://doi.org/10.1126/science.228.4704.1147

Gorelick, N., Hancher, M., Dixon, M., Ilyushchenko, S., Thau, D., & Moore, R. (2017). Google earth engine: Planetary-scale geospatial analysis for everyone. *Remote Sensing of Environment, 202*, 18–27. https://doi.org/10.1016/j.rse.2017.06.031

Guisan, A., & Zimmermann, N. E. (2000). Predictive habitat distribution models in ecology. *Ecological Modelling, 135*(2–3), 147–186. https://doi.org/10.1016/s0304-3800(00)00354-9

Hall, D. K., & Riggs, G. A. (2007). Accuracy assessment of the MODIS snow products. *Hydrological Processes: An International Journal, 21*(12), 1534–1547. https://doi.org/10.1002/hyp.6715

Hantson, S., & Chuvieco, E. (2011). Evaluation of different topographic correction methods for Landsat imagery. *International Journal of Applied Earth Observation and Geoinformation, 13*(5), 691–700. https://doi.org/10.1016/j.jag.2011.05.001

Holling, C. S. (Ed). (1978). *Adaptive environmental assessment and management (International Series on Applied Systems Analysis, Vol. 3)*. Wiley.

Huete, A. R. (1988). A soil-adjusted vegetation index (SAVI). *Remote Sensing of Environment, 25*(3), 295–309. https://doi.org/10.1016/0034-4257(88)90106-X

Huete, A., Didan, K., Miura, T., Rodriguez, E. P., Gao, X., & Ferreira, L. G. (2002). Overview of the radiometric and biophysical performance of the MODIS vegetation indices. *Remote Sensing of Environment, 83*(1–2), 195–213. https://doi.org/10.1016/s0034-4257(02)00096-2

Immerzeel, W. W., Van Beek, L. P., & Bierkens, M. F. (2010). Climate change will affect the Asian water towers. *Science, 328*(5984), 1382–1385. https://doi.org/10.1126/science.1183188

Jun, W. (2022). A multiscale assessment of snow leopard distribution, habitat-use and landscape connectivity in a new national park in China (Doctoral dissertation, Manchester Metropolitan University).

Kääb, A., Berthier, E., Nuth, C., Gardelle, J., & Arnaud, Y. (2012). Contrasting patterns of early twenty-first-century glacier mass change in the Himalayas. *Nature, 488*(7412), 495–498. https://doi.org/10.1038/nature11324

Karnieli, A., Agam, N., Pinker, R. T., Anderson, M., Imhoff, M. L., Gutman, G. G., et al. (2010). Use of NDVI and land surface temperature for drought assessment: Merits and limitations. *Journal of Climate, 23*(3), 618–633. https://doi.org/10.1175/2009jcli2900.1

Kellndorfer, J., Cartus, O., & Helfrich, S. (2025). Meeting the big synthetic aperture radar data processing challenge: Introducing the software for earth big data processing, prediction modeling, and organization cloud-scaling solution. *IEEE Geoscience and Remote Sensing Magazine*. https://doi.org/10.1109/mgrs.2024.3519549

Khiry, M. A. (2007). Spectral mixture analysis for monitoring and mapping desertification processes in semi-arid areas in North Kordofan State, Sudan. https://e-docs.geo-leo.de/bitstreams/b5f09586-12c2-4304-8145-e67bc2e0cfea/download

Kogan, F. N. (1997). Global drought watch from space. *Bulletin of the American Meteorological Society, 78*(4), 621–636. https://doi.org/10.1175/1520-0477(1997)078%3c0621:gdwfs%3e2.0.co;2

Kokhanovsky, A., Box, J. E., Vandecrux, B., Mankoff, K. D., Lamare, M., Smirnov, A., & Kern, M. (2020). The determination of snow albedo from satellite measurements using fast atmospheric correction technique. *Remote Sensing, 12*(2), 234. https://doi.org/10.3390/rs12020234

Kumar, V., Sharma, K. V., Pham, Q. B., Srivastava, A. K., Bogireddy, C., & Yadav, S. M. (2024). Advancements in drought using remote sensing: Assessing progress, overcoming challenges, and exploring future opportunities. *Theoretical and Applied Climatology, 155*(6), 4251–4288. https://doi.org/10.1007/s00704-024-04914-w

Kushwaha, A., Bhatnagar, R., Kumar, P., Zucca, C., Srivastava, S., & Ajai. (2024). Long Term Monitoring of Ecological Status of Major Deserts of the World. *Journal of the Indian Society of Remote Sensing, 52*(8), 1819–1839. https://doi.org/10.1007/s12524-024-01915-0

Lahat, D., Adali, T., & Jutten, C. (2015). Multimodal data fusion: An overview of methods, challenges, and prospects. *Proceedings of the IEEE, 103*(9), 1449–1477. https://doi.org/10.1109/JPROC.2015.2460697

Li, P., Tariq, A., Li, Q., Ghaffar, B., Farhan, M., Jamil, A., et al. (2023). Soil erosion assessment by RUSLE model using remote sensing and GIS in an arid zone. *International Journal of Digital Earth, 16*(1), 3105–3124. https://doi.org/10.1080/17538947.2023.2243916

Li, X., Yang, D., Wang, J., & Pan, G. (2025). Prediction of the change in suitable growth area of *Sabina tibetica* on the Qinghai-Tibetan plateau using MaxEnt model. *Frontiers in Forests and Global Change, 8*, 1465416. https://doi.org/10.3389/ffgc.2025.1465416

Liang, S., Fang, H., Chen, M., Shuey, C. J., Walthall, C., Daughtry, C., et al. (2002). Validating MODIS land surface reflectance and albedo products: Methods and preliminary results. *Remote Sensing of Environment, 83*(1–2), 149–162. https://doi.org/10.1016/s0034-4257(02)00092-5

Lillesand, T., Kiefer, R. W., & Chipman, J. (2015). *Remote sensing and image interpretation*. Wiley.

Lissovsky, A. A., & Dudov, S. V. (2021). Species-distribution modeling: advantages and limitations of its application. 2. MaxEnt. *Biology Bulletin Reviews, 11*(3), 265–275. https://doi.org/10.1134/S2079086421030087

Longley, P. A., Goodchild, M. F., Maguire, D. J., & Rhind, D. W. (2015). *Geographic information science and systems*. Wiley.

Lu, D., Mausel, P., Brondizio, E., & Moran, E. (2004). Change detection techniques. *International Journal of Remote Sensing, 25*(12), 2365–2401. https://doi.org/10.1080/0143116031000139863

Ma, L., Liu, Y., Zhang, X., Ye, Y., Yin, G., & Johnson, B. A. (2019). Deep learning in remote sensing applications: A meta-analysis and review. *ISPRS Journal of Photogrammetry and Remote Sensing, 152*, 166–177. https://doi.org/10.1016/j.isprsjprs.2019.04.015

Manfreda, S., McCabe, M. F., Miller, P. E., Lucas, R., Pajuelo-Madrigal, V., Mallinis, G., et al. (2018). On the use of unmanned aerial systems for environmental monitoring. *Remote Sensing, 10*(4), 641. https://doi.org/10.3390/rs10040641

Maxar Technologies. (2025). Satellite imagery products. Retrieved September 21, 2025, from https://www.maxar.com/products/satellite-imagery

McFeeters, S. K. (1996). The use of the normalized difference water index (NDWI) in the delineation of open water features. *International Journal of Remote Sensing, 17*(7), 1425–1432. https://doi.org/10.1080/01431169608948714

Myneni, R. B., Hoffman, S., Knyazikhin, Y., Privette, J. L., Glassy, J., Tian, Y., et al. (2002). Global products of vegetation leaf area and fraction absorbed PAR from year one of MODIS data. *Remote Sensing of Environment, 83*(1–2), 214–231. https://doi.org/10.1016/S0034-4257(02)00074-3

Nagler, T., & Rott, H. (2002). Retrieval of wet snow by means of multitemporal SAR data. *IEEE Transactions on Geoscience and Remote Sensing, 38*(2), 754–765. https://doi.org/10.1109/36.842004

NASA. (2025a). Moderate resolution imaging spectroradiometer (MODIS). Goddard Space Flight Center.Retrieved September 21, 2025, from https://modis.gsfc.nasa.gov/

Niraj, K. C., Gupta, S. K., & Shukla, D. P. (2022). A Comparison of image-based and physics-based atmospheric correction methods for extracting snow and vegetation cover in Nepal Himalayas using landsat 8 oli images. *Journal of the Indian Society of Remote Sensing, 50*(12), 2503–2521. https://doi.org/10.1007/s12524-022-01616-6

Ostrom, E. (2009). A general framework for analyzing sustainability of social-ecological systems. *Science, 325*(5939), 419–422. https://doi.org/10.1126/science.1172133

Padró, J. C., Carabassa, V., Balagué, J., Brotons, L., Alcañiz, J. M., & Pons, X. (2019). Monitoring opencast mine restorations using unmanned aerial system (UAS) imagery. *Science of the Total Environment, 657*, 1602–1614. https://doi.org/10.1016/j.scitotenv.2018.12.156

Panigada, C., Tagliabue, G., Zaady, E., Rozenstein, O., Garzonio, R., Di Mauro, B., et al. (2019). A new approach for biocrust and vegetation monitoring in drylands using multi-temporal Sentinel-2 images. *Progress in Physical Geography: Earth and Environment, 43*(4), 496–520. https://doi.org/10.1177/0309133319841903

Park, H. C., Lim, J. C., Lee, J. H., & Lee, G. G. (2017). Predicting the potential distributions of invasive species using the Landsat imagery and MaxEnt: Focused on. *Journal of the Korean Society of Environmental Restoration Technology, 20*(1), 1–12. https://doi.org/10.13087/kosert.2017.20.1.1

Penjor, U. (2022). Biodiversity and global land-use change in Bhutan: A model for the Himalayas (Doctoral dissertation, University of Oxford).

Pettorelli, N., Schulte to Bühne, H., Tulloch, A., Dubois, G., Macinnis-Ng, C., Queirós, A. M., et al. (2018). Satellite remote sensing of ecosystem functions: opportunities, challenges and way forward. *Remote Sensing in Ecology and Conservation, 4*(2), 71–93. https://doi.org/10.1002/rse2.59

Phillips, S. J., Anderson, R. P., & Schapire, R. E. (2006). Maximum entropy modeling of species geographic distributions. *Ecological Modelling, 190*(3–4), 231–259. https://doi.org/10.1016/j.ecolmodel.2005.03.026

Piao, S., Ciais, P., Huang, Y., Shen, Z., Peng, S., Li, J., et al. (2010). The impacts of climate change on water resources and agriculture in China. *Nature, 467*(7311), 43–51.

Poggio, L., De Sousa, L. M., Batjes, N. H., Heuvelink, G. B., Kempen, B., Ribeiro, E., & Rossiter, D. (2021). SoilGrids 2.0: producing soil information for the globe with quantified spatial uncertainty. *Soil, 7*(1), 217–240. https://doi.org/10.5194/soil-7-217-2021

Preston, T. M., Johnston, A. N., Ebenhoch, K. G., & Diehl, R. H. (2023). Beyond presence mapping: Predicting fractional cover of non-native vegetation in Sentinel-2 imagery using an ensemble of MaxEnt models. *Remote Sensing in Ecology and Conservation, 9*(4), 512–526. https://doi.org/10.1002/rse2.325

Reichle, R. H. (2008). Data assimilation methods in the earth sciences. *Advances in Water Resources, 31*(11), 1411–1418. https://doi.org/10.1016/j.advwatres.2008.01.001

Ren, H., & Zhou, G. (2014). Determination of green aboveground biomass in desert steppe using litter-soil-adjusted vegetation index. *European Journal of Remote Sensing, 47*(1), 611–625. https://doi.org/10.5721/EuJRS20144734

Riaño, D., Chuvieco, E., Salas, J., & Aguado, I. (2003). Assessment of different topographic corrections in Landsat-TM data for mapping vegetation types. *IEEE Transactions on Geoscience and Remote Sensing, 41*(5), 1056–1061. https://doi.org/10.1109/tgrs.2003.811693

Rosenqvist, A., Shimada, M., Suzuki, S., Ohgushi, F., Tadono, T., Watanabe, M., et al. (2014). Operational performance of the ALOS global systematic acquisition strategy and observation plans for ALOS-2 PALSAR-2. *Remote Sensing of Environment, 155*, 3–12. https://doi.org/10.1016/j.rse.2014.04.011

Rouse, Jr, J. W., Haas, R. H., Schell, J. A., & Deering, D. W. (1973). Monitoring vegetation systems in the great plains with ERTS. In *Third NASA Earth Resources Technology Satellite Symposium* (Vol. 1, pp. 309–317).

Rouse Jr, J. W., Haas, R. H., Schell, J. A., & Deering, D. W. (1974). Paper a 20. In *Third Earth Resources Technology Satellite-1 Symposium: The Proceedings of a Symposium Held by Goddard Space Flight Center at Washington, DC on December 10-14, 1973: Prepared at Goddard Space Flight Center* (Vol. 351, p. 309).

Salih, A. A., Ganawa, E. T., & Elmahl, A. A. (2017). Spectral mixture analysis (SMA) and change vector analysis (CVA) methods for monitoring and mapping land degradation/desertification in arid and semiarid areas (Sudan), using Landsat imagery. *The Egyptian Journal of Remote Sensing and Space Science, 20*, S21–S29. https://doi.org/10.1016/j.ejrs.2016.12.008

Schmitt, M., & Zhu, X. X. (2016). Data fusion and remote sensing: An ever-growing relationship. *IEEE Geoscience and Remote Sensing Magazine, 4*(4), 6–23. https://doi.org/10.1109/mgrs.2016.2561021

Shen, H., Li, X., Cheng, Q., Zeng, C., Yang, G., Li, H., & Zhang, L. (2015a). Missing information reconstruction of remote sensing data: A technical review. *IEEE Geoscience and Remote Sensing Magazine, 3*(3), 61–85. https://doi.org/10.1109/mgrs.2015.2441912

Shen, M., Piao, S., Jeong, S. J., Zhou, L., Zeng, Z., Ciais, P., et al. (2015b). Evaporative cooling over the Tibetan Plateau induced by vegetation growth. *Proceedings of the National Academy of Sciences, 112*(30), 9299–9304. https://doi.org/10.1073/pnas.1504418112

Shupe, S. M., & Marsh, S. E. (2004). Cover-and density-based vegetation classifications of the Sonoran Desert using Landsat TM and ERS-1 SAR imagery. *Remote Sensing of Environment, 93*(1–2), 131–149. https://doi.org/10.1016/j.rse.2004.07.002

Snyder, K. A., Wehan, B. L., Filippa, G., Huntington, J. L., Stringham, T. K., & Snyder, D. K. (2016). Extracting plant phenology metrics in a great basin watershed: Methods and considerations for quantifying phenophases in a cold desert. *Sensors, 16*(11), 1948. https://doi.org/10.3390/s16111948

Tadono, T., Ishida, H., Oda, F., Naito, S., Minakawa, K., & Iwamoto, H. (2014). Precise global DEM generation by ALOS PRISM. *ISPRS Annals of the Photogrammetry, Remote Sensing and Spatial Information Sciences, 2*, 71–76. https://doi.org/10.5194/isprsannals-ii-4-71-2014

Tmušić, G., Manfreda, S., Aasen, H., James, M. R., Gonçalves, G., Ben-Dor, E., et al. (2020). Current practices in UAS-based environmental monitoring. *Remote Sensing, 12*(6), 1001. https://doi.org/10.3390/rs12061001

Torres, R., Snoeij, P., Geudtner, D., Bibby, D., Davidson, M., Attema, E., et al. (2012). GMES Sentinel-1 mission. *Remote Sensing of Environment, 120*, 9–24. https://doi.org/10.1016/j.rse.2011.05.028

Turner, B. L., Kasperson, R. E., Matson, P. A., McCarthy, J. J., Corell, R. W., Christensen, L., et al. (2003a). A framework for vulnerability analysis in sustainability science. *Proceedings of the National Academy of Sciences, 100*(14), 8074–8079. https://doi.org/10.1073/pnas.1231335100

Turner, W., Spector, S., Gardiner, N., Fladeland, M., Sterling, E., & Steininger, M. (2003b). Remote sensing for biodiversity science and conservation. *Trends in Ecology & Evolution, 18*(6), 306–314. https://doi.org/10.1016/s0169-5347(03)00070-3

U.S. Geological Survey (USGS). (2025). Landsat missions. Retrieved September 21, 2025, from https://www.usgs.gov/landsat-missions

Ulaby, F. T., Moore, R. K., & Fung, A. K. (1981). Microwave remote sensing: Active and passive. Volume 1-microwave remote sensing fundamentals and radiometry.

Urbazaev, M., Hess, L. L., Hancock, S., Sato, L. Y., Ometto, J. P., & Thiel, C. (2022). Assessment of terrain elevation estimates from ICESat-2 and GEDI spaceborne LiDAR missions across different land cover and forest types. *Science of Remote Sensing, 6*, Article 100067. https://doi.org/10.1016/j.srs.2022.100067

USGS. (2023). What are the advantages of satellite imagery? U.S. Geological Survey. Retrieved September 21, 2025, from https://www.usgs.gov/faqs/what-are-advantages-satellite-imagery

Ustin, S. L., & Gamon, J. A. (2010). Remote sensing of plant functional types. *New Phytologist, 186*(4), 795–816. https://doi.org/10.1111/j.1469-8137.2010.03284.x

Wagner, W., Sabel, D., Doubkova, M., Bartsch, A., & Pathe, C. (2009). The potential of Sentinel-1 for monitoring soil moisture with a high spatial resolution at global scale. In *Symposium of Earth Observation and Water Cycle Science* (Vol. 3, p. 60).

Wan, Z., Hook, S., & Hulley, G. (2015). MYD11A1 MODIS/aqua land surface temperature/emissivity daily L3 global 1km SIN grid V006. In *NASA EOSDIS Land Processes Distributed Active Archive Center (DAAC) data set, MYD11A1-006*. https://doi.org/10.3133/fs20163070
Wang, C., Guo, H., Zhang, L., Qiu, Y., Sun, Z., Liao, J., et al. (2015). Improved alpine grassland mapping in the Tibetan Plateau with MODIS time series: A phenology perspective. *International Journal of Digital Earth, 8*(2), 133–152. https://doi.org/10.1080/17538947.2013.860198
Wang, Z., Wu, B., Zhang, M., & Ma, Z. (2023). An advanced coverage estimation method to quantify biological soil crust coverage using Sentinel-2 imagery in desert and sandy land of China. *Giscience & Remote Sensing, 60*(1), 2257470. https://doi.org/10.1080/15481603.2023.2257470
Wang, Q., Cui, G., Liu, H., Huang, X., Xiao, X., Wang, M., et al. (2025). Spatiotemporal dynamics and potential distribution prediction of *Spartina alterniflora* Invasion in Bohai Bay Based on Sentinel time-series data and MaxEnt modeling. *Remote Sensing, 17*(6), 975. https://doi.org/10.3390/rs17060975
Weinstein, B. G. (2018). A computer vision for animal ecology. *Journal of Animal Ecology, 87*(3), 533–545. https://doi.org/10.1111/1365-2656.12780
Willis, K. S. (2015). Remote sensing change detection for ecological monitoring in United States protected areas. *Biological Conservation, 182*, 233–242. https://doi.org/10.1016/j.biocon.2014.12.006
Wu, Z., Lei, S., Bian, Z., Huang, J., & Zhang, Y. (2019). Study of the desertification index based on the albedo-MSAVI feature space for semi-arid steppe region. *Environmental Earth Sciences, 78*(6), 232. https://doi.org/10.1007/s12665-019-8111-9
Xu, P., Tsendbazar, N. E., Herold, M., De Bruin, S., Koopmans, M., Birch, T., et al. (2024). Comparative validation of recent 10 m-resolution global land cover maps. *Remote Sensing of Environment, 311*, Article 114316. https://doi.org/10.1016/j.rse.2024.114316
Xue, J., & Su, B. (2017). Significant remote sensing vegetation indices: A review of developments and applications. *Journal of Sensors, 2017*, 1353691. https://doi.org/10.1155/2017/1353691
Zanaga, D., Van De Kerchove, R., Daems, D., De Keersmaecker, W., Brockmann, C., Kirches, G., et al. (2022). ESA World Cover 10 m 2021 v200. https://doi.org/10.1002/essoar.10509476.1
Zhang, T. (2005). Influence of the seasonal snow cover on the ground thermal regime: An overview. *Reviews of Geophysics, 43*(4). https://doi.org/10.1029/2004rg000157
Zhu, X. X., Tuia, D., Mou, L., Xia, G. S., Zhang, L., Xu, F., & Fraundorfer, F. (2017). Deep learning in remote sensing: A comprehensive review and list of resources. *IEEE Geoscience and Remote Sensing Magazine, 5*(4), 8–36. https://doi.org/10.1109/mgrs.2017.2762307
Zhu, X., Chen, J., Gao, F., Chen, X., & Masek, J. G. (2010). An enhanced spatial and temporal adaptive reflectance fusion model for complex heterogeneous regions. *Remote Sensing of Environment, 114*(11), 2610–2623. https://doi.org/10.1016/j.rse.2010.05.032
Zhu, Z., Piao, S., Myneni, R. B., Huang, M., Zeng, Z., Canadell, J. G., et al. (2016). Greening of the earth and its drivers. *Nature Climate Change, 6*(8), 791–795. https://doi.org/10.1038/nclimate3004
Zribi, M., Le Hégarat-Mascle, S., Taconet, O., Ciarletti, V., Vidal-Madjar, D., & Boussema, M. R. (2003). Derivation of wild vegetation cover density in semi-arid regions: ERS2/SAR evaluation. *International Journal of Remote Sensing, 24*(6), 1335–1352. https://doi.org/10.1080/01431160210146668

Effects of Primary and Secondary Umbels in the Germination Behavior of *Heraclium candicans* Wall. ex. DC. and *Selinum tenuifolium* Wall ex C.B. Clarke, the High-Altitude Umbellifers

Anil Kumar Bisht, Harsh Kumar Chauhan, and Sheetal Oli

1 Introduction

Apiaceae (umbellifers/Umbelliferae) is one of the important families of aromatic flowering plants consisting of 4089 species (Catalogue of Life, 2019). The family is named after type genus *Apium* and commonly known as celery, carrot, or parsley family. Members of the family are distributed throughout the globe, mostly in northern temperate regions and high altitudes of tropics (Aktürk et al., 2022). The family is characterized by aromatic herbaceous nature, alternate leaves, hollow stems, small flowers, and inflorescence with flowers mostly aggregating in terminal umbels (Christensen & Brandt, 2006). Members of this family are used for various purposes, including nutrition, medicine, beverages, spices, repellents, staining, cosmetics, fragrances, and industrial uses (Das et al., 2023; Thiviya et al., 2022). They have several documented ethnobotanical uses including the treatment of digestive, endocrine, reproductive, and respiratory disorders (Khoja et al., 2024). The family has gained importance as the potential source of natural agro-chemicals and for several pharmacological activities such as anti-tumor, antimicrobial, anti-inflammatory, radical scavenging, diuretic, gastrointestinal, and anti-obesity (Das et al., 2023; Sayed-Ahmad et al., 2017).

Indian Himalayan Region (IHR) harbors a great diversity of medicinal plants, of which many species belong to Apiaceae (Bisht et al., 2008; Mehta et al., 2021). In IHR, the cold deserts of northwestern Himalayas (Ladakh, Lahul Spiti, and high-altitude regions of Uttarakhand and Himachal Pradesh) are highly fragile (Rawat et al., 2023). These regions are characterized by extreme climatic conditions such as low temperature, limited precipitation, nutrient poor soil, and short growing seasons.

A. K. Bisht (✉) · H. K. Chauhan · S. Oli
Department of Botany, D.S.B. Campus, Kumaun University, Nainital, India
e-mail: bishtakb@rediffmail.com

Y. S. Rawat and C. P. Kala (eds.), *Cold Deserts*,
https://doi.org/10.1007/978-3-032-29119-6_7

Climate change is further exacerbating the fragile ecosystems of the cold desert by disrupting the phenology and seed germination (Kanwar & Kuniyak, 2022). Studying seed germination ecology in such ecosystems is important for dealing with anthropogenic climate change and developing conservation strategies of the native plant species. To test the germination ecology, we selected two species, namely *Selinum tenuifolium* Wall ex C.B. Clarke (Syn *Selinum wallichianum* Raizada & H.O. Saxena) and *Heracleum candicans* Wall. ex. DC. (Syn *Heracleum lanatum* Michx) *of Apiaceae.* Interestingly, 25% of the total representative of Apiaceae in IHR has medicinal importance, and 70% are confined to the subalpine–alpine zone (Bisht et al., 2008). Among others, *S. tenuifolium* and *H. candicans* are economically important species of Apiaceae having good potential of cultivation in IHR. *S. tenuifolium* is nearly endemic in the Himalayas and distributed at an altitude between 2400 and 4800 m asl (Chauhan, 1999). The aerial parts of the species are aromatic and used as antispasmodic, diuretic, stimulants, and carminatives (Mathela & Sinha, 1976; Shrestha et al., 2025; Tewari & Mathela, 2003). Roots act as an insect repellent, and root essential oil has antimicrobial and antibacterial properties (Mathela & Sinha, 1976). *H. candicans* is a native Himalayan species and distributed in temperate and alpine forest of Afghanistan, Bhutan, south-west China, North India, Nepal, and West Pakistan at 2000–4300 m asl (Butola & Samant, 2010; Jan et al., 2018). It is used as a spice, carminative, stimulant, aphrodisiac, and nerve tonic and also used in the treatment of leucoderma, abdominal colic, and other digestive disorders (Chauhan, 1999; Dar et al., 2022; Sarin & Jamwal, 1975). It possess antibacterial, antifungal, anti-inflammatory activities (Bal-Tembe et al., 1996; Dar et al., 2022; Kaur et al., 2003). Although several active constituents are reported from this species, Xanthotoxin is most valuable and in constant demand in the pharmaceutical industry (Chauhan et al., 2014).

High-altitude medicinal plants are often slow to reach maturity and to produce the tradable components (Chauhan et al., 2018). Overexploitation coupled with erratic seed germination due to dormancy has threatened the existence of *S. tenuifolium* and *H. candicans* in the Himalayas (Joshi & Dhar, 2003; Joshi et al., 2006). Previous reports confirm the significant positive effects of chilling (4 °C for 14 days) highlighting chilling as the best treatment to improve seed germination in both species (Joshi & Dhar, 2003; Joshi et al., 2006). Positional effect (i.e., presence of seeds on primary or secondary umbels) on seed mass/size has been reported from many species. Seed mass/size can affect germination time, germination percentage, dispersal, seed water relation thereby affecting seedling establishment, growth, and survival of the species (Kolodziejek, 2017). Both the selected species bore seeds on primary and secondary umbels, therefore it becomes imperative to test the chilling responses in seeds from primary and secondary umbels to develop better germination protocols for the regeneration and existence in the nature. Hence, the present study has dealt; (i) to assess the positional effect (i.e., presence of seeds on primary or secondary umbels) on seed germination, (ii) to investigate the effects of chilling on seeds collected from primary and secondary umbels, and (iii) to assess the variations in seed germination response of selected species to positional and chilling effects.

2 Material and Methods

2.1 Selected Species

The present study was conducted on two Himalayan Apiaceae: *Heracleum candicans* and *Selium tenuifolium*. *H. candicans* is distributed from the temperate to alpine region, and its range is extended to cold desert areas while *S. tenuifolium* is restricted to alpine and subalpine zones. Both species share similar ecological constraints such as low temperature, short growing season, and poor nutrient soil. Study on seed germination of these two important species provides a comparative account to understand whether cold deserts impose unique germination challenges or such traits are common among the high-altitude Apiaceae.

2.2 Seed Collection

Mature seeds of both species were collected at the time of their natural dispersal (i.e., September–October) from Marhi (Himachal Pradesh) and Dwali (Uttarakhand), West Himalayan state of India. At each population, seeds were collected randomly from the primary and secondary umbels of about 100 plants, cleaned and pooled to represent the genetic diversity. After collection, seeds were stored at room temperature (20–25 °C) for 3 weeks until the beginning of the experiment.

2.3 Viability Test

Seeds viability was tested using 1% (w/v) solution of 2, 3, 5-triphenyltetrazolium chloride (TTC) (Sigma) (Hendry & Grime, 1993). Forty seeds in replicate of three from each seed type (collected from primary and secondary umbel) of both species were soaked in TTC solution for 24 h and stored in dark at 25 ± 2 °C. Seeds were bisected longitudinally and observed under a simple microscope, seeds showing red-stained embryo represented viable seeds.

2.4 Seed Germination Test and Statistical Analysis

In the previous reports, both species showed the best germination by pre-chilling at (4 °C) (Joshi & Dhar, 2003; Joshi et al., 2006). In this experiment, pre-chilling treatment (4 °C) for 7, 14, 21 and 28 days was used to compare its effects on seed germination and mean germination time (MGT). Stored seeds were sterilized by % (v/v) Molyclean (Molychem, India) for 15 min, 1% Bavistin solution (BAFS India

Ltd., India) for 10 min, 0.1% $HgCl_2$ for 3 min, followed by three rinses of sterile double distilled water after each treatment. To evaluate the chilling effects, three replicates of 30 seeds were germinated in an incubator and monitored for 28 days. The experiment was terminated after 33 days, since no new seeds germinated after 5 days (Bhatt et al., 2019).

Germination was conducted in 90 mm-diameter petri-dishes containing moistened Whatman No. 1 filter paper. Petri-dishes were sealed with parafilm, to avoid contamination and incubated at 25 ± 2 °C. Petri-dishes were labeled to represent species, seeds collected from primary and secondary umbels, and pre-chilling treatments. A set of seeds without pre-sowing treatment was also included as a control. Germinated seeds were counted (seeds with ≥2 mm radicle considered germinated) and removed daily. The germination time was calculated as

$$\mathrm{MGT} = \sum (ni \times di)/N,$$

where n = number of seeds germinated at day ni, di = incubation period in days, N = total number of seeds germinated in the test. The experiment was conducted in a completely randomized design. The standard error and one-way Anova followed by post hoc Duncan test were performed by using software SPSS 21.0 version. The significance level of types of umbels and species on the mean germination percentage and MGT was analyzed by independent sample t-test using the same software.

3 Results

3.1 *Viability Test*

Both the species have shown significant seed viability. The seed viability was 95 ± 0.83% and 93.3 ± 0.48% for the seeds collected from primary and secondary umbels of *S. tenuifolium* and 91.66 ± 1.27% and 88.33 ± 1.27% for the seeds collected from primary and secondary umbels of *H. candicans*.

3.2 *Effects of Pre-chilling Treatment on the Seed Germination in* Selinum tenuifolium

In *S. tenuifolium*, the germination percentage of the seeds collected from primary and secondary umbels varied from 17.73 to 94.40% and 7.53 to 82.16%, respectively. The maximum germination was reported by the pre-chilling treatment for 28 days (Table 1). The mean germination time has reduced drastically to 3.09 days by the pre-chilling of seeds collected from primary umbels for 14 days. The minimum MGT

(1.58 days) in case of seeds collected from secondary Umbel was obtained by pre-chilling treatment of seeds for 7 days. The effects of pre-chilling treatment on the germination percentage as well as MGT were significant ($P \leq 0.05$), despite the source of collection of seeds i.e., primary and secondary umbels (Table 1).

3.3 *Effects of Pre-chilling Treatment on the Seed Germination in* Heracleum candicans

In *H. candicans*, the germination percentage of the seeds collected from primary and secondary umbels varied from 37.73 to 88.86% and 22.10 to 59.93%, respectively. The maximum germination was reported by the pre-chilling treatment for 28 days in case of seeds collected from primary umbels and 21 days in case of seeds collected from secondary umbels (Table 1). The minimum MGT was obtained by pre-chilling treatment of seeds for 7 days. The effects of pre-chilling treatment on the germination percentage as well as MGT were significant ($P \leq 0.05$) in seeds collected from primary and secondary umbels (Table 1).

3.4 *Difference Between the GP and the MGT of Seeds on the Basis of the Source of the Seeds Collection (i.e., Primary and Secondary Umbels) and Species*

Independent sample paired t-test revealed no significant difference between the germination percentage and the MGT of the seeds collected from primary and secondary umbels in both the species. Even there was no significant difference between the germination percentage and the MGT among the species (Table 2).

4 Discussion

A prominent seed viability is the indicator of the capability of seed germination under favorable conditions. However, considerable low germination of seeds in control and some treatments in both the species predicts the presence of the seed dormancy. Different seed dormancy types have been reported from several alpine/subalpine plants (Amen, 1966; Bernareggi et al., 2016; Peng et al., 2023; Schwienbacher et al., 2011). Pre-chilling treatments for longer duration (i.e., 28 and 21 days) appeared the best treatment for the improvement in the germination percentage of seeds of *S. tenuifolium* and *H. candicans* (Table 1). However, in both the species pre-chilling for shorter duration, i.e., 7 and 14 days have reduced the MGT (Table 1). Chilling is well-known to induce cold acclimation, which leads to acquisition of freezing

Table 1 Effects of pre-chilling treatment on the seed germination in *Selinum tenuifolium* and *Heracleum candicans*

	Selinum tenuifolium				*Heracleum candicans*			
	Germination percentage		Mean germination time (days)		Germination percentage		Mean germination time (days)	
Treatment	Primary umbels	Secondary umbels	Primary umbels	Secondary umbels	Primary umbels	Secondary umbels	Primary umbels	Secondary umbels
Control	$63.26^{b} \pm 13.33$	$74.43^{b} \pm 5.56$	$17.10^{c} \pm 3.60$	$20.10^{c} \pm 1.50$	$42.16^{a,b} \pm 8.02$	$35.53^{a,b} \pm 3.99$	$11.40^{a} \pm 2.16$	$9.61^{b,c} \pm 1.08$
Chilling for 7 days	$22.20^{a} \pm 4.85$	$7.53^{a} \pm 4.64$	$4.06^{a} \pm 1.17$	$1.58^{a} \pm 0.98$	$37.73^{a} \pm 11.30$	$22.10^{a} \pm 6.74$	$6.47^{a} \pm 2.57$	$3.78^{a} \pm 1.19$
Chilling for 14 days	$17.73^{a} \pm 1.13$	$9.96^{a} \pm 1.93$	$3.09^{a} \pm 0.33$	$1.62^{a} \pm 0.32$	$37.73^{a} \pm 8.02$	$41.10^{a,b,c} \pm 4.85$	$6.66^{a} \pm 1.92$	$7.08^{a,b} \pm 0.59$
Chilling for 21 days	$91.06^{c} \pm 5.53$	$72.00^{b} \pm 6.65$	$10.58^{b} \pm 0.86$	$7.73^{b} \pm 1.00$	$65.50^{b,c} \pm 6.74$	$59.93^{c} \pm 8.81$	$9.14^{a} \pm 0.99$	$12.27^{c} \pm 1.60$
Chilling for 28 days	$94.40^{c} \pm 2.20$	$82.16^{b} \pm 14.43$	$16.16^{b,c} \pm 1.33$	$15.07^{c} \pm 3.15$	$88.86^{c} \pm 4.43$	$54.30^{b,c} \pm 2.20$	$17.57^{b} \pm 1.29$	$12.18^{c} \pm 0.94$
F-value	28.08*	22.33*	12.61*	23.73*	7.83*	6.79*	5.88*	10.13*

*Significant ($P \leq 0.05$); means values followed by the same letter(s), i.e., a, b, and c in a column are not significantly different ($P \leq 0.05$) based on Duncan test

Table 2 Summary of independent sample *t*-test carried among different parameters

S. No.	Paired parameters	Significant level at 0.05
1	GP of seeds collected from primary umbel and secondary umbels in *Selinum tenuifolium*	Non-significant
2	MGT of seeds collected from primary umbels and secondary umbels in *Selinum tenuifolium*	Non-significant
3	GP of seeds collected from primary umbels and secondary umbels in *Heracleum candicans*	Non-significant
4	MGT of seeds collected from primary umbels and secondary umbels in *Heracleum candicans*	Non-significant
5	GP of seeds collected from primary umbels of *Selinum tenuifolium* and *Heracleum candicans*	Non-significant
6	GP of seeds collected secondary umbels of *Selinum tenuifolium* and *Heracleum candicans*	Non-significant
7	MGT of seeds collected from primary umbels of *Selinum tenuifolium* and *Heracleum candicans*	Non-significant
8	MGT of seeds collected from secondary umbels of *Selinum tenuifolium* and *Heracleum candicans*	Non-significant

temperature (Thomashow, 1999). MGT considerably increased with an increase in pre-chilling duration. The observations may be related to slacken metabolic activity of the embryo under prolonged chilling treatment.

In the present study, the seeds collected from primary umbels of *S. tenuifolium* and *H. candicans* have shown the germination percentage of 94.40 and 88.86% after the pre-chilling treatment for 28 days. The observations are in agreement that the members of Apiaceae have good germination efficiency; however, the optimization of suitable conditions for seed germination is crucial. Extensive germination trials on 88 members of family Apiaceae reported that seventeen (i.e., 19% of those tested) exhibited 90–100% germination under optimum conditions (Moravcova et al., 2005).

A better germination in both species after a pre-chilling of longer duration is attributable to the natural adaption mechanism of the alpine plants and cold deserts in which the seeds are compelled to tolerate cold and chilling environment for at least six months. These findings are highly relevant to cold deserts where plants experience extreme winter chilling and short growing seasons. Improved germination parameters with chilling suggest that both the species are well adapted to cold desert conditions where seeds lie dormant under snow cover for months before germination in spring. This adaption mechanism ensures that the seedling emerges at the time of spring, when temperature and soil conditions are favorable for seedling emergence. In winter, an accumulation of the memory of prolonged chilling is needed for controlling the loss of bud dormancy, bud burst, and germination in spring (Penfield, 2008). The pre-chilling may change the hormonal balance of seed and improve germination by enhancing the level of gibberellic acid and cytokinin activity and/or a decline in abscisic acid (Copeland & McDonald, 2001). Prolonged pre-chilling treatment requirement indicates that both the selected species may have deep physiological dormancy. This type of dormancy is also observed in other members of Apiaceae such as *Aegopodium podagraria* (Vandelook et al., 2009). Pre-chilling and cold stratification are reported to improve seed germination in numerous alpine/subalpine plants (Baskin et al., 1998; Vandelook et al., 2009) including Himalayan plants (Gashi et al., 2019; Joshi et al., 2006; Yaqoob & Nawchoo, 2017). Overexploitation, habitat degradation, and climate change are already impacting the cold deserts; therefore, understanding such dormancy breaking mechanisms is vital for developing conservation strategies in cold deserts.

Members of the Apiaceae consist of the inflorescence in which series of umbels (producing seeds) develop in a sequential manner. The position of seed is known to affect the seed mass, morphology, germination, and dormancy (Baskin et al., 1998; Kolodziejek, 2017). The non-significant variances in the seed germination percentage and MGT from primary and secondary umbels are interesting (Table 1) for the development of agro-techniques/mass cultivation. Seeds can be collected for mass cultivation from primary as well as secondary umbels. In other members of the Apiaceae family such as *Angelica archangelica* and *Peucedanum oreoselinum,* the seeds produced by primary umbels have shown significant competitive advantage over the seeds produced by secondary umbels, and the germination percentage was higher in seeds obtained from primary umbels (Kolodziejek, 2017; Ojala, 1985). In other species (*Petroselinium crispum, Daucus carota,* and *Apium graveolens*) as well, the germination was poor in seeds produced on secondary, tertiary, and quaternary umbels as compared to primary umbels (Thomas, 1978, 1996; Thomas et al., 1979). However, in other umbellifers like *Heracleum mantegazzianum* and *Angelica acutiloba,* the position effect does not mark its appearance, i.e., germination of seeds obtained from different umbel orders does not differ significantly (Phip et al., 2007). The positional effect of seeds in Apiaceae is species specific (Kołodziejek, 2017).

5 Conclusion

The present study concludes that the pre-chilling of seeds for 28 and 21 days is best treatment for improving the germination percentage in *H. candicans* and *S. tenuifolium* for regeneration in the alpine and cold desert regions. Seeds germination and the MGT did not differ significantly among the seeds of primary and secondary umbels, so seeds can be collected for mass cultivation from primary as well as secondary umbels. This research study presents a simple and reliable method for the large-scale cultivation of *H. candicans* and *S. tenuifolium.* Further, it highlights that the germination ecology of *H. candicans* and *S. tenuifolium* illustrates the adaptive strategies in cold desert ecosystems, where prolonged chilling serves as a cue for breaking dormancy.

References

Aktürk, C., Özerkan, M., Yılmaz, A., Kaynak, G., & Yılmaz, Ö. (2022). A new distribution area of *Angelica archangelica* L. (Apiaceae) around Uludağ (Bursa) and its micromorphological, anatomical, palynological properties. *Journal of Biological and Environmental Sciences, 16*(47), 22–32.

Amen, R. D. (1966). The extent and role of seed dormancy in alpine plants. *The Quarterly Review of Biology, 41*(3), 271–281.

Bal-Tembe, S., Joshi, D. D., & Lakdawala, A. D. (1996). Isolation of the anti-inflammatory principle of *Heracleum candicans. Indian Journal of Chemistry, 35*(5), 518–519.

Baskin, J. M., Nan, X., & Baskin, C. C. (1998). A comparative study of seed dormancy and germination in an annual and a perennial species of Senna (Fabaceae). *Seed Science Research, 8*(4), 501–512.

Bernareggi, G., Carbognani, M., Mondoni, A., & Petraglia, A. (2016). Seed dormancy and germination changes of snowbed species under climate warming: The role of pre-and post-dispersal temperatures. *Annals of Botany, 118*(3), 529–539.

Bhatt, A., Bhat, N. R., Lozano-Isla, F., Gallacher, D., Santo, A., Batista-Silva, W., et al. (2019). Germination asynchrony is increased by dual seed bank presence in two desert perennial halophytes. *Botany, 97*(11), 639–649.

Bisht, A. K., Bhatt, A., Rawal, R. S., & Dhar, U. (2008). Assessment of reproductive potential of different populations of *Angelica glauca* Edgew., a critically endangered Himalayan medicinal herb. *Journal of Mountain Science, 5*(1), 84–90.

Butola, J. S., & Samant, S. S. (2010). *Saussurea* species in Indian Himalayan region: Diversity, distribution and indigenous uses. *International Journal of Plant Biology, 1*(1), e9–e9.

Catalogue of Life. (2019). Annual checklist. Retrieved August 08, 2025, from http://www.catalogueoflife.org/annual-checklist/2019/browse/tree/id/25de583791577526fa6112aba66f51cc

Chauhan, H. K., Bisht, A. K., Bhatt, I. D., Bhatt, A., Gallacher, D., & Santo, A. (2018). Population change of *trillium govanianum* (Melanthiaceae) amid altered indigenous harvesting practices in the Indian Himalayas. *Journal of Ethnopharmacology, 213*, 302–310.

Chauhan, N. S. (1999). *Medicinal and aromatic plants of Himachal Pradesh.* Indus Publishing Company.

Chauhan, R. S., Nautiyal, M. C., Tava, A., & Cecotti, R. (2014). Essential oil composition from leaves of *Heracleum candicans* wall.: A sustainable method for extraction. *Journal of Essential Oil Research, 26*(2), 130–132.

Christensen, L. P., & Brandt, K. (2006). Bioactive polyacetylenes in food plants of the Apiaceae family: Occurrence, bioactivity and analysis. *Journal of Pharmaceutical and Biomedical Analysis, 41*(3), 683–693.

Copeland, L. O., & McDonald, M. (2001). *Principles of seed science and technology.* Kluwer Academic Publisher. https://doi.org/10.1007/978-1-4615-1619-4

Dar, N. A., Raja, W. Y., Tewari, D., & Bhat, Z. A. (2022). Pharmacognostic study of roots and aerial parts of less explored *Heracleum candicans* wall. Ex DC. From Betaab Valley, Pahalgam, Kashmir, India. *Indian Journal of Natural Products and Resources, 13*(3), 362–373.

Das, G., Das, S., Talukdar, A. D., Venil, C. K., Bose, S., Banerjee, S., Shin, H. S., Gutierrez-Grijalva, E. P., Heredia, J. B., & Patra, J. K. (2023). Pharmacology and ethnomedicinal potential of selected plants species from Apiaceae (Umbelliferae). *Combinatorial Chemistry and High Throughput Screening, 26*(2), 256–288.

Gashi, B., Osmani, M., & Sali, A. L. I. U. (2019). Breaking seed dormancy of *Tulipa scardica* Bornm. and Tulipa kosovarica Kit Tan, Shuka & Krasniqi by pre-chilling, plant growth regulators and some chemical treatments. *Acta agriculturae Slovenica, 113*(2), 203–210.

Hendry, G. A., & Grime, J. P. (Eds.). (1993). *Methods in comparative plant ecology: A laboratory manual.* Springer Science & Business Media.

Jan, M., Singh, S., Maqbool, F., & Nawchoo, I. A. (2018). Direct shoot regeneration from hypocotyl explants of *Heracleum candicans* wall: A vulnerable high value medicinal herb of Kashmir Himalaya. *African Journal of Agricultural Research, 13*(28), 1419–1424.

Joshi, M., & Dhar, U. (2003). Effect of various presowing treatments on seed germination of *Heracleum candicans* wall. ex DC.: A high value medicinal plant. *Seed Science and Technology, 31*(3), 737–743.

Joshi, M., Rawal, R. S., & Dhar, U. (2006). Breaking seed dormancy of *Selinum wallichianum*, a medicinal plant of the Himalaya. *Seed Technology, 28*(1), 87–92.

Kanwar, N., & Kuniyal, J. C. (2022). Vulnerability assessment of forest ecosystems focusing on climate change, hazards and anthropogenic pressures in the cold desert of Kinnaur district, northwestern Indian Himalaya. *Journal of Earth System Science, 131*(1), 51.

Kaur, M., Thakur, Y., Thakur, M., & Rana, R. C. (2003). Antimicrobial effect of hydrophilic and hydrophobic extracts of *Achyranthes aspera* and *Heracleum candicans*. In *Proceedings of the IUFRO-ISTS-UHF international conference on world perspective on short rotation forestry for industrial and rural development, Nauni, Solan, India* (pp. 399–404).

Khoja, A. A., Hamid, M., Hamid, B., Waheed, M., Jameel, M. A., Bussmann, R. W., & Haq, S. M. (2024). Himalayan Apiaceae-a comprehensive ecological and ethnobotanical evaluation. *Ethnobotany Research and Applications, 27*, 1–8.

Kołodziejek, J. (2017). Effect of seed position and soil nutrients on seed mass, germination and seedling growth in *Peucedanum oreoselinum* (Apiaceae). *Scientific Reports, 7*(1), 1959.

Mathela, C. S., & Sinha, G. K. (1976). Study on some essential oils from aromatic plants of Nainital. *Indian Perfume, 20*, 23–24.

Mehta, P., Bisht, K., & Sekar, K. C. (2021). Diversity of threatened medicinal plants of Indian Himalayan region. *Plant Biosystems—An International Journal Dealing with all Aspects of Plant Biology, 155*(6), 1121–1132.

Moravcová, L., Perglová, I., Pyšek, P., Jarošík, V., & Pergl, J. (2005). Effects of fruit position on fruit mass and seed germination in the alien species *Heracleum mantegazzianum* (Apiaceae) and the implications for its invasion. *Acta Oecologica, 28*(1), 1–10.

Ojala, A. (1985). Seed dormancy and germination in Angelica archangelica subsp. archangelica (Apiaceae). In *Annales Botanici Fennici* (pp. 53–62). The Finnish Botanical Publishing Board.

Penfield, S. (2008). Temperature perception and signal transduction in plants. *New Phytologist, 179*, 615–628. https://doi.org/10.1111/j.1469-8137.2008.02478.x

Peng, D. L., Geng, B. Y., Qin, Y. B., Yang, L. E., Baskin, C. C., & Baskin, J. M. (2023). Ecophysiology of seed dormancy and germination in the alpine-subalpine medicinal plant species *Sinopodophyllum hexandrum* (Royle) TS Ying. *Journal of Applied Research on Medicinal and Aromatic Plants, 32*, 100448.

Phip, N. T., Nojima, H., & Tashiro, T. (2007). Effect of umbel order and umbellet position on the production and characteristics of seeds and on the development and growth of seedlings in *Angelica acutiloba* Kitagawa. *Japanese Journal of Tropical Agriculture, 51*(2), 46–53.

Rawat, Y. S., Negi, V. S., Moussa, I. M., Zaman, W., & Elansary, H. O. (2023). Diversity, distribution and vegetation assessment of woody plant species in the cold desert environment, North-Western Himalaya, India. *Sustainability, 15*(13), 10429.

Sarin, Y. K., & Jamwal, P. S. (1975). A note on *Heracleum candicans* roots; a new source of xanthotoxin. *Indian Journal of Pharmacy, 37*(6), 145–147.

Sayed-Ahmad, B., Talou, T., Saad, Z., Hijazi, A., & Merah, O. (2017). The Apiaceae: Ethnomedicinal family as source for industrial uses. *Industrial Crops and Products, 109*, 661–671.

Schwienbacher, E., Navarro-Cano, J. A., Neuner, G., & Erschbamer, B. (2011). Seed dormancy in alpine species. *Flora-Morphology, Distribution, Functional Ecology of Plants, 206*(10), 845–856.

Shrestha, R. K., Maharjan, P. S., & Shrestha, B. G. (2025). Phytochemical analysis and bioactivity evaluation of root extract of *Selinum tenuifolium* wall. From Nepal. *Journal of Nepal Chemical Society, 45*(2), 135–146.

Tewari, M., & Mathela, C. S. (2003). Chemical compositions of the leaf and root oils of *Selinum tenuifolium* wall. *Indian Perfumer, 47*(4), 343–345.

Thiviya, P., Gunawardena, N., Gamage, A., Madhujith, T., & Merah, O. (2022). Apiaceae family as a valuable source of biocidal components and their potential uses in agriculture. *Horticulturae, 8*(7), 614.

Thomas, T. H. (1978). Relationship between bolting-resistance and seed dormancy of different celery cultivars. *Scientia Horticulturae, 9*(4), 311–316.

Thomas, T. H. (1996). Relationships between position on the parent plant and germination characteristics of seeds of parsley (*Petroselinium crispum* Nym.). *Plant Growth Regulation, 18*(3), 175–181.

Thomas, T. H., Biddington, N. L., & O'toole, D. F. (1979). Relationship between position on the parent plant and dormancy characteristics of seeds of three cultivars of celery (*Apium graveolens*). *Physiologia Plantarum, 45*(4), 492–496.

Thomashow, M. F. (1999). Plant cold acclimation: Freezing tolerance genes and regulatory mechanisms. *Annual Review of Plant Biology, 50*(1), 571–599.

Vandelook, F., Bolle, N., & Van Assche, J. A. (2009). Morphological and physiological dormancy in seeds of *Aegopodium podagraria* (Apiaceae) broken successively during cold stratification. *Seed Science Research, 19*(2), 115–123.

Yaqoob, U., & Nawchoo, I. A. (2017). Conservation and cultivation of *ferula jaeschkeana* Vatke: A species with deep complex morphophysiological dormancy. *Proceedings of the National Academy of Sciences, India Section B: Biological Sciences, 87*(2), 315–325.

Sustaining Livelihoods and Mitigating Climate Change Impacts Through Cold Desert Agroforestry Systems in the Lahaul Valley of Northwestern Himalaya, India

Yashwant S. Rawat and Chandra Prakash Kala

1 Introduction

Cold deserts occupy in a petty part of the Earth's land surface worldwide, are extensive arid regions characterized by scanty rainfall, prolonged snowfall, and limited vegetation, which is sparsely dispersed across the mountain ranges (Aswal & Mehrotra, 1994; G.B. Pant Institute of Himalayan Environment and Development, 1995–96; Kala, 2002; Saxena et al., 2011; Negi et al., 2018; Behera et al., 2014 Rawat et al., 2023). The cold desert regions remain completely snowbound for months and exhibit extreme climatic conditions, i.e., a temperature range of −45 °C in winter to 40 °C in summer, rainfall <60 mm, etc. (Behera et al., 2014; Rawat et al., 2023). There are many high-altitude cold deserts located in the north and south arid principle belts including North-American deserts and the Patagonian desert of Argentina. While Asia encompasses the largest area of cold desert in the world, extending through Afghanistan, India, Russia, Mongolia, China, Kyrgyz Republic, Iran, Tajikistan, and Pakistan (Negi et al., 2018; Saxena et al., 2011).

Cold deserts are critical in offering many ecosystem services, including biodiversity, and unique habitats. These regions are the source of glaciers and water bodies; however, the rapid reduction in the glaciers has future implications for sustaining agro-ecosystems and downstream water resources (Rawat et al., 2023). The socio-economic activities are limited to the short summer period due to prolong snowy

Y. S. Rawat (✉)
Department of Wood Technology Management, Faculty of Civil Technology, Federal Technical and Vocational Training Institute (TVTI), Addis Ababa, Ethiopia
e-mail: yasrawat@gmail.com

C. P. Kala
Ecosystems and Environment Management, Indian Institute of Forest Management (IIFM), Bhopal, India

Y. S. Rawat and C. P. Kala (eds.), *Cold Deserts*,
https://doi.org/10.1007/978-3-032-29119-6_8

winter in many parts of the cold deserts. Local communities residing in the high-altitude cold deserts perform all the rituals, agriculture, and forestry practices during the short summer season to sustain their life (Rawat, 2006; Rawat et al., 2006, 2009). Thus, the ecological understanding of the bio-resources is critical for the broader socio-economic development, biodiversity conservation, landscape restoration, and ecological development of the region. The vegetation found in these areas is broadly segregated into alpine forests (extension of temperate forests), dry alpine scrub, alpine meadows, and alpine stony deserts (Aswal & Mehrotra, 1994; Sood et al., 2001). Nonetheless, high-altitude cold deserts have rarely been at the forefront of ecosystem-based approach and sustainability research due to their inherently scanty and disperse nature of vegetation patterns, inhospitable climate, and socio-economic marginality (Behera et al., 2014; Rawat et al., 2023; Saxena et al., 2011). High-altitude cold deserts are vulnerable ecosystems due to their extreme social-ecological setup and climate change impacts, where vegetation and crops have to withstand the effects of drought, low temperature, wind, and erosion (Kuniyal et al., 2004; Rawat et al., 2023). Due to low temperature and large snow deposits, the sapling establishment and growth are poor, and the establishment of plantations has usually been unsuccessful except in some areas (Rawat & Everson, 2013; Rawat et al., 2006). The social-ecological systems including agro-ecosystem, biodiversity, and vegetation patterns in cold desert regions across the world are poorly understood because of their inaccessibility and remoteness (Rawat et al., 2023).

In India, a significant geographical area of North-western states, particularly Ladakh (Leh-Kargil districts) and Himachal Pradesh is occupied by cold deserts (Anonymous, 1993). Cold desert in Himachal Pradesh is bound by the Karakoram mountains in the north, Rudok and Chummurti district of Tibet in the East and South-East, Kullu and Kinnaur district of Himachal Pradesh in the South, and Jammu and Kashmir in the west (Anonymous, 1993). The cold deserts extend to Ladakh regions, Lahual and Spiti district with pooh subdivision of Kinnaur, and some parts of Chamba district of the Himachal Pradesh. The cold desert of Western Himalayas represents characteristic vegetation pattern, biome of alpine grasslands, and fascinating endemic medicinal and aromatic plants associations (Aswal & Mehrotra, 1994; Kala, 2002; Singh et al., 1997; Sood et al., 2001). Cold desert is a vast reservoir of high-value plant taxa, but ruthlessly divested due to over exploitation and illegal harvesting (Kuniyal et al., 2005; Rawat & Everson, 2012). This unique ecological system embodies a diverse range of socio-cultural and religious beliefs that reflect the integrity of its socio-cultural framework. The total area of the Lahaul and Spiti constitutes one-fourth of the entire Himachal state; however, Lahaul valley merely occupies 3.9% of the total revenue area of the state and 15.6% revenue area of the Lahaul and Spiti district (Anonymous, 1993; Vishvakarma et al., 2005).

Agroforestry and forestry have improved the economic conditions of farmers in the cold desert regions by providing food, fruits, fuelwood, timber, and fodder (Rawat et al., 2009). For instance, willow is a single most important multipurpose agroforestry tree species grown widely in cold desert regions of India for fuelwood, land restoration, and combating desertification (Oinam et al., 2005; Rawat & Everson, 2013; Rawat et al., 2006). Agroforestry and forestry practices play a pivotal role in reducing the pressure on scanty and sparsely distributed forest resources, while also meeting the subsistence needs of the people (Rawat et al., 2006, 2009). Owing to the high velocity of winds, poor soil nutrients, and harsh climatic conditions, plants in cold deserts tend to become prostrate, thick, woody, cushion-forming and bushy, with long, deep roots and small leaves (Sekar & Srivastava, 2010; Srivastava & Shukla, 2015; Shukla & Srivastava, 2020; Sekar et al., 2022). The cold desert has an ultra-varied topography, a factor that fosters species diversity and endemism (Rawat, 2005; Rawat & Adhikari, 2005). In this region, along the altitudinal gradient, distinct spatial and temporal changes in the distribution, density, and species composition are apparent in the vegetation (Aswal & Mehrotra, 1994; Kala & Mathur, 2002; Negi et al., 2018; Rawat et al., 2023; Sekar et al., 2022). The floristic altitudinal zonation found along the altitudinal gradient corresponds to the influence of physiographic factors that determine a characteristic type of vegetation (Behera et al., 2001, 2014).

The cold arid regions of the Himalayas represent unique challenges and opportunities for agroforestry systems, particularly in areas like the Lahaul valley of Himachal Pradesh, India, where harsh climates and limited resources necessitate innovative and sustainable agricultural practices (Tewari, 2016; Rawat et al., 2009, 2023). The climatic conditions of the cold deserts are due to their location in the interior of the continent and high-altitude environmental conditions. Cold deserts are well-recognized phyto-geographical region of the Indian Himalaya due to herbaceous vegetation composition having immense economic value and sparsely distributed tree species (Behera et al., 2001, 2014; Rawat, 2005; Rawat & Adhikari, 2005). *Betula utilis*, *Pinus wallichiana*, *Abies pindrow*, *Picea smithiana*, and *Cedrus deodara* are some of the tree species found on the north-facing slopes in small patches; south-facing dry slopes comprised of *Rosa moschata*, *R. webbiana*, and *Juniperus macropoda* (Rawat et al., 2023). The agroforestry system is considered as more appropriate than the traditional cropping practices for the conservation of natural resources in the cold desert ecosystem. Similarly, agroforestry systems of humid tropics have more potential to produce valuable timber with comparatively shorter cycles (Somarriba et al., 2001) as compared to temperate agroforestry. Agroforestry is one of the best possible land management practices for the proper utilization of space both above and below ground and for nutrients and moisture supply in different soil layers and over head space competing for sunlight and other uses (Khybri et al., 1992). Agroforestry systems play an important role in alleviating pressure on forest resources by providing fodder and fuelwood, as well as enhancing site conditions (Kala, 2025; Maikhuri & Semwal, 1997). Agroforestry is integrated land-use system of woody and non-woody component with raring of animals in the same management unit. Matching tree species to appropriate site conditions and stand manage is important for a sound agroforestry system (Somarriba et al., 2001). Agroforestry practices in the cold desert regions have been known since times immemorial and many indigenous and exotic species of trees are grown under such a system to fulfill the needs of local communities (Rawat et al., 2023). Irrigated agriculture along with agroforestry is one of the positive indicators to maintain food, fodder, and fuelwood security in the cold deserts (Kuniyal et al., 2004; Rawat & Vishvakarma, 2011; Vishvakarma et al., 1998a, 1998b).

The integration of trees into agricultural landscapes can provide several benefits, including soil stabilization, soil protection, microclimate regulation, and diversification of income sources for local communities as well as beautification benefits, including several recreational services (Hamadi et al., 2019; Chauhan et al., 2021; Kala, 2025). Besides, agroforestry systems are considered vital for achieving the goal of increasing forest cover from the current level of less than 25–33% (Deshmukh et al., 2020). In cold desert environments, agroforestry plays a pivotal role in mitigating the impacts of extreme temperatures, wind erosion, and limited precipitation, thereby promoting agricultural productivity and ecological resilience (Rawat et al., 2023). Traditional ecological knowledge plays a crucial role in shaping agroforestry practices in the Himalayan region, where local communities have developed time-tested strategies for managing natural resources sustainably (Maikhuri & Semwal, 1997; Rawat et al., 2023; Salick et al., 2023). These strategies often involve the selection of hardy tree species that can withstand the harsh climatic conditions and provide valuable products such as fuelwood, fodder, and fruits.

In the typically xeric cold desert conditions, only two cultivated species each of willow and poplar have successfully adapted to the climate of Lahaul, flourishing more effectively than all other species. These species are *Salix fragilis* L., *S. alba* L., *Populus balsamifera* L., and *P. nigra* L. *S. fragilis,* in particular, is cultivated extensively under the agroforestry and forestry systems around the settlements (Aswal & Mehrotra, 1994) for past 150 years (Rawat et al., 2006). While the seabuckthorn grows naturally in and around the cold desert agroforestry systems. In this backdrop, the present study has focused on (1) understanding the cold desert agroforestry systems in relation to land use/land cover and cropping patterns (including traditional, cash, and horticulture crops as well as fodder production); (2) examining the density and diversity of plant species within cold desert agroforestry; (3) conducting research on fuelwood and fodder production under cold desert agroforestry systems.

2 Materials and Methods

2.1 Study Area and Climate

The present study was conducted in four villages of the Lahaul valley, namely Kuthar (2600 m), Hinsa (2700 m), Jahlma (3000 m), and Khoksar (3200 m), located along an altitudinal gradient in the cold desert district Lahaul–Spiti of Himachal Pradesh, India (Fig. 1). The district is located between 31° 44′ 34″ N and 32° 59′ 57″ N latitude and 76° 46′ 29″ E and 78° 41′ 34″ E longitude, characterized by extreme cold desert climatic conditions with a short summer and prolonged winter. These villages were selected based on variations in altitude, climatic conditions, and existing land-use systems, serving as representative examples of cold desert regions. The region experiences a cold arid climate with low annual precipitation, and irrigation is essential for sustaining agriculture. The terrain is predominantly mountainous, with elevations ranging from 2,400 to 6,400 m, resulting in complex topography. Lahaul valley is a landlocked region, accessible only via the Rohtang Pass (3,978 m) and the Rohtang Tunnel (now Atal Tunnel) from the neighboring Kullu district of the Himachal Pradesh state of India, primarily during summer months. The area is dissected by glacier-fed streams and rivers, which sustain its natural drainage systems. The Lahual valley is divided into three distinct sub-valleys: Chandra valley, which serves as the catchment area for the Chandra River; Bhaga valley, the catchment area for the Bhaga River; and Pattan valley, which extends from Tandi (the confluence of the Chandra and Bhaga rivers) to a few kilometers beyond Kuthar, forming the Chandra–Bhaga river (known as the Chenab in Jammu and Kashmir state of India). The region exhibits a mosaic of microclimate due to its varied topography and altitudinal gradients, fostering high floral endemism through speciation. The landscape comprises of natural ecosystems like forests and alpine pastures on steep slopes, which is grazing grounds for migratory transhumance communities, while human modified systems are agroforestry and cultivated areas on gentle slopes. The Pattan valley hosts numerous small villages where local communities and livestock depend heavily on limited bio-resources for sustenance in the cold desert ecosystems (Rawat et al., 2006, 2023).

Climatically, the Lahaul valley is cold and arid, with low rainfall, high snowfall, and severe winters. The valley has two distinguished seasons, a short-lived summer and a prolonged severe winter. The soils are skeletal due to the hilly terrain and high snowfall. The Lahaul valley has an extremely harsh climate, with annual rain and snowfall ranging from 241.5 to 272.4 mm and 466.2 to 693.2 mm, respectively. The temperature reaches to a maximum of 27.8 °C in July and a minimum of −13.1 °C during the month of January (Rawat et al., 2009). Climatically, Khoksar and Hinsa are relatively humid and wet areas as compared to Jahlma and experience rain for a few days during the rainy season, whereas, Jahlma and adjacent areas remain dry even during the rainy months. Winter starts in the middle of October and continues until the end of April of the subsequent year. In general, January is the coldest month, and July is the warmest month for the entire valley. Two vegetation zones occur in the Lahaul valley, the temperate zone at an altitude between 2400 and 3300 m and the alpine zone above 3300 m (Aswal & Mehrotra, 1994). Shrubs are among the dominant vegetation in the valley. Juniper species mainly grow on the dry south-facing slopes, except for *J. communis*, which occurs on the moist north-facing slopes. Sparsely distributed *Pinus wallichiana, Abies pindrow, Picea smithiana,* and *Cedrus deodara* are also found on north-facing slopes in scattered patches.

2.2 Methods

2.2.1 Land-Use Systems and Irrigated Area

Primary data on land use and irrigated area was collected through field surveys, focused group discussions (FGDs), and participatory rural appraisal (PRA) protocols

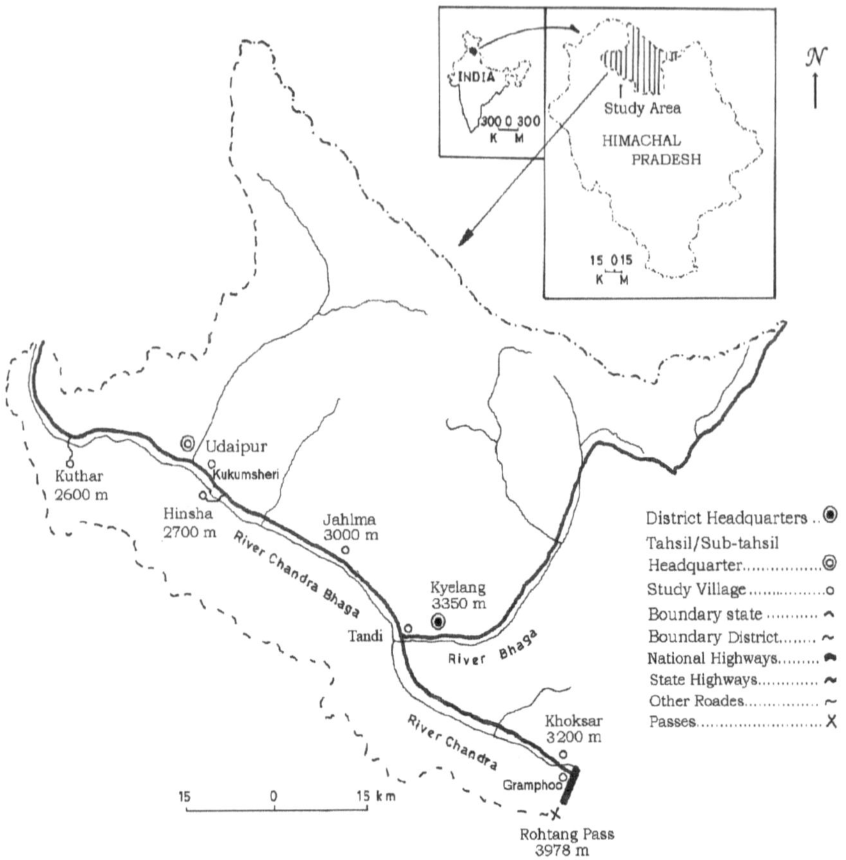

Fig. 1 Locations of the study sites in cold desert of the Lahaul valley

during the cropping season. Information was obtained from structured questionnaires and interviews with farmers in each village (Kuthar = 17 households; Hinsa = 52; Jahlma = 41; Khoksar = 12) (Table 1). The total irrigated land in each village was categorized into following land-use system: arable land (net sown area), kitchen gardens, cultivated grasslands, and forestry (*Salix* plantation). The data illustrates spatial variability in irrigation-based land utilization patterns along the altitudinal gradient of the Lahaul valley.

2.2.2 Crops Grown Under Willow-Based Agroforestry Systems

Willow (*Salix fragilis*)-based agroforestry systems are widely practiced in the Lahaul valley. The region lies at altitudes ranging from 2400 to 3200 m above mean sea level

Table 1 Land-use system and irrigated area in four studied villages in the Lahaul valley

Irrigated land	Villages				Total villages under study
	Kuthar (2600 m) n = 17	Hinsa (2700 m) n = 52	Jahlma (3000 m) n = 41	Khoksar (3200 m) n = 12	
1. Arable land as net area sown	6.3 (61.2)[a]	20.4 (63.4)	36.3 (63.6)	6.6 (59.5)	69.6 (62.9)
2. Kitchen garden	0.1 (1.0)	0.3 (0.9)	0.6 (1.1)	0.1 (0.9)	1.1 (1.0)
3. Cultivated grassland	3.9 (37.9)	9.6 (29.8)	13.2 (23.1)	3.6 (32.4)	30.3 (27.4)
4. Forestry (Salix plantation)	–	1.9 (5.9)	7.0 (12.3)	0.8 (7.2)	9.7 (8.8)
Gross irrigated area	10.3 (100)	32.2 (100)	57.1 (100)	11.1 (100)	110.7 (100)

n Indicates number of households interviewed
[a]Values in parentheses are in percentage of the total land area

and experiences a short-growing season (April–October). The willow-based agroforestry systems typically integrate agricultural crops, vegetables, fodder, medicinal, and cash crops within and around *Salix* plantations, utilizing available irrigated land. Data was collected through the field surveys, direct crop observations, and farmer interviews in the selected villages. The data was also collected on crop species cultivated, cropping patterns, and their respective yields under willow-based agroforestry systems, which were recorded for each crop. The crops were categorized into traditional crops, vegetables, fodder crops, medicinal plants, and cash crops, respectively. Sowing and harvesting periods were recorded based on local practices and verified on ground in the year around. The sowing and harvesting months refer to planting and annual harvest periods, respectively.

2.2.3 Vegetation Structure Under Willow-Based Agroforestry Systems

The survey of vegetation (trees and shrubs) and sampling was carried out in agroforestry of the selected villages during the growing season. The site selection was based on the physical character and dominance of vegetation. The four villages were selected to understand the effects of an altitudinal gradient and slope aspect on the composition and structure of the vegetation. Additionally, it could be inferred that altitudinal differences and slope aspects are sufficient to anticipate differences in vegetation structure and composition. The aim was to explore how the unique cold desert agroforestry system is interlinked and functions in order to achieve farmers' livelihoods, sustainable forest management, and environmental sustainability. It has been observed that agroforestry and forestry systems are floristically simpler than natural forest fragments (Rawat et al., 2023).

A line transect method was used to assess the vegetation patterns of trees and shrubs along an altitudinal gradient and marked topographic features in the agroforestry systems. The line transect method was used to make sure that the quadrats laid out were distributed evenly throughout the agroforestry systems studied. Quadrats ($10 \times 10\ m^2$) were laid down at every 10 m of intervals using a line running method (Kershaw, 1973). At each selected site, three transects were laid down along an altitudinal gradient, starting at the base of the slopes and ending at the agroforestry system. The first transect was aligned randomly on one side of the agroforestry system by using a compass. Thereafter, the other transects were laid at the intervals of 200 m from each other in each management unit. Thus, the distance between transects was 200 m. In total, 60 quadrats were laid out and analyzed, with 20 quadrats in each of transect in the agroforestry systems, to obtain detailed information on the structure, composition and vegetation pattern of woody plant species (e.g., trees and shrubs). The diameter at breast height (DBH) of tree species was measured at 1.37 m from the ground level. However, in case of shrubs, the diameter was measured at 5 cm above the ground level. Individuals more than 5.0 cm in diameter were considered trees, and individuals between 1.7 and 5.0 cm in diameter were considered saplings or shrubs (as described by Knight, 1963).

The density, defined as the total number of individuals per unit area, was calculated following the method described by Curtis and McIntosh (1950). The abundance-frequency (A/F) ratio was used to interpret the distribution patterns of species. This ratio indicates regular (<0.025), random (0.025–0.05), and contagious (>0.05) distribution patterns (Curtis & Cottom, 1956). Species richness (i.e., trees and shrubs) was determined as the number of species per unit area, following the methods of Mac Arthur (1965) and Whittaker (1975). The relative density, relative frequency, and relative dominance (basal area) were calculated according to Phillips (1959). The Importance Value Index (IVI), which is defined as the sum of the relative density, relative frequency, and relative dominance, was calculated for the various species. Fresh samples of the species found in each sample unit were collected and identified with the help of local floras (Aswal & Mehrotra, 1994; Sood et al., 2001). The nomenclature and nativity of the species were identified according to Anonymous (1886–1970), and information on endemism based on species distribution was obtained from the literature and from local floras (Aswal & Mehrotra, 1994; Sood et al., 2001). The distributional range and elevation limits of plants were also studied to determine the species turnover along the gradients. It has been observed that due to scarce forest resources and a prolonged winter period in the study area, the dependency of local communities on agroforestry systems is critical for sustaining life in the cold desert environment. Hence, the ecological understanding of the structure and composition of agroforestry systems is important for the broader socio-economic and ecological development of the region (Rawat et al., 2023).

2.2.4 Fuelwood and Fodder Production

Fuelwood and fodder data were collected through household surveys, field measurements, and community resource assessments. The total trees found in agroforestry systems were enumerated, and number of trees lopped annually for fuelwood and fodder were recorded. The internal and external sources of fuelwood and fodder were evaluated. The internal sources include the extraction of fuelwood and fodder from the land area acquired by the farmers in the landscapes while the external sources include the purchase of the fuelwood and fodder. Sources of fodder (including agroforestry, willow plantations, kitchen garden waste, forest grasses, purchased grass, feed, and oilcake residues) were examined. The structured questionnaires and focused group discussions (FGDs) were held with the farmers to obtain and verify the data. For each source of the category (e.g., willow under agroforestry, willow plantation, poplar, seabuckthorn, shrubs, agricultural waste, dung, natural forest, and purchased fuelwood and fodder) and annual production (MT) was estimated. Where possible, field sampling was conducted to measure fresh and dry fodder yields from representative plots within willow-based agroforestry systems. For sources such as purchased feed or oilcake, quantities were recorded from farmer interviews and local trade records.

3 Results and Discussion

3.1 Land-Use System and Irrigated Area

Land-use types in the region are categorized into agriculture, plantations, settlement, kitchen garden, barren land, and forests. The land use and land cover have changed drastically in the cold deserts of the Lahaul valley (Kuniyal et al., 2004; Oinam et al., 2005) as a consequence of climate change and developmental activities, leading to changes in cropping pattern, as well as agriculture and forestry practices (Rawat et al., 2023). Local communities perform all the rituals, agriculture, and forestry practices during a short summer season to sustain the life because of prolong winters in cold desert conditions (Rawat & Everson, 2013). The information derived from household surveys delineates the distribution of land among arable farming, kitchen gardens, cultivated grassland, and forestry, providing insight into the agricultural structure and resource allocation in this high-altitude cold desert region. Total gross land area across all studied villages was 110.7 ha (Table 1). The present study reveals a clear hierarchy in land use, consistent across all villages: Arable land is the dominant category, followed by cultivated grassland, forestry, and finally, kitchen gardens. This structural composition underscores a mixed agropastoral economy, heavily reliant on crop production but supported by significant investment in livestock fodder while maintaining the naturally growing trees and planted *Salix* on the margins of the lands (Rawat et al., 2006, 2023). There is a substantial disparity in the scale of agricultural operations among the villages: Jahlma (3000 m) is the predominant agricultural center, accounting for over half of the total land area. Hinsa (2700 m) is the second-largest contributor, while Kuthar (2600 m) and Khoksar (3200 m) are similar in scale, respectively.

Arable land, representing the net area sown with crops, is the cornerstone of the irrigated system, constituting 62.9% of the total land. The share of arable land is remarkably consistent across villages, ranging from 59.5% in Khoksar to 63.6% in Jahlma (Table 1). This consistency highlights the primary role of staple crop production for subsistence and cash, irrespective of altitude or village size (Kuniyal et al., 2004). Cultivated grassland is the second most critical land use, comprising 27.4% of the total area. This reflects the integral role of livestock in the local economy, providing dung, draught power, dairy, meat, and wool products. The relative importance of this land type, however, shows a notable inverse relationship with the presence of forestry (Rawat & Vishvakarma, 2011; Saxena et al., 2011). For instance, Kuthar, which has no *Salix* plantations, dedicates the highest proportion of its land (37.9%) to grassland. In contrast, Jahlma, with the most extensive forestry, allocates the least (23.1%). Forestry (Salix plantation), accounting for 8.8% of the total land, willow plantations are a specialized and important land use, particularly in Jahlma (12.3%) and Hinsa (5.9%). *Salix* is a multipurpose tree species vital for soil stabilization, fuelwood, fodder, and basket weaving, indicating a strategic investment in long-term ecological and economic security (Rawat & Everson, 2013). Kitchen gardens category is minimal, consistently representing about 1.0% of land across all villages. These small plots are essential for household nutrition, providing a diverse range of vegetables but do not constitute a significant land use in terms of area (Oinam et al., 2005). The land-use systems in the studied villages of the Lahaul valley are characterized by a stable, irrigated agro pastoral model (Saxena et al., 2011). The primary division between arable land and cultivated grassland demonstrates a balanced focus on crop and livestock production (Rawat & Vishvakarma, 2011). Variations between villages, particularly the significant investment in forestry in Jahlma and Hinsa, suggest adaptive strategies that may be influenced by a combination of altitude, local ecology, and socio-economic factors (Rawat et al., 2006). The present study is the foundation of the rural economy in this high-altitude cold desert region, where efficient water management is critical for sustaining both food and fodder security (Saxena et al., 2011; Vishvakarma et al., 2005), as majority of the land area is irrigated by snow melt irrigation system through water canals (*kuhls*).

3.2 Crops Grown Under Willow-Based Agroforestry Systems

Agroforestry systems in the Himalayan cold deserts represent an ecologically adaptive and socio-economically vital land-use practice (Rawat et al., 2023). In regions such as the Lahaul valley, where climatic conditions are characterized by short-growing seasons, low precipitation, and harsh winters, conventional mono cropping is often unsustainable (Kuniyal et al., 2004). Willow (*Salix* spp.)-based agroforestry systems have been traditionally adopted to protection against these limitations (Rawat, 2006; Rawat et al., 2006). Willows provide fuelwood, timber, and microclimatic amelioration, while intercropped agricultural species contribute to food, fodder, and cash income (Rawat et al., 2009; Vishvakarma et al., 2011). A wide range of crops including cereals, vegetables, fodder, medicinal, and cash crops are cultivated within this agroforestry systems, ensuring both livelihood security and ecological sustainability (Sharma et al., 2022; Quandt et al., 2023).

3.3 Traditional Food Crops

Traditional staple cereals such as barley (*Hordeum vulgare*), wheat (*Triticum aestivum*), and maize (*Zea mays*) continue to be the backbone of subsistence farming in the Lahaul valley (Kuniyal et al., 2004). Among these, maize records the highest productivity (3.26 t ha^{-1}), followed by wheat (2.28 t ha^{-1}) and barley (1.91 t ha^{-1}) (Table 2). Pseudocereals such as buckwheat (*Fagopyrum esculentum* and *F. tataricum*), well adapted to marginal soils, contribute significantly to both household nutrition and fodder resources (Vishvakarma et al., 2011). Oilseed crops such as mustard (*Brassica nigra*) are also integrated, providing edible oil and crop residues for livestock. The dual purpose of these crops ensures a balance between human food security and livestock feeding requirements. The majority of the crops are cultivated during the short summer period, contributing significantly to household food and fodder security (Kuniyal et al., 2004).

3.4 Vegetable Crops

Vegetables occupy a prominent position in willow-based agroforestry systems. Vegetable crops play a pivotal role in enhancing household nutrition and providing marketable surplus (Kuniyal et al., 2004; Vishvakarma et al., 2011). Cabbage (*Brassica oleracea var. capitata*; 17.90 t ha^{-1}), beetroot (*Beta vulgaris*; 16.39 t ha^{-1}), and potato (*Solanum tuberosum*; 13.01 t ha^{-1}) show particularly high productivity under willow-based agroforestry systems (Table 2). Root crops such as carrot (*Daucus carota*) and radish (*Raphanus sativus*) demonstrate a vital adaptability to the valley's environmental conditions, producing yields above 10 t ha^{-1}. The inclusion of tomato (*Lycopersicon lycopersicum*; 12.26 t ha^{-1}) and capsicum (*Capsicum annuum*; 2.66 t ha^{-1}) diversifies agriculture produce and contributes to household income through local and regional markets (Vishvakarma et al., 2011). These crops are cultivated successfully during April–May sowings with harvesting extending up to October and also show excellent adaptation to the cold desert conditions and provide a great

Table 2 Crops grown under willow-based agroforestry systems in cold desert of the Lahaul valley

Botanical name	Vernacular name	Family	Production t ha^{-1}			Cropping season
(i) *Traditional crops*			Yield	Fodder	Sowing time	Harvesting time
Amaranthus paniculatus L	Amaranths	Amaranthaceae	1.10	1.3	July–Aug	Sept–Oct
Brassica nigra Koch	Mustard	Cruciferae	3.01	2.8	July–Aug	Sept–Oct
Fagopyrum esculentum Moench	Buckwheat-I (Kathu)	Polygonaceae	1.25	2.4	July–Aug	Sept–Oct
F. tataricum Gaertn	Buckwheat-II (Phulan)	Polygonaceae	1.44	2.0	July–Aug	Sept–Oct
Hordeum vulgare L	Barley	Gramineae	1.91	2.1	Apl–May	July
Phaseolus spp.	Kidney bean	Leguminoceae	2.24	3.6	Apl–May	Aug–Sept
Triticum aestivum L	Wheat	Gramineae	2.28	2.0	Apl–May	Aug–Sept
Zea mays L	Maize	Gramineae	3.26	3.4	Apl–May	Aug–Sept
(ii) *Vegetables*						
Beta vulgaris L	Beet Root	Chinopodiaceae	16.39	1.9	Apl–May	June–Oct
Brassica oleracea botrytis L	Cauliflower	Cruciferae	12.03	1.1	Apl–May	June–Oct
Brassica oleracea var. capitata L	Cabbage	Cruciferae	17.90	1.3	Apl–May	Aug–Sept
Brassica rapa L	China sarson	Cruciferae	7.09	0.1	Apl–May	June–Oct
Brassica tournefortii Gouan	Turnip	Cruciferae	12.63	1.4	Apl–May	June–Oct
Capsicum annuum L	Chilly	Solanaceae	2.66	1.6	Apl–May	June–Oct
Daucus carota L	Carrot	Umbelliferae	11.16	1.2	Apl–May	Aug–Sept
Lycopersicon lycopersicum L	Tomato	Solanaceae	12.26	2.8	Apl–May	June–Oct
Raphanus sativus L	Radish	Cruciferae	10.62	1.1	Apl–May	June–Oct
Raphanus sp.	Labuk	Cruciferae	13.81	1.6	Apl–May	June–Oct
(iii) *Fodder crop*						
Avena sativa L	Oat (jai)	Gramineae	–	2.5	–	–
(iv) *Medicinal crops*						
Inula recemosa Hook.f	Manu*	Asteraceae	4.26	1.6	Nov	Sept–Oct
Saussurea costus (Falc.) Lipsch	Kuth*	Compositae	3.92	1.7	Nov	Sept–Oct
(v) *Cash crops*						
Humulus lupulus L	Hops*	Cannabaceae	1.65	2.4	Oct–Nov	July–Aug

(continued)

Table 2 (continued)

Botanical name	Vernacular name	Family	Production t ha^{-1}			Cropping season
Pisum sativum L	Pea	Leguminoseae	5.53	2.5	Apl–May	Sept–Oct
Solanum tuberosum L	Potato	Solanaceae	13.01	2.0	Apl–May	Sept–Oct

* Perennial crop

taste. The vegetables provide both high caloric value and marketable surpluses, contributing to cash income for farming households (Oinam et al., 2004–05).

3.5 *Fodder Crops*

Livestock rearing forms an essential livelihood activity in the Lahaul valley, necessitating reliable fodder supply (Rawat & Vishvakarma, 2011). Therefore, fodder production is an integral component of the agroforestry system due to the dependence of local communities on livestock (Rawat et al., 2006). Oat (*Avena sativa*), cultivated primarily as a fodder crop, produces 2.5 t ha^{-1} of biomass (Table 2). Additionally, by-products from cereals and pulses (e.g., wheat, barley, maize, and kidney beans) contribute significantly to fodder availability, ensuring year round livestock feeding (Kuniyal et al., 2004; Rawat & Vishvakarma, 2011; Vishvakarma et al., 2011). This integration of food-fodder nexus enhances resilience against feed scarcity during long winters for livestock.

3.6 *Medicinal and Aromatic Crops*

The cold desert ecosystem of the Lahaul is rich in medicinal plant biodiversity (Aswal & Mehrotra, 1994; Sood et al., 2001). Medicinal plants are of both cultural and commercial significance in the cold deserts (Kuniyal et al., 2005). Two important perennial medicinal crops, *Inula racemosa* (Manu) and *Saussurea costus* (Kuth), are cultivated within the willow-based agroforestry systems (Rawat & Everson, 2011; Rawat et al., 2004). *Inula racemosa* (Manu) and *Saussurea costus* (Kuth), cultivated as perennial crops, yield 4.26 and 3.92 t ha^{-1}, respectively (Table 2). These species are sown in November and harvested in September–October after 3 years later (Kuniyal et al., 2005; Rawat & Everson, 2011; Rawat et al., 2004). These species are highly valued in traditional medicine and pharmaceutical industries for their root-based bioactive compounds. Their integration into agroforestry systems not only diversifies income sources but also supports the conservation of native medicinal flora of the region (Kuniyal et al., 2005).

3.7 *Cash Crops*

Cash crops provide critical market linked income opportunities to the farmers in the Lahaul valley (Table 2). Cash crops such as hops (*Humulus lupulus*), pea (*Pisum sativum*), and potato (*Solanum tuberosum*) are cultivated to meet market demands in the Lahaul valley (Kuniyal et al., 2004; Vishvakarma et al., 2011). Among these, pea (*Pisum sativum*) records a yield of 5.53 t ha^{-1}, while potato (*Solanum tuberosum*) remains a dominant cash crop due to its high productivity (13.01 t ha^{-1}) and established market demand. Hops (*Humulus lupulus*), cultivated as a perennial crop, is harvested between July and August with an average yield of 1.65 t ha^{-1}. Their commercial value in brewing industries highlights the potential of willow-based agroforestry to integrate high-value niche crops (Rawat, 2006; Rawat et al., 2006; Vishvakarma et al., 2011). However, the cultivation of the hops has declined sharply due to processing and market problems.

Willow-based agroforestry systems in the Lahaul valley represent a climate resilient and ecologically sustainable land-use strategy for the Himalayan cold desert (Rawat et al., 2006, 2009; Singh et al., 1997). The integration of cereals, vegetables, fodder, medicinal, and cash crops ensures food and nutritional security, provides sustainable fodder resources, and enhances household income diversification (Kuniyal et al., 2004; Vishvakarma et al., 2011). This diversified cropping pattern also safeguards farming communities against climatic uncertainties and market fluctuations, thereby offering a viable strategy for climate resilient agriculture in Himalayan cold deserts (Vishvakarma et al., 2011). Given their ecological and socio-economic importance, willow-based agroforestry systems offer a replicable pathway for sustainable agricultural intensification in fragile mountain ecosystems (Rawat et al., 2006, 2009). However, currently, the horticulture crops have been inducted in the willow-based agroforestry systems. The environmental conditions have become conducive for horticulture crops like apple, apricot, cherry, plum, pear, almond, and walnuts may be due to global environment change (Rawat et al., 2023). However, the preference of raising apple remains in the Lahual valley.

3.8 *Vegetation Structure Under Willow-Based Agroforestry Systems*

The woody vegetation under willow-based agroforestry systems in the Lahaul valley exhibited considerable heterogeneity across sites (Khoksar, Jahlma, Hinsa, and Kuthar) (Table 3). A total of 42 species were recorded, distributed among trees, small trees, and shrubs, with variations in density, basal area, and IVI (Rawat et al., 2023). At the site level, species richness and abundance were highest at Hinsa and Kuthar, compared with Khoksar and Jahlma. The higher density and basal area values at Hinsa and Jahlma reflect relatively more diverse and structurally complex communities, likely supported by site-specific microclimates and management interventions

(Rawat et al., 2006, 2023). Willow-based agroforestry systems are developed by local communities to meet fuelwood, fodder, and timber requirements in the Lahaul valley (Rawat et al., 2023; Singh et al., 1997). The growth conditions, the use of farmyard manure (FYM), and fertilizer in the agricultural system improve the density, basal area, and IVI of the species as the altitude declines and vice versa in the study area (Kuniyal et al., 2004; Rawat et al., 2023).

Among tree species, *Salix fragilis* dominated across all sites, attaining the highest IVI values (214.54 at Khoksar; 123.89 at Jahlma; 37.14 at Hinsa; and 20.17 at Kuthar) (Table 3). The dominance of *S. fragilis* is attributed to its deliberate planting within agroforestry systems for fuelwood, fodder, and microclimatic regulation (Rawat et al., 2023; Sharma et al, 2022). Other tree species such as *Juglans regia*, *Juniperus polycarpos*, and *Corylus jaquemontii* have contributed substantially to basal area at certain sites. For instance, *J. polycarpos* at Hinsa has achieved a density of 378 ha^{-1} with a basal area of 2,199.28 $cm^2/100\ m^2$ and an IVI of 97.92, indicating its ecological importance at this site (Rawat & Everson, 2012). Small tree species were represented more prominently at Jahlma and Kuthar, with total densities of 150 ha^{-1} and 147 ha^{-1}, respectively. *Pyrus malus* was the most abundant small tree species across both sites, recorded 67 ha^{-1} at Jahlma and Hinsa. *Fraxinus xanthoxyloides* (Hinsa: 300 ha^{-1}, IVI: 32.38) and *Salix acmophylla* (Jahlma: 34 ha^{-1}, IVI: 7.65) were also important contributors. These species enhances the ecological and economic value of the agroforestry systems by providing fruits, fodder, and wood resources (Rawat & Everson, 2013; Rawat et al., 2006).

Shrubs exhibited the greatest species richness, with site-specific variations (Table 3). At Khoksar, shrubs were sparsely distributed, dominated by *Rosa webbiana* (33 ha^{-1}, IVI: 26.64). In contrast, Jahlma and Hinsa recorded much higher shrub densities (788 and 1,044 ha^{-1}, respectively), with *Hippophae rhamnoides* (440 ha-1 at Jahlma, IVI: 55.36; 378 ha^{-1} at Hinsa, IVI: 26.63) being the most dominant shrub species. At the Kuthar site, the shrub density was higher 1,628 ha^{-1}, with notable contributions from *Lonicera hypoleuca* (365 ha^{-1}, IVI: 32.74) and *Spiraea canescens* (247 ha^{-1}, IVI: 20.08). These shrubs are ecologically important, functioning as soil binders, providing fodder, and supporting biodiversity through habitat provision (Behera et al., 2001, 2014; Negi et al., 2018; Rawat et al., 2023; Sekar et al., 2022).

3.9 Diversity and Ecological Significance

The distribution of IVI values across tree, small tree, and shrub layers indicates the ecological dominance of *Salix fragilis* and *Hippophae rhamnoides* within these systems (Table 3). However, the presence of multiple associated species such as *Juglans regia*, *Pyrus malus*, and *Lonicera hypoleuca* suggests a structurally and functionally diverse community (Behera et al., 2001, 2014; Rawat et al., 2023). Diversity Indices (A/F ratios) generally indicated regular to contagious distribution patterns, with some species (e.g., *Robinia pseudoacacia*, *Prunus communis*) showing clumped

Table 3 Density, basal area, IVI, and diversity of plant species under agroforestry system in cold desert of the Lahaul valley

Species	Khoksar				Jahlma				Hinsa				Kuthar			
Scientific name	Density ha^{-1}	Basal area (cm^2/100 m^2)	IVI	A/F	Density ha^{-1}	Basal area (cm^2/100 m^2)	IVI	A/F	Density ha^{-1}	Basal area (cm^2/100 m^2)	IVI	A/F	Density ha^{-1}	Basal area (cm^2/100 m^2)	IVI	A/F
Tree																
Acer accuminatum L	–[a]	–	–	–	–	–	–	–	–	–	–	–	75	77.42	12.53	0.19
Celtis australis L	–	–	–	–	–	–	–	–	–	–	–	–	2	11.26	1.21	0.51
Corylus jaquemontii L	–	–	–	–	–	–	–	–	–	–	–	–	73	218.09	22.60	0.10
Euonymous fimbriatus Wall	–	–	–	–	–	–	–	–	–	–	–	–	61	14.03	6.79	0.32
Fraxinus floribanda Wall. ex Roxb	–	–	–	–	–	–	–	–	–	–	–	–	41	69.91	10.56	0.11
Juglans regia L	–	–	–	–	4	17.38	2.35	0.41	56	281.62	15.50	0.11	29	691.45	46.43	0.09
Juniperus polycarpos C. Koch	–	–	–	–	17	16.90	5.06	0.30	378	2199.28	97.92	0.06	–	–	–	–
Malus baccata (L.) Borkh	–	–	–	–	12	22.73	4.45	0.28	22	61.58	5.19	0.18	8	5.28	1.58	0.51
Populus nigra L	7	36.84	32.12	0.34	37	266.87	19.85	0.14	11	29.75	3.76	0.09	20	54.10	5.47	0.57
Prunus cornuta (Wall. ex royle) Stued	–	–	–	–	20	18.10	4.88	0.49	–	–	–	–	20	25.96	3.80	0.57
Robinia pseudoacacia L	–	–	–	–	5	5.20	1.49	1.16	56	17.33	7.82	0.11	20	8.67	2.32	1.28
Salix fragilis L	102	197.98	214.54	0.17	365	1862.36	123.89	0.10	244	564.25	37.14	0.12	57	219.80	20.17	0.15
Total	109	234.82	246.66	–	460	2209.54	161.97	–	767	3153.81	167.33	–	406	1395.97	133.46	–
Small tree																

(continued)

Table 3 (continued)

Species	Khoksar				Jahlma				Hinsa				Kuthar			
Scientific name	Density ha^{-1}	Basal area (cm^2/100 m^2)	IVI	A/F	Density ha^{-1}	Basal area (cm^2/100 m^2)	IVI	A/F	Density ha^{-1}	Basal area (cm^2/100 m^2)	IVI	A/F	Density ha^{-1}	Basal area (cm^2/100 m^2)	IVI	A/F
Crataegus songarica C. Koch	–	–	–	–	6	6.89	1.64	1.40	–	–	–	–	–	–	–	
Fraxinus xanthooxiloides (G.Don.) DC	–	–	–	–	–	–	–	–	300	148.60	32.38	0.07	82	13.93	8.22	0.33
Prunus armeniaca L	–	–	–	–	6	15.52	2.43	0.62	–	–	–	–	6	24.03	2.60	0.38
Prunus avium L	–	–	–	–	3	0.55	1.16	0.70	–	–	–	–	–	–	–	
Prunus communis Huds	–	–	–	–	18	38.44	5.53	0.44	–	–	–	–	12	15.33	2.35	0.77
Prunus sp.	–	–	–	–	16	1.40	1.66	13.95	–	–	–	–	–	–	–	
Pyrus malus L	–	–	–	–	67	61.10	14.40	0.23	67	38.74	6.57	0.54	18	14.48	2.57	1.15
Salix acmophylla Boiss	–	–	–	–	34	40.24	7.65	0.47	–	–	–	–	29	16.23	4.12	0.48
Total	–	–	–	–	150	164.14	34.47	–	367	187.34	38.95	–	147	84.00	19.86	–
Shrub																
Berberis jaeschkeana Schneid	9	0.06	11.68	1.80	18	0.52	2.23	3.95	44	1.16	4.46	0.36	–	–	–	–
Berberis pseudumbellata Parker	–	–	–	–	–	–	–	–	–	–	–	–	12	0.46	1.02	3.06
Cotoneaster obtusus Wall. ex Lindl	–	–	–	–	–	–	–	–	144	1.69	13.82	0.13	118	17.57	10.51	0.38

(continued)

Table 3 (continued)

Species	Khoksar				Jahlma				Hinsa				Kuthar			
Scientific name	Density ha^{-1}	Basal area (cm^2/100 m^2)	IVI	A/F	Density ha^{-1}	Basal area (cm^2/100 m^2)	IVI	A/F	Density ha^{-1}	Basal area (cm^2/100 m^2)	IVI	A/F	Density ha^{-1}	Basal area (cm^2/100 m^2)	IVI	A/F
Euonymus monbeigii W.W. Sm													25	9.04	3.06	0.74
Hippophae rhamnoides L. ssp. turkestanica Rousi	–	–	–	–	440	172.23	55.36	0.26	378	73.61	26.63	0.34	92	1.08	5.20	5.99
Juniperus communis L. var saxatilis Pallas	–	–	–	–	–	–	–	–	22	0.22	3.41	0.18	31	8.99	3.33	0.91
Jasminum humile L	–	–	–	–	–	–	–	–	–	–	–	–	29	14.06	4.45	0.31
Lonicera hypoleuca Decne	–	–	–	–	24	1.16	4.00	0.82	167	21.80	17.81	0.08	365	33.52	32.74	0.10
Parrotopsis jacquemontiana Render	–	–	–	–	–	–	–	–	–	–	–	–	84	14.76	8.81	0.27
Prunus jacquemontii Hook f. Fl. Brit.Ind	–	–	–	–	–	–	–	–	–	–	–	–	37	22.77	4.87	0.61
Prunus sp.	–	–	–	–	–	–	–	–	–	–	–	–	33	2.45	3.48	0.54
Prunus sp.	–	–	–	–	–	–	–	–	–	–	–	–	10	0.83	0.95	2.55
Rhamnus prostrata Jacq. ex Parker	–	–	–	–	–	–	–	–	–	–	–	–	49	7.39	5.40	0.35

(continued)

Table 3 (continued)

Species	Khoksar				Jahlma				Hinsa				Kuthar			
Scientific name	Density ha^{-1}	Basal area (cm^2/100 m^2)	IVI	A/F	Density ha^{-1}	Basal area (cm^2/100 m^2)	IVI	A/F	Density ha^{-1}	Basal area (cm^2/100 m^2)	IVI	A/F	Density ha^{-1}	Basal area (cm^2/100 m^2)	IVI	A/F
Ribes glaciale Wall	–	–	–	–	–	–	–	–	–	–	–	–	35	1.76	3.08	1.02
Ribes grossularia auct.non L	–	–	–	–	31	0.61	4.06	169	–	–	–	–	137	2.43	10.06	0.56
Rhamnus sp.	–	–	–	–	–	–	–	–	–	–	–	–	71	4.54	5.32	1.15
Rosa webbiana Wall. ex Royle	33	0.27	26.64	6.75	275	4.12	37.92	15	289	2.01	27.61	0.07	84	2.05	7.16	0.45
Sorbus aucuparia L	13	1.58	15.02	2.70	–	–	–	–	–	–	–	–	–	–	–	–
Sorbus lanata (D.Don) S.Schauer	–	–	–	–	–	–	–	–	–	–	–	–	51	46.38	8.26	0.27
Sorbaria tomentosa (Lindl.) Rehder	–	–	–	–	–	–	–	–	–	–	–	–	102	4.09	7.63	0.74
Spiraea canescens D.Don	–	–	–	–	–	–	–	–	–	–	–	–	247	10.27	20.08	0.20
Viburnum continifolium D.Don	–	–	–	–	–	–	–	–	–	–	–	–	16	1.51	1.26	4.08
Total	55	1.91	53.34	–	788	178.64	103.57	–	1044	100.49	93.74	–	1628	205.95	146.67	–
Gross total	164	236.73	300.00	–	1398	2552.32	300.00	–	2178	3441.64	300.00	–	2181	1685.92	300.00	–

[a] – absent

Note Khoksar, Jahlma, study sites are located on the South-facing slopes while Hinsa and Kuthar sites are located on North-facing slopes distribution due to localized planting or natural regeneration niches (Masoodi et al., 2021). Overall, the willow-based agroforestry systems of the Lahaul valley support a multi-strata vegetation structure, with willows forming the dominant upper canopy, complemented by small trees, and a rich shrub layer (Negi et al., 2018; Rawat & Everson, 2012, 2013; Rawat et al., 2006; Sekar et al., 2022). This structural diversity enhances ecological resilience, provides multiple livelihood benefits, and contributes to the sustainability of these systems in fragile cold desert environments (Rawat et al., 2023; Saxena et al., 2011).

3.10 Fuelwood Production

Fuelwood constitutes a primary household energy source in the cold desert of the Lahaul valley (Rawat et al., 2009). Willow-based agroforestry systems are the principal fuelwood providers, contributing between 18.19 and 20.1% of household fuelwood supply depending on the village (Figs. 2 and 3). Willow (*Salix fragilis*) trees, grown both under agroforestry and plantations, are central to this provision, contributing 18.90–69.51% of fuelwood demand at the village level. Fuelwood supply in the Lahaul valley was derived from multiple sources, with farmer managed willow-based agroforestry systems constituting the largest share (17.41% of the total, 61.28 MT). *Salix fragilis* grown under agroforestry provided consistent production across sites, ranging from 4.64 MT in Khoksar to 26.14 MT in Jahlma. Willow plantations further contributed 29.94% (105.38 MT), indicating their role as a deliberate and managed source of wood energy, especially at Jahlma (64.68 MT) and Hinsa (38.06 MT). The land holding of the farmers determines the number of trees in agroforestry and forestry systems in the present study (Acharya, 2006; Acharya et al., 2011). The influence of social and ecological factors is considered when determining the use of certain species in agroforestry and forestry systems (Rawat et al., 2006, 2009, 2023).

Poplar (*Populus* spp.) plantations contributed only marginally (1.04%; 3.65 MT), reflecting their low prevalence in these systems. Seabuckthorn (*Hippophae rhamnoides*) stands also contributed minimally (0.36%; 1.25 MT), although their role as a multipurpose species (fuelwood, fodder, soil stabilization) remains ecologically important (Behera et al., 2001, 2014; Negi et al., 2018; Rawat et al., 2023; Sekar et al., 2022). Agricultural waste and shrubs/grassland vegetation were additional household level sources (Vishvakarma et al., 2011). Shrubs around agriculture fields and grasslands supplied 18.40 MT (5.23%), with the highest extraction recorded at Kuthar (6.70 MT). Dung was another significant energy source found in the Lahaul valley, contributing 5% (17.24 MT), with the highest dependency observed in Jahlma (11.35 MT).

Natural forests remain a critical source of domestic energy, accounting for 31.38% of the total fuelwood production (110.43 MT) (Rawat & Everson, 2012; Rawat et al., 2009, 2023). Forest-derived fuelwood accounted for 1.30–68.42% of supply, varying by village, with Kuthar showing the highest forest dependence (68.42%) and Jahlma

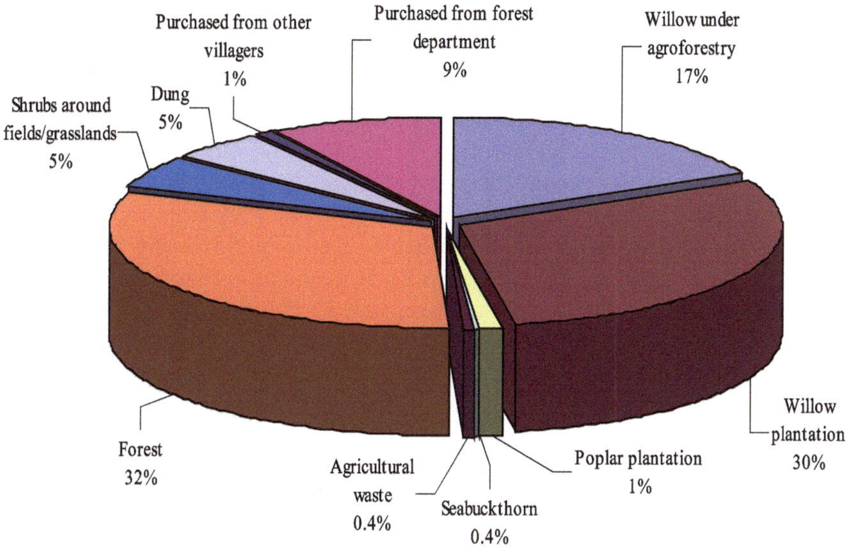

Fig. 2 Contribution of various sources in meeting total fuelwood requirement at the valley level

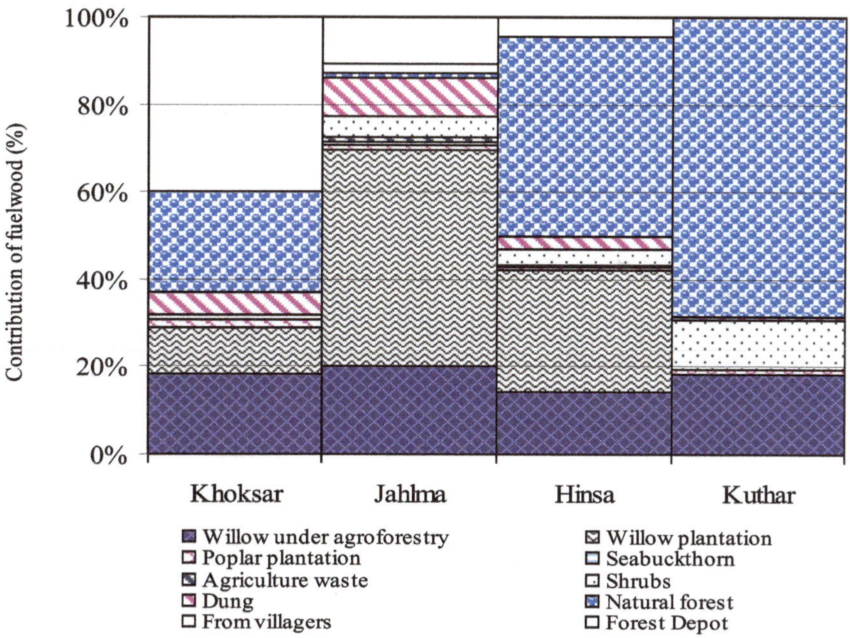

Fig. 3 Contribution of various sources in meeting total fuelwood requirement at the village level

the lowest (1.30). Site wise, forest dependence was highest at Kuthar (39.87 MT; 68.42% of site level fuelwood supply) and Hinsa (63.14 MT; 45.79%). In contrast, Jahlma (1.7 MT; 1.3%) and Khoksar (5.72 MT; 22.79%) showed relatively lower reliance, reflecting either restricted access or greater availability of on farm resources. Across the valley, *Salix fragilis* alone supplied 17.41% of fuelwood from agroforestry and 29.94% from forestry, cumulatively meeting nearly 47.35% of total household energy requirements from the species itself. Agroforestry systems and plantation provided a major share of fuelwood highlighted their role in reducing dependence on natural forests and ensuring energy security in a remote and harsh environmental conditions (Rawat et al., 2009; Rawat & Eversion, 2012; Rawat et al., 2023). Jahlma exhibiting the highest productivity due to greater adoption of willow plantations. This confirms the multifunctional role of agroforestry in cold desert energy provisioning (Rawat et al., 2013). External purchases were negligible and restricted to Khoksar and Jahlma, further indicating that the region is largely self-sufficient in fuelwood, despite illicit harvesting and collection of fuelwood from natural forests (Rawat & Eversion, 2012).

At Hinsa, the majority of fuelwood was obtained from natural forests (45.79%), while at Kuthar, forest dependence was even higher (68.42%). In contrast, Jahlma exhibited the highest reliance on willow plantations (49.50%) and willow-based agroforestry (20.01%), reflecting the success of farm-based energy production interventions in reducing forest dependence (Rawat et al., 2006; Rawat & Eversion, 2012; Rawat et al., 2023). Khoksar exhibited a mixed strategy, with contributions from agroforestry (18.49%), forest (22.79%), and purchased sources (40.04%), suggesting limited on farm resources and greater market reliance (Rawat & Eversion, 2012). At the site level, the total annual fuelwood availability was 351.93 MT, with farmer managed sources (59.29%) contributing more than natural forests (31.38%) and purchased wood (9.33%), respectively. This indicates that willow-based agroforestry systems have become the backbone of rural household energy security in the Lahaul valley (Rawat et al., 2006, 2009; Rawat & Eversion, 2013; Rawat et al., 2023). However, the persistent reliance on natural forests, especially in Hinsa and Kuthar, highlights continued extraction pressure on fragile cold desert ecosystems (Rawat et al., 2023). The results underscore the importance of expanding willow and poplar plantations, promoting multipurpose shrubs like seabuckthorn, and strengthening community managed fuelwood sources (Rawat et al., 2006, 2009). Such interventions could reduce forest degradation, enhance household energy self-sufficiency, and improve the ecological sustainability of fuelwood use in the cold desert environments of the Northwestern Himalaya (Rawat et al., 2023).

3.11 Fodder Production and Household Livelihoods

Willow-based agroforestry systems were the single most important contributor, providing 77–89% of the total fodder depending on the village (Table 4). The highest share from agroforestry was recorded at Kuthar (88.94%), followed by Khoksar

(84.56%), while Jahlma and Hinsa reported slightly lower proportions at 77.41% and 78.01%, respectively. Agroforestry contributed 2.92 MT/year per household basis, clearly underscoring its role in livestock sustenance and productivity, while site-specific differences were emerged clearly in the study. Khoksar (37.30 MT) and Jahlma (233.78 MT) reported the highest aggregate fodder availability per household, meeting 100% of estimated livestock needs. Hinsa (145.59 MT) and Kuthar (34.00 MT) similarly achieved full self-sufficiency in the study area (Vishvakarma et al., 2011). Agroforestry remained the dominant source across all sites, but Jahlma's stronger dependence on willow plantations (16.91%) and Hinsa's higher external purchases reflect contrasting local resource dynamics (Rawat, 2006). A highly developed willow-based agroforestry systems were found in Jahlma, a middle-altitude site in the study area. In this area, natural forest species were very scanty and sparse (Rawat et al., 2006, 2023). Farmers have maintained high densities of willow trees in agroforestry and forestry (plantations) management units to meet fodder, fuelwood, and timber requirements (Rawat & Vishvakarma, 2011; Rawat et al., 2023). Trees serve as essential sources of fodder for livestock that subsequently provide food products such as milk and meat and maintain soil productivity through compost and manure in the Northwestern Himalaya of India (Vishvakarma et al., 1998a, 1998b) and in Nepal Himalaya (Acharya, 2006; Acharya et al., 2011).

Willow plantations alone contributed up to 16.91% of the total fodder in Jahlma, underlining their importance in livestock-based farming systems (Rawat et al., 2023). Willow plantations represented the second important internal source, contributing 16.91% (39.54 MT) at Jahlma and 16.05% (23.36 MT) at Hinsa, but only a minor share in Khoksar (4.80%) and absent in Kuthar. This reflects both differential adoption of willow plantations and varying land availability (Rawat & Everson, 2013; Rawat et al., 2006). Kitchen garden residues (vegetable waste) played only a marginal role, contributing less than 1% of total fodder across all sites (Vishvakarma et al., 2011).

Kitchen garden residues provided a minor share (<1%), while forest grasses were rarely used, with only Kuthar reporting a small contribution (3.58 MT; 10.53%). This minimal dependence on forest fodder contrasts with the heavy reliance on forests for fuelwood (Figs. 2 and 3), indicating site-specific pressures and possibly stricter protection of grazing rights. The overall contribution of internal sources was substantial, ranging between 89.26 and 94.64% of total fodder supply, with an average per household provision of 3.46 MT/year (94.58% of total). Willow plantations contributing an additional 4.80% (Khoksar) to 16.91% (Jahlma). These results highlight the efficiency of agroforestry systems in ensuring fodder security in the resource scarce cold desert (Rawat et al., 2006, 2023).

External fodder sources (purchased grasses, oil cakes, and feed) contributed less than 6% of total supply, underscoring the self-reliance of households. The total fodder production equaled the actual livestock requirements, indicating a condition of near fodder self-sufficiency (Rawat & Vishvakarma, 2011). External sources, including purchased grass, commercial feed, and oilseed cakes, supplemented internal supplies, but accounted for only 0.21–10.29% of total fodder across villages. Among these, purchased grass was significant in Khoksar (4.93%) and Hinsa (1.32%), while

Table 4 Fodder production (MT) from various sources in cold desert of the Lahaul valley (values in parentheses % of the total)

	Khoksar (12)	Jahlma (41)	Hinsa (52)	Kuthar (17)	Per HH/MT/Y	Kg/day/HH
A. *Internal sources*						
1. Agroforestry	31.54 (84.56)	180.96 (77.41)	113.58 (78.01)	30.24 (88.94)	2.92	8.00
2. Willow under plantation	1.79 (4.80)	39.54 (16.91)	23.36 (16.05)	–	0.53	1.45
3. Kitchen garden (vegetables waste)	0.13 (0.35)	0.75 (0.32)	0.45 (0.31)	0.11 (0.32)	0.01	0.03
Total	33.46 (89.71)	221.25 (94.64)	137.39 (94.37)	30.35 (89.26)	3.46 (94.58)	9.48
Per ha. aggregate yield	2.95	3.68	3.66	2.91	–	–
B. *Grasses from forest*	–	–	–	3.58(10.53)	0.03	0.08
C. *External sources*						
1. Purchased grass from other villagers	1.84 (4.93)	0.95 (0.41)	1.92 (1.32)	0.07 (0.21)	0.04	0.11
2. Feed	2.00 (5.36)	11.55 (4.94)	6.20 (4.26)	–	0.16	0.44
3. Oil (mustard)	–	0.03 (0.01)	0.08 (0.05)	0.001 (0.003)	0.001	0.002
Total	3.84 (10.29)	12.53 (5.36)	8.20 (5.63)	0.07 (0.21)	0.20 (5.42)	0.55
Grand total	37.30 (100.00)	233.78 (100.00)	145.59 (100.00)	34.00 (100.00)	3.69	10.12

Animal feed (1bag = 50 kg), – data not available

purchased feed dominated at Jahlma (11.55 MT; 4.94%) and Hinsa (6.20 MT; 4.26%). Oilseed cakes (primarily mustard) were negligible across sites. On a per household basis, external sources contributed only 0.20 MT/year (5.42% of total), suggesting that households are largely self-reliant in fodder production.

These results demonstrate that willow-based agroforestry systems are the backbone of fodder security, reducing dependence on external feed inputs and conserving fragile alpine pastures (Tewari et al., 2016; Vishvakarma et al., 2011). The adoption of *Salix* plantation in Jahlma and Hinsa reflects both cultural acceptance and greater availability of suitable land for willow plantation (Rawat & Everson, 2013; Rawat et al., 2023; Rawat et al., 2006, 2009). *Salix fragilis* mitigates pressure on natural forests, supports year round household needs, and stabilizes local ecosystems (Rawat & Everson, 2013; Rawat et al., 2006). On average, agroforestry contributed largely of household needs for both fuelwood and fodder (Rawat & Vishvakarma,

2011; Rawat et al., 2009). This dual provisioning service is particularly critical in the context of the Lahaul valley, where remoteness, harsh climate, and short-growing seasons limit access to external resources (Rawat & Everson, 2013; Rawat et al., 2023). By ensuring steady supplies of energy and livestock feed, agroforestry underpins household survival and sustains the mixed farming system typical of the region (Rawat et al., 2009).

3.12 Implications for Sustainable Livelihoods and Resource Management

The findings collectively underscore that agroforestry in the cold desert is multifunctional and indispensable for rural livelihoods (Rawat et al., 2006, 2009, 2023). Households meet most of their fuelwood and fodder needs internally, reducing dependence on relict sparse forests. Meeting energy and fodder demands through on farm sources, and extraction pressure on relict natural forests and pastures is reduced (Rawat & Vishvakarma, 2011). Willow-based agroforestry systems ensure year round availability of critical resources under short-growing seasons and extreme climatic conditions (Rawat & Everson, 2013). Prioritizing *Salix fragilis* in future agroforestry expansion can maximize both energy and fodder production while enhancing ecological services, such as soil stabilization and biodiversity conservation (Hamadi et al., 2019; Rawat et al., 2009, 2023). Additionally, willow plantations in the region are also vital for cricket bat industry of India (Rawat & Everson, 2013).

Willow-based agroforestry shows a strong adaptability to cold arid conditions and provides both ecological and socio-economic benefits to local communities (Rawat et al., 2009, 2023). Scaling up such the prototypes, coupled with scientific management of plantations, can further enhance resource availability, reduce forest degradation, and improve resilience of mountain farming systems in the cold deserts of the western Himalaya to external forces, including anthropogenic climate change (Rawat & Everson, 2013; Rawat et al., 2006, 2023). The results highlights that willow-based agroforestry systems not only serve as a critical source of fuelwood but also provide the majority of household fodder needs (Rawat & Everson, 2012; Rawat et al., 2009). Agroforestry systems is contributing nearly 95% of the total fodder, significantly reduces pressure on fragile alpine pastures and forests, while supporting year round livestock feeding. Limited dependence on external purchases suggests economic savings for households, although reliance on feed inputs at Jahlma and Hinsa may increase if agroforestry productivity declines (Vishvakarma et al., 2011). Willow-based agroforestry systems play a central role in ensuring fodder security, sustaining livestock dependent livelihoods, and maintaining ecological balance in the cold desert of the Lahaul valley (Rawat & Vishvakarma, 2011). Further, the willow-based agroforestry is a cornerstone of sustainable mountain farming in the Lahaul valley, delivering multiple ecosystem services and supporting resilient livelihoods in cold arid landscapes (Rawat & Everson, 2013; Rawat et al., 2006, 2009, 2023).

Strategic promotion and scientific management of willow plantations will be critical for sustaining household energy, income, and fodder security while conserving fragile alpine ecosystems (Quandt et al., 2023; Rawat et al., 2023).

4 Conclusions

The Lahaul valley supports limited but ecologically and economically significant vegetation patches that meet the subsistence needs of local communities. Vegetation composition and density vary markedly with altitude and slope aspect, reflecting the influence of climatic and edaphic gradients. However, increased anthropogenic pressures, limited natural regeneration, and changing land use and climate are contributing to the degradation of relict natural forest patches. Sustainable management and expansion of willow-based agroforestry systems can mitigate these pressures by providing fuelwood, fodder, and timber, thereby reducing dependence on natural forests. Conservation planning should prioritize ecosystem restoration through community-based management, slope stabilization, and the use of native and multipurpose species such as *Salix fragilis* and *Hippophae rhamnoides*. Strengthening agroforestry practices offers a viable strategy for climate change adaptation, biodiversity conservation, and livelihood improvement in the fragile cold desert ecosystems of the Himalaya.

Acknowledgements We thank the G. B. Pant National Institute of Himalayan Environment (NIHE), Kosi-Katarmal, Almora, Uttarakhand, for providing necessary facilities and the Technical and Vocational Training Institute, Addis Ababa, Ethiopia, while finalizing the manuscript. The first author also acknowledges the Council of Scientific and Industrial Research (CSIR), Govt. of India, New Delhi, for awarding the Senior Research Fellowship to the first author. The help and support provided during the study by the various sources and people are also gratefully acknowledged.

References

Acharya, K. P. (2006). Linking trees on farms with biodiversity conservation in subsistence farming systems in Nepal. *Biodiversity and Conservation, 15*, 631–646.

Acharya, B. K., Chettri, B., & Vijayan, L. (2011). Distribution pattern of trees along an elevation gradient of Eastern Himalaya. *India Acta Oecology, 37*, 329–336.

Anonymous. (1886–1970). *Index Kewensis Plantarum Hanerogamarum* (vol. 1–2) (1883–1885) and 15 Suppl. Oxford, UK: Clarendron Press.

Anonymous. (1993). *Action plan on cold desert: an integrated approach for sustainable development* (pp. 1–40). Solan: Regional Center, National Afforestation and Ecodevelopment Board (NAEB), Dr.Y.S. Parmar University of Horticulture & Forestry (UHF).

Aswal, B. S., & Mehrotra, B. N. (1994). *Flora of Lahual-Spiti. A cold desert in North-West Himalaya* (pp. 10–15). Dehradun, India: Bishan Singh Mahendra Pal Sigh.

Behera, M. D., Kushwaha, S. P. S., & Roy, P. S. (2001). Forest vegetation characterization and mapping using IRS-1C satellite images in Eastern Himalayan Region. *Geocarto International, 16*, 53–62.

Behera, M. D., Matin, S., & Roy, P. S. (2014). Biodiversity of Kargil Cold Desert in the Ladakh Himalaya. In S.-I. Nakano, T. Yahara & T. Nakashizuka (Eds.), *Integrative observations and assessments. Ecological research monographs* (pp. 253–274). https://doi.org/10.1007/978-4-431-54783-9.

Chauhan, D. S., Bisht, D. S., Deorai, M., Rawat, D. S., & Sundriyal, R. C. (2021). A sustainable approach for livelihood improvement and integrated natural resource management in Central Himalaya, India. *Current Science, 120*(5). https://doi.org/10.18520/cs/v120/i5/825-834

Curtis, J. T., & McIntosh, R. R. (1950). The interrelations of certain analytic and synthetic phytosocialogical charaters. *Ecology, 46*, 84–89.

Curtis, J. T., & Cottom, G. (1956). Plant ecology workbook. Laboratory field production in Andropogan gerardi. *Ecology, 31*, 488–489.

Deshmukh, H. K., Tandel, M. B., Gunaga, R., Thakur, N., Dobriyal, M., Singh, N., Chhatrola, H. N., & Mevada, R. J. (2020). Three decades of review on existing agroforestry systems and practices in South Gujarat. *International Journal of Current Microbiology and Applied Sciences, 8*(8), 2973–2978. https://doi.org/10.20546/ijcmas.2020.908.334

G.B. Pant Institute of Himalayan Environment and Development. (1995–1996). *Growth problems of tree species in the Lahaul valley in highland of North-West Himalaya* (Annual Report (1995–1996)). Retrieved September 8, 2025, fromhttps://www.gbpihed.gov.in/PDF/Annual%20Reports/Annual_Report_2018_19_English.pdf.

Hamadi, A., Larid, M., Gómez, J. A., & Ouabel, H. (2019). Potential impacts of agroforestry on controlling soil degradation by water erosion in the agricultural lands of foothills North-West of Dahra (Mostaganem, Algeria). *Proceedings, 30*, 50. https://doi.org/10.3390/proceedings2019030050

Kala, C. P. (2002). *Medicinal plants of Indian Trans-Himalaya*. Dehradun, India: Bishen Singh Mahendra Pal Singh.

Kala, C. P., & Mathur, V. B. (2002). Patterns of plant species distribution in the Trans-Himalayan region of Ladakh, India. *Journal of Vegetation Science, 13*(6), 751–754.

Kala, C. P. (2025). Agroforestry in a changing climate: Challenges, opportunities and solutions. *Ecological Frontiers, 45*(2), 269–276. https://doi.org/10.1016/j.ecofro.2024.11.003

Kershaw, K. K. (1973). *Quantitative and dynamic plant ecology* (2nd ed.) (p. 308). London, UK: ELBS and Edward Arnold Ltd.

Khybri, M. L., Gupta, R. K., Ram, S., & Tomar, H. P. S. (1992). Crop yield of rice and wheat grown in rotation as intercrop with tree species in the outer hills of Western Himalaya. *Agroforestry System, 17*, 193–204.

Knight, D. H. (1963). A distance method for constructing forest profile diagram and obtaining structural data. *Tropical Ecology, 4*, 89–94.

Kuniyal, J. C., Vishvakarma, S. C. R., & Singh, G. S. (2004). Changing Crop Biodiversity and resource use efficiency of traditional versus introduced crops in cold desert of north-western Indian Himalaya: A case of Lahaul valley. *Biodiversity and Conservation, 13*, 1271–1304.

Kuniyal, C. P., Rawat, Y. S., Oinam, S. S., Kuniyal, J. C., & Vishvakarma, S. C. R. (2005). Kuth (*Saussurea lappa*) cultivation in the cold desert environment of the Lahaul valley, northwestern Himalaya, India: Arising threats and need to revive socio-economic values. *Biodiversity and Conservation, 14*(5), 1035–1045.

Maikhuri, R. K., & Semwal, R. L. (1997). Agroforestry for rehabilitation of degraded community lands: A case study in the Garhwal Himalaya, India. *International Tree Crops Journal, 9*, 89–99.

Mac Arthur, R. H. (1965). Pattern of species diversity. *Biological Reviews, 40*, 510–533.

Masoodi, T.H., Shah, I.A., Dar, M.U.D., Sofi, P.A., & Mugloo, J.A. (2021). Evaluation of probability distribution functions applied to tree diameter in a mixed uneven Kiker (Robinia pseudoacacia) stand of Kashmir Himalaya, India. *Current Science, 121*(4), 573–577. https://doi.org/10.18520/cs/v121/i4/573-577

Negi, V. S., Joshi, B. C., Pathak, R., Rawal, R. S., & Sekar, K. C. (2018). Assessment of fuelwood diversity and consumption patterns in cold desert part of Indian Himalaya: Implication for conservation and quality of life. *Journal of Cleaner Production, 196*, 23–31.

Oinam, S. S., Rawat, Y. S., Khwairakpam, G. S., Khoiyangbam, R. S., Kuniyal, J. C., & Vishvakarma, S. C. R. (2005). Land use and land cover changes in Jahlma watershed of the Lahaul valley, cold desert region of the northwestern Himalaya, India. *Journal of Mountain Science, 2*(2), 129–136.

Oinam, S. S., Kuniyal, J. C., Rawat, Y. S., Vishvakarma, S. C. R., & Pandey, D. C. (2004–05). Food habits and dietary system of the local communities in cold desert of the Lahaul valley, northwestern Himalaya, India. *Journal of Regional Science and Development, 3–4*(1–2), 22–42.

Phillips, E. A. (1959). *Methods of vegetation study*. New York, NY, USA: Henry Holt and Co. Inc.

Quandt, A., Neufeldt, H., & Kayla Gorman, K. (2023). Climate change adaptation through agroforestry: Opportunities and gaps. *Current Opinion in Environmental Sustainability, 60*, Article 101244. https://doi.org/10.1016/j.cosust.2022.101244

Rawat, G. S. (2005). *Alpine Meadows of Uttaranchal: Ecology, land use and status of medicinal and aromatic plants*. Dehra Dun, India: Bishen Singh Mahendrapal Singh.

Rawat, G. S., & Adhikari, B. S. (2005). Floristics and distribution of plant communities across moisture and topographic gradients in Tso Kar Basin, Changthang Plateau, Eastern Ladakh. *Arctic, Antarctic, and Alpine Research, 37*, 539–544.

Rawat, Y. S., Oinam, S. S., Vishvakarma, S. C. R., & Kuniyal, J. C. (2004). *Saussurea costus* (Falc.) Lipsch.: A promising medicinal crops under cold desert agro-ecosystem in North Western Himalaya. *Indian Journal of Forestry, 24*(3), 297–303.

Rawat, Y. S. (2006). *Vegetational analysis, socio-economic and cultural aspects of Salix sp. in cold desert environment of the Lahaul valley, north-western Himalaya (HP)*. Ph.D. Thesis, Garhwal University, Srinagar Garhwal, U.K. India. p 270.

Rawat, Y. S., Oinam, S. S., Vishvakarma, S. C. R., Kuniyal, C. P., & Kuniyal, J. C. (2006). Willow (*Salix fragilis* Linn.): A multipurpose tree species under pest attack in the cold desert of Lahaul valley, north-western Himalaya, India. *Ambio, 35*, 43–48.

Rawat, Y. S., Vishvakarma, S. C. R., & Todaria, N. P. (2009). Fuelwood consumption pattern of tribal communities in cold desert of the Lahaul valley, North-Western Himalaya, India. *Biomass and Bioenergy, 33*, 1547–1557.

Rawat, Y. S., & Everson C. S. (2011). *Inula racemosa* Hook. f.: A potential medicinal crop in cold desert agroecosystem North Western Himalaya, India. *Journal of Medicinal Plant Research, 5*(26), 6218–6223.

Rawat, Y. S., & Vishvakarma, S. C. R. (2011). Ecological analysis of indigenous agroforestry systems in Kullu and Lahaul valleys, Himachal Pradesh, India. *Proceedings of the National Academy of Science, India (Sect. B), 81*, 160–170.

Rawat, Y. S., & Everson, C. S. (2012). Ecological status and uses of juniper species in the cold desert environment of the Lahaul Valley, north-western Himalaya, India. *Journal of Mountian Science, 9*, 676–686.

Rawat, Y. S., & Everson, C. S. (2013). Availability and use of willow species in representative cold desert areas of Northwestern Himalaya, India. *Journal of Mountian Science, 10*(3), 472–481.

Rawat, Y. S., Negi, V. S., Moussa, I. M., Zaman, W., & Elansary, H. O. (2023). Diversity, distribution and vegetation assessment of woody plant species in the cold desert environment, north-western Himalaya, India. *Sustainability, 15*, 10429. https://doi.org/10.3390/su151310429

Salick, J., Ghimire, S. K., Fang, Z., Dema, S., Konchar, K. M., Byg, A., Ross, N., Bauer, K., Hart, R., Lloyd, A., Staver, B., Na, A., Dorje, S., Qizhu, Xian, H., Aifang, X., Maorong, X., Xiaowu, S., et al. (2023). Himalayan Alpine vegetation, climate change and mitigation. *Journal of Ethnobiology, 34*(3), 276–293. https://doi.org/10.2993/0278-0771-34.3.276 (Original work published 2014)

Saxena, K. G., Liang, L., & Xue, X. (2011). *Global change, biodiversity and livelihood in cold desert region of Asia* (vol. 322). Dehradun, India: Bishen Singh Mahendra Pal Singh. ISBN 978-81-211-0780-8

Sekar, K. C., & Srivastava, S. K. (2010). A supplement to the flora of Lahaul-Spiti. *Journal of Non-Timber Forest Products, 17*, 233–258.

Sekar, K. C., Pandey, A., Joshi, B. C., Bhatt, I. D., Bhojak, P., Dey, D., Thapliyal, N., Bisht, K., Bisht, M., Negi, V. S., et al. (2022). *Floristic diversity in cold desert regions of Uttarakhand Himalaya* (p. 537). Phytotaxa.

Shukla, A. N., & Srivastava, S. K. (2020). Flora of Ladakh: An annotated inventory of flowering plants. *Biodiversity of the Himalaya: Jammu and Kashmir State* (pp. 673–730). Springer.

Srivastava, S. K., & Shukla, A. N. (2015). Flora of cold desert-Western Himalaya, India (vol. 2). Kolkata, India: Botanical Survey of India.

Sharma, R., Pant, S. K., Bishist, R., Sharma, P. P., & Gautam, K. L. (2022). Agroforestry systems in north-western Himalayas, India: An overview. *Agrobiodiversity and Agroecology, 2*(1), 1–20.

Singh, G. S., Ram, S. C., & Kuniyal, J. C. (1997). Changing traditional land use pattern in Great Himalaya: A case study of Lahaul valley. *Journal of Environmental System, 25*(2), 195–211.

Sood, S. K., Nath, R., Kalia, D. C., et al. (2001). *Ethnobotany of cold desert tribes of Lahoul-Spiti (N.W. Himalaya)* (p. 228). New Delhi: Deep Publication.

Somarriba, E., Valdivieso, Vasquez, W., & Galloway, G. (2001). Survival, growth, timber productivity and site index of *Cordia alliodora*, in forestry and agroforestry systems. *Agroforestry Systems, 51*, 111–118.

Tewari, J. C., Pareek, K., Raghuvanshi, M. S., et al. (2016). Fodder production system—A major challenge in cold arid region of Ladakh, India. *MOJ Ecology and Environmental Science, 1*(1), 22–28. https://doi.org/10.15406/mojes.2016.01.00005

Vishvakarma, S. C. R, Kuniyal, J. C., & Singh, G. S. (1998). *Indigenous agroforestry system of north western Himalaya. Research for mountain development. Some initiatives and accomplishments* (pp. 99–118) Gyanodaya Prakashan, Nainital.

Vishvakarma, S. C. R., Kuniyal, J. C., & Singh, G. S. (1998). Indigenous agroforestry system of North Western Himalaya. In *Research for mountain development: Some initiative and accomplishment* (pp. 99–118). Gyanodaya prakashan: Nainital, India.

Vishvakarma, S. C. R., Rawat, Y. S., Oinam, S. S., & Kuniyal, J. C. (2005). Irrigation (*Kuhls*): A community management system in cold desert of the Lahaul valley, northwestern Himalaya, India. In P. Pradhan & U. Gautam (Eds.), *Proceedings: Third International Seminar on Farmer Managed Irrigation Systems and Governance Alternatives* (pp. 425–444). Kathmandu, Nepal.

Vishvakarma, S. C. R., Kuniyal, J. C., Rawat, Y. S., & Oinam, S. S. (2011). Cold desert agroecosystems: A case study from Lahaul Valley, India. In K. G. Saxena, L. Liang, & X. Xue (Eds.), *Global change, biodiversity and livelihoods in cold desert Region of Asia* (pp. 107–126). United Nations University.

Whittaker, R. H. (1975). *Communities and ecosystem* (2nd ed., p. 385). Macmillan Publishing Co.

Community-Led Forestry and Landscape Restoration for Climate Change Mitigation and Adaption in Cold Desert, Northwestern Himalaya, India

Yashwant S. Rawat

1 Introduction

Cold deserts typically found in high mountain ranges, serve as sources of many ecosystem services including glaciers and water bodies (Behera et al., 2014). These unique high-altitude mountain ecosystems significantly contribute in environmental sustainability by sustaining biodiversity conservation, sustainable forest management, and broader socio-economic activities in many parts of the cold deserts (Rawat et al., 2023; Saxena et al., 2011). In view of global climate change impacts, the rapid decline of glaciers and environmental degradation poses significant challenges for sustaining the cold desert social–ecological systems and downstream water resources (Rawat et al., 2023). Cold deserts are among the most vulnerable regions to global climate change due to their limited biological resources, lack of technological support in many areas, rising temperatures, desertification, and land degradation (Rawat & Everson, 2012; Sekar et al., 2010; Srivastava & Shukla, 2015; Shukla & Srivastava, 2020; Sekar et al., 2022). Furthermore, the social–ecological systems of cold desert regions—including agro-ecosystems, biodiversity, and vegetation patterns—remain poorly understood due to their inaccessibility and remoteness (Rawat et al., 2023). These landscapes are increasingly impacted by climate change and anthropogenic activities, posing significant challenges for their sustainability and conservation (Kuniyal et al., 2004; Oinam et al., 2005). These regions remain snowbound for months and experience extreme climatic conditions, with temperatures ranging from −45 °C in winter to 40 °C in summer and annual rainfall of less than 60 mm (Behera et al., 2014; Rawat et al., 2023). Cold desert ecosystems have already crossed critical climate change thresholds, leading to rapid transformations in cropping patterns, eco-tourism, horticulture, biodiversity, and forestry (Rawat et al., 2023). Given their

Y. S. Rawat (✉)
Department of Wood Technology Management, Faculty of Civil Technology, Federal Technical and Vocational Training Institute (TVTI), Addis Ababa, Ethiopia
e-mail: yasrawat@gmail.com

Y. S. Rawat and C. P. Kala (eds.), *Cold Deserts*,
https://doi.org/10.1007/978-3-032-29119-6_9

ecological significance and sustainability, cold deserts are vital for biodiversity conservation, climate change mitigation, and ecosystem restoration efforts.

The cold deserts are extended all over the world while a large portion of the land in Asia particularly in the high mountains comes under cold deserts (Negi et al., 2018; Saxena et al., 2011). In India, the Cold Desert Zone of the Indian Trans-Himalaya (ITH) spans across three distinct biogeographic regions: (i) the Ladakh Mountains, encompassing Kargil, Nubra, and Zanskar in Jammu and Kashmir, as well as Lahaul–Spiti and Kinnaur in Himachal Pradesh; (ii) the Tibetan Plateau, including the Changthang region of Ladakh and portions of Uttarakhand's northern territories; and (iii) the Sikkim Plateau (Srivastava, 2010; Srivastava & Singh, 2005). The cold deserts of the Trans-Himalayan region, particularly Jammu and Kashmir, Ladakh, and Himachal Pradesh, cover about 2.3% (74,809 km^2) of the total geographic area of India, which include the Leh and Kargil districts of Ladakh, along with the Lahaul–Spiti district and parts of Kinnaur and Chamba districts in Himachal Pradesh (Anonymous, 1993). In Himachal Pradesh, cold deserts encompass approximately 35% of the total land area, with the majority lying in the Lahaul–Spiti district, which accounts for 25% of the cold desert landscape of the state (Sekar & Srivastava, 2010).

Cold deserts are particularly vulnerable ecosystems, shaped by extreme environmental and socio-ecological conditions (Sekar et al., 2010; Srivastava & Shukla, 2015; Shukla & Srivastava, 2020; Sekar et al., 2022). Vegetation and crops in these areas endure severe drought, low temperatures, strong winds, and soil erosion (Kuniyal et al., 2004; Oinam et al., 2005). Due to harsh climate and substantial snow deposits, sapling establishment and growth are generally poor, making afforestation efforts largely unsuccessful except in the selected locations (Rawat et al., 2006; Rawat & Eversion, 2013; Sekar et al., 2010; Srivastava & Shukla, 2015; Shukla & Srivastava, 2020; Sekar et al., 2022). Cold deserts play a crucial role in providing diverse ecosystem services, including biodiversity conservation and maintenance of unique habitats (Rawat et al., 2023). However, socio-economic activities in these regions are largely confined to the short summer period due to the prolonged and harsh winters (Aswal & Mehrotra, 1994; G.B. Pant National Institute of Himalayan Environment (NIHE), 1995–96; Kala, 2002; Rawat & Eversion, 2013). Local communities residing in these landscapes carry out agricultural, forestry, and cultural practices during this season to sustain their livelihoods (Rawat & Everson, 2013). Therefore, a comprehensive ecological understanding of bio-resources is essential for promoting socio-economic development, biodiversity conservation, landscape restoration, and overall ecological resilience in these fragile environments (Rawat & Everson, 2012; Rawat et al., 2023).

Vegetation in the cold deserts of the Lahaul valley is broadly categorized into alpine forests (extensions of temperate forests), dry alpine scrub, alpine meadows, and alpine stony deserts (Aswal & Mehrotra, 1994; G.B. Pant National Institute of Himalayan Environment (NIHE), 1995–96; Kala, 2002). Despite their ecological significance, these regions have rarely been at the forefront of ecosystem-based approaches and sustainability research (Rawat et al., 2023). This is primarily due to their sparse and scattered vegetation, inhospitable climate, and socio-economic marginality. In the Lahaul valley, farmers actively participate in reforestation, community-based resource management and ecological restoration activities for sustenance (e.g., fuelwood, fodder, and soil and water conservation), biodiversity, livelihoods, and sustainable forest management (Rawat et al., 2006, 2009, 2023). Although agro-ecosystems—particularly agroforestry—help meet local needs by providing food, fruits, firewood, and fodder in certain areas, the demand for their subsistence resources continues to exert significant pressure on bio-resources (Rawat et al., 2006, 2023). Therefore, a strong focus on responsible resource use and sustainability is essential in the context of natural resource extraction in cold deserts. In the Lahaul valley, agroforestry systems and plantations (hereafter called "forestry" in the text) serve as vital sources of food, fruits, firewood, fodder, and raw materials, contributing to both social and environmental sustainability (Rawat et al., 2009, 2023). Due to their adaptability to diverse climatic conditions, including harsh environments, these systems play a crucial role in supporting the socio-ecological resilience of the Lahaul valley. Agro-ecosystems, agroforestry, and forestry systems need to embrace best practices to adapt and minimize the impacts of climate change and reduce the pressure on remnant natural forest patches (Rawat et al., 2023). Therefore, the present study deals with assessing the community-led forestry and landscape restoration in the cold desert of Lahaul valley in the northwestern Himalaya, India.

2 Materials and Methods

2.1 Study Area

The present study was conducted in the Lahaul valley within Lahaul–Spiti district of Himachal Pradesh, India. It is located between latitudes 31° 44′ 34″ N and 32° 59′ 57″ N and longitudes 76° 46′ 29″ E and 78° 41′ 34″ E. The district experiences extreme cold desert climatic conditions characterized by prolonged winters and short summers. The topography of the study area is dominated by mountain outcrops, intersected by glacier-fed rivers and streams that form its natural drainage system called Chandra-Bhaga catchments which after flowing further are named as River Chenab in Jammu and Kashmir.

The Lahaul valley, a landlocked area, remains accessible only by a recently opened Rohtang tunnel and occasionally via the Rohtang Pass (3,978 m), connecting it to Kullu district during summer and winter months. The Lahaul valley elevation ranges from 2,400 to 6,400 m, creating a complex temperate and alpine terrain marked by diverse microclimates due to geographical and topographical variations. These harsh climatic conditions have driven speciation in flora, contributing to the high endemicity of the region.

The Lahual valley is divided into three sub-valleys: Chandra, Bhaga, and Pattan (Fig. 1). The Chandra and Bhaga valleys are drained by the Rivers Chandra and Bhaga, respectively, which meet at Tandi to form the River Chandra-Bhaga (known as the Chenab in Jammu and Kashmir). The Pattan valley extends from Tandi to areas beyond Kuthar and hosts several small villages along the River Chandra-Bhaga. Land use in the region reflects in its topography: steep slopes retain natural forests and alpine pastures, while gentler terrains are cultivated, supporting a mix of agroforestry systems including horticulture and human-dominated landscapes.

The Lahaul valley exhibits sparse forest cover, which is 7.9% of the total geographical area of the Lahaul–Spiti district. These forests exist as fragmented, discontinuous patches, predominantly in the Bhaga and Pattan valleys (Aswal & Mehrotra, 1994). Cold Desert is sparsely visible in the north-facing slopes, while south-facing slopes of the Chandra valley are nearly barren. Sampling sites were located near settlements such as Khoksar (3,200 m), Jahlma (3,000 m), Hinsa (2,700 m), and Kuthar (2,600 m). Land use patterns vary significantly by slope aspect and elevation; agroforestry systems dominate between 2,400 and 3,000 m, forestry (2,610–3,010 m) and natural forests (2,700–4,300 m) on the north-facing slopes. While on the south-facing slopes, agroforestry (2,650–3,200 m) and forestry (2,660–3,200 m) occur

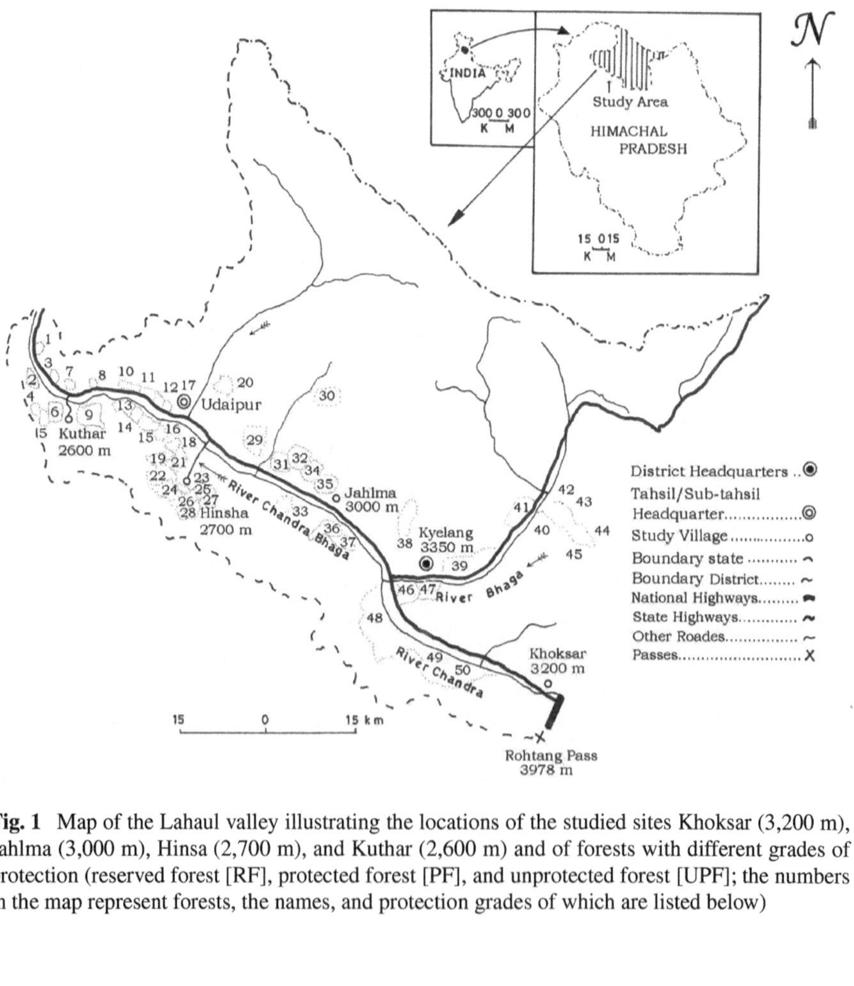

Fig. 1 Map of the Lahaul valley illustrating the locations of the studied sites Khoksar (3,200 m), Jahlma (3,000 m), Hinsa (2,700 m), and Kuthar (2,600 m) and of forests with different grades of protection (reserved forest [RF], protected forest [PF], and unprotected forest [UPF]; the numbers in the map represent forests, the names, and protection grades of which are listed below)

at higher elevations compared to the north-facing, with natural forests spanning 2,400–4,600 m. Notably, *J. indica* populations extend at elevations up to 5,000 m.

1. Krakun RF	11. Khurea RF	21. Bardang PF	31. Duling RF	41. Jispa RF
2. Chhaned RF	12. Suthal RF	22. Bardang UPF	32. Lingche RF	42. Sunshi RF
3. Surnali RF	13. Salrot RF	23. Trilokinath PF	33. Nalda PF	43. Milang RF
4. Bharaur PF	14. Ratoli RF	24. Khangar UPF	34. Chewar RF	44. Vochi RF
5. Bhujund PF	15. Salpat RF	25. Bhiyari PF	35. Othang RF	45. Yoche RF
6. Kain RF	16. Salpat UPF	26. Bhingar UPF	36. Rital RF	46. Guru Ghantal RF
7. Lohni P F	17. Riodhan RF	27. Gilding PF	37. Shangrang RF	47. Kardang RF
8. Dhandal RF	18. Lobar RF	28. Gilding UPF	38. Kyelang RF	48. Gushal RF
9. Romas RF	19. Lobar UPF	29. Sindwari PF	39. Sitingri RF	49. Mulling RF
10. Dareda RF	20. Phatgahar PF	30. Lot PF	40. Shang Kolong RF	50. Dalingcha RF

Forests 1–37 are situated in the Pattan valley, 38–47 in the Bhaga valley, and 48–50 in the Chandra valley

2.2 *Climate and Vegetation Zone*

Lahaul valley features a cold, arid climate marked by minimal rainfall, heavy snowfall, and extreme winters. It is broadly divided into two distinct seasons: a short summer and a prolonged severe winter, while the skeletal soils of the region reflect its rugged terrain and climatic extremes. Annual precipitation ranges from 241.5 to 272.4 mm as rainfall and 466.2 to 693.2 mm as snowfall, with maximum rise at 27.8 °C temperature in July and minimum at -3.1 °C in January (Rawat et al., 2009). Micro-climatic variations exist across the valley; Khoksar and Hinsa experience marginally higher humidity and sporadic rainfall during monsoon months, while Jahlma and surrounding areas remain arid even in the rainy season. Winters span mid-October to late April, with January and July representing the coldest and warmest months, respectively. Vegetation is stratified into two zones: a temperate zone (2,400–3,300 m) and an alpine zone (>3,300 m) (Aswal & Mehrotra, 1994). Shrubs dominate the landscape, with Juniper species primarily occupying dry south-facing slopes. J. *communis* is an exception, thriving on moist north-facing slopes. Scattered patches of *Pinus wallichiana*, *Abies pindrow*, *Picea smithiana*, and *Cedrus deodara* also occur on north-facing slopes.

2.3 Crop Farming

Crop farming in the Lahaul valley has short summer (May–September) growing season. Historically, agriculture relied on traditional practices with diverse crops such as *Fagopyrum* spp., *Hordeum himalayense*, *Saussurea costus*, and *Inula racemosa* (Kuniyal et al., 2004). However, recent shifts toward cash crops like potatoes, peas, hops and horticulture crops supported by irrigation via *kuhls* (traditional water channels) have transformed local economies. Cultivated land is concentrated between 2,400 and 3,600 m asl in the villages, though landholdings remain small: ~ 0.12 ha/capita in Khoksar and Jahlma, and 0.08 ha/capita in Hinsa and Kuthar (Rawat & Vishvakarma, 2011). While cash crops are exported, staples, fuelwood, and timber are imported from the adjoining district Kullu.

2.4 Vegetation and Land Use

Agroforestry systems prioritize *Salix* spp. and *Populus* spp. for fuelwood and fodder, while *Hippophae rhamnoides* thrives densely on slopes and river banks naturally. Sub-valleys vegetation varies; sparse *Betula utilis*, *Pinus wallichiana*, *Juniperus communis*, and *J. indica* on north-facing slopes in the Chandra sub-valley. While Bhaga sub-valley, dominated by *Juniperus polycarpos* on south-facing slopes (Aswal & Mehrotra, 1994; Rawat & Everson, 2012). Scattered *Pinus wallichiana*, *Abies pindrow*, and *Cedrus deodara* at lower elevations, with *Betula utilis* at higher north-facing slopes in the Pattan valley. Scrublands-comprising grasslands interspersed with *Rosa webbiana*, *Berberis pseudumbelata*, and *Ribes grossularia* form critical habitats for wildlife and livestock (Rawat et al., 2006). Beyond these zones, stony outcrops and snowbound, devoid of shrubs, areas dominate.

Forest and scrubland degradation, driven by overharvesting of fuelwood, fodder, and timber (Rawat & Everson, 2012; Rawat et al., 2009), is projected to intensify with population growth and development, exacerbating risks of resource depletion, soil erosion, and biodiversity loss (Rawat et al., 2006). Soils are predominantly sandy loam or moraine, with localized variations including clayey loam, gravelly sand, or silt. Alluvial deposits are along riverbanks, while brown-earth soils (brown-yellowish) occur at lower elevations. Saline and alkaline soils, scattered widely, arise from aridity, saline water, or seasonal salt deposition through streams. Traditional *kuhls* sustain irrigation for agriculture and plantations, though soils remain vulnerable to fluvial, avalanche, and wind transport (Vishvakarma et al., 2005).

Table 1 Land use system and irrigated area in four studied villages in the Lahaul valley

Irrigated land	Villages				Total villages under study
	Kuthar (2600 m) n = 17	Hinsa (2700 m) n = 52	Jahlma (3000 m) n = 41	Khoksar (3200 m) n = 12	
1. Arable land as net area sown	6.3 (61.2)[a]	20.4 (63.4)	36.3 (63.6)	6.6 (59.5)	69.6 (62.9)
2. Kitchen garden	0.1 (1.0)	0.3 (0.9)	0.6 (1.1)	0.1 (0.9)	1.1 (1.0)
3. Cultivated grassland	3.9 (37.9)	9.6 (29.8)	13.2 (23.1)	3.6 (32.4)	30.3 (27.4)
4. Forestry (Salix plantation)	-	1.9 (5.9)	7.0 (12.3)	0.8 (7.2)	9.7 (8.8)
Gross irrigated area	10.3 (100)	32.2 (100)	57.1 (100)	11.1 (100)	110.7 (100)

n Indicates number of households interviewed
[a] Values in parentheses are in percentage of the total land area

2.5 *Survey and Sampling*

The present study was carried out in four villages, namely, Kuthar, Hinsa, Jahlma, and Khoksar, located at different altitudinal gradients in the cold desert of the Lahaul valley. The study employed a stratified sampling approach across four villages in the Lahaul valley for land-use systems. Villages were selected based on elevation, microclimatic variations, and land-use patterns (e.g., agroforestry systems, natural forests). A total of 122 households were surveyed, with sample sizes stratified by village: Kuthar (n = 17), Hinsa (n = 52), Jahlma (n = 41), and Khoksar (n = 12) (Table 1). This stratification ensured representation of diverse agro-ecological conditions and socio-economic practices. Data on land-use categories (e.g., agriculture, kitchen gardens, forestry) were collected through structured household interviews, with percentages calculated relative to total irrigated land. Environmental challenges and *Salix* (willow) species distribution were assessed via field observations and checklists, recording presence (√) or absence (–) across villages. For preferential site ranking, a scoring system based on 122 respondents was utilized, ensuring community perspectives informed restoration priorities.

2.6 *Data Collection, Processing, and Analysis*

This study adopted a mixed-methods approach, combining qualitative and quantitative analysis. The survey was carried out to collect the data based on questionnaires and structured interviews. The questionnaires were administered and field interviews were conducted. Closed-ended and Open-ended questionnaires with fixed response questions were used to collect the data from the respondents. Household interviews were conducted to collect numerical data on land-use (hectares, percentages) and restoration program metrics. Qualitative assessments of environmental issues and associated species were documented using categorical (presence/absence) and descriptive methods. To monitor the association of plant species, a transect study and inventory of plant resources were developed. Plots of 10 × 10 m^2 were laid down using a transect line from the base of the landscape to the end during the growing season, when evergreen species and other associated plant species were easier to identify. The selection of plots along the transect line was guided by both practical and statistical considerations (Rawat et al., 2023). This approach aimed to reduce bias by distributing plots across key gradients of environmental variation present at each of the sites (e.g., topography, vegetation). The selected plots collectively captured the species-specific patterns and association of the plant species to adequately understand the ecological process and characteristics. Government records from the Forest Department in Keylong, which serves as the headquarter of the Lahual, were consulted to collect restoration area statistics. Quantitative data were standardized into percentages and aggregated totals (e.g., gross irrigated area). Environmental problems were categorized by village incidence, with remedial measures thematically coded. *Salix* (Willow) species presence was mapped by slope aspects the north- and south-facing slopes, and associated flora were taxonomically classified. Preferential ranks were derived from scored marks, normalized into ordinal rankings (1–5). Descriptive statistics summarized land-use patterns, and restoration trends were analyzed. Comparative analysis highlighted inter-village disparities in land-use percentages and environmental challenges. Thematic analysis synthesized qualitative remedial measures, emphasizing adaptive strategies (e.g., terracing, organic farming). Species distribution data were analyzed ecologically to identify slope-specific afforestation preferences. The methodologies were used for robust environmental and socio-economic analyses in fragile ecosystems.

3 Results and Discussion

3.1 Land Use System and Irrigated Area

In the study area of Lahual, the arable land as net area sown constitutes the largest proportion of irrigated land in all the villages, ranging from 59.5% (Khoksar) to 63.6% (Jahlma) (Table 1). Jahlma has the highest absolute agricultural area (36.3 ha), likely due to being located in a wider part of the valley and water availability. Kitchen gardens represent <1.1% of irrigated land across all the villages, with negligible spatial variation. This reflects their supplementary role in household food security and income. Cultivated grasslands account for 23.1–37.9% of the total irrigated land, declining with elevation (highest in Kuthar at 2,600 m). This inverse relationship with altitude may reflect greater reliance on fodder production at lower elevations, where livestock integration is more feasible. *Salix* plantations are absent in Kuthar but contribute 5.9–12.3% of the total irrigated land in other villages. Jahlma has the highest *Salix* coverage (7.0 ha), suggesting targeted agroforestry practices for soil stabilization, fuelwood, fodder, and other (e.g., wood products, basketry). Jahlma has the largest gross irrigated area (57.1 ha), while Khoksar (3,200 m) has the smallest (11.1 ha), likely due to relatively harsher climatic constraints and narrower valley at higher elevations. The number of households correlates with agricultural land use (e.g., Hinsa, with 52 households, has 20.4 ha agriculture; Khoksar 12 households and 6.6 ha). This underscores a central role of agriculture to livelihoods of the local people.

The decline in total irrigated area with elevation (e.g., Khoksar at 3,200 m; Jahlma at 3,000 m) aligns with known constraints of high-altitude agro-ecosystems, including shorter growing seasons, colder temperatures, and limited water availability. Cultivated grasslands' higher prevalence at lower elevations (e.g., Kuthar) may reflect greater reliance on livestock, consistent with studies in similar Himalayan regions (Oinam et al., 2005).

The scattered trees of *Salix* at Kuthar and its prominence at Jahlma and Hinsa suggest site-specific agroforestry adaptations. Cold tolerance of *Salix* and utility for soil conservation, fuelwood, fodder, and artisanry make it a strategic crop in the valley. Its limited adoption at Khoksar may reflect extreme climatic thresholds and prioritization of staple cash crops like pea and potato (Rawat et al., 2006). Agricultural crops' overwhelming share >59% of irrigated land highlights dependency on staple cash crops (e.g., pea and potato) and traditional food crops (e.g., barley, buckwheat) for subsistence requirements. However, this mono-cultural leaning may increase vulnerability to climate shifts or pest outbreaks, emphasizing the need for diversification (e.g., integrating drought-resistant crops or agroforestry). The correlation between household numbers and agricultural area (e.g., Hinsa, Jahlma) implies that larger communities manage more land collectively, possibly through shared irrigation systems. This mirrors community-based resource management patterns observed in the trans-Himalayan villages (Vishvakarma et al., 2005). The minimal allocation to kitchen gardens (<1.1%) suggests underutilized potential for enhancing dietary diversity and food security. Promoting nutrient-dense, high-value crops in these spaces could improve resilience. Similarly, expanding *Salix* plantations in higher elevations (with climate-smart species) could mitigate erosion and diversify incomes. This suggests prioritizing mixed-cropping systems, agroforestry, and climate-resilient *Salix* varieties to balance productivity and sustainability. This study underscores the interplay of altitude, agro-ecology, and socio-economics in shaping land-use strategies in the Lahaul valley, offering insights for sustainable mountain development mainly in the cold desert areas.

3.2 Ecological Restoration Program

The spatial and temporal trends in ecological restoration efforts led by the Forest Department of Himachal Pradesh in the cold desert of the Lahaul valley, focused on *Salix* (willow), seabuckthorn, and poplar species between 1992–93 and 2002–03. The maximum land area was planted at Udaipur (886.4 ha) followed by Keylong (737.4 ha), Tindi (417.7 ha) and Jahlma (316.5 ha). The restoration program in the Lahaul valley aligns with their ecological vulnerability as cold desert zones, where soil erosion, glacial melt, and desertification necessitate urgent intervention (Vishvakarma et al., 2005). The early to mid-1990s surge probably represents the initial emphasis of government on large-scale afforestation to stabilize delicate ecosystems, which aligns with national policies linking biodiversity protection with desertification control (Rawat et al., 2006). Plantation in Lahaul valley is a balance between leveraging modern horticulture system (i.e., apple, apricot, cherry, plum, pear, almond, and walnuts) for economic growth and practicing ecological restoration with hardy native species (i.e., *Salix*, seabuckthorn, poplar, junipers, and birch) (Rawat et al., 2023). The focus is increasingly on climate-resilient, multipurpose species that can secure both the environment and the livelihoods of the local communities in this fragile, high-altitude landscape. This is considered an ecosystem-based approach for soil and water conservation, adaptation to climate change, biodiversity conservation, job creation and sustainable forest management. Restoration activities are increasingly framed within state and national climate change action plans, recognizing the vulnerability and sustainability in the Himalayan region.

3.3 *Preferences of Farmers for Planting of* Salix *(Willow)*

The efforts of the Forest Department of Lahual for the ecological restoration were supported by the local communities, especially the farmers. The species of Salix being one of the preferential species for ecological restoration, the ranking of its planting sites was determined by interviewing farmers (n = 122) in the cold desert region of the Lahaul valley. Farmers prioritized planting of Salix on personal land (rank 1), followed by own wasteland (rank 2), along water channels (rank 3), and erosion-prone land (rank 4). Lower preferences were observed for agroforestry premises (rank 5), government land (rank 6), and community land (rank 7). This hierarchy highlights a strong inclination toward utilizing privately owned or managed land for Salix plantation, with marginal interest in shared or public spaces. The agroforestry premises are exclusively used for agriculture (i.e., cash and traditional crops) and horticulture crops to avoid the competition, and maintain the higher productivity (Rawat et al., 2023).

The prioritization of personal land and own wasteland underscores farmers' preference for sites where they exercise direct control over resource management and benefit from economic returns (e.g., fodder, fuelwood). Role of Salix in stabilizing soil and providing ecosystem services likely drives its adoption on erosion-prone lands (rank 4), aligning with studies emphasizing Salix species' utility in combating desertification in fragile cold deserts (Rawat & Everson, 2013; Rawat et al., 2006). The preference for planting along water channels (rank 3) reflects high water requirement of Salix and strategic use of irrigation infrastructure of farmers to enhance survival rates in arid conditions. Lower rankings for government and community lands (ranks 6–7) suggest institutional and socio-cultural barriers, such as ambiguous land tenure policies, limited incentives for collective action, or historical distrust in shared resource management. The modest ranking of agroforestry premises (rank 5) may indicate competing land-use priorities (e.g., crop production) or a lack of awareness about agro-ecological benefits of *Salix* (e.g., microclimate moderation) (Rawat & Everson, 2013). These findings mirror global trends where land tenure security and immediate economic returns strongly influence agroforestry adoption (Mbow et al., 2014). However, the low interest in communal lands highlights a critical gap in leveraging Salix planting for large-scale ecological restoration in cold deserts (Rawat et al., 2023). Policymakers must address tenure issues and incentivize collaborative planting on erosion-prone or degraded public lands to align individual farmer priorities with broader environmental goals. It has been suggested that promotion of Salix (willows) planting on private/wastelands through subsidies or technical support enhances farmer livelihoods and soil health. Integrate willows into government-led watershed management programs, particularly along water channels and erosion zones in the cold deserts would be appropriate solutions for a long run. Strengthening community-based initiatives to overcome barriers to collective land use would foster ecological resilience in the Lahaul valley.

3.4 Why **Salix** *Is so Significant in Lahual Valley?*

The cultivated *Salix* species like *Salix fragilis* occurred in all the studied villages. The study reveals that north-facing slopes had the greater diversity of *Salix kerelinii* (Khoksar), *S. daphnoides* (Hinsa), and *S. acmophylla* (Jahlma, Kuthar), which are exclusively distributed in the north-facing slopes. Highest *Salix* diversity was in agroforestry systems of Jahlma and Hinsa, reflecting intentional cultivation for soil stabilization, fulfilling day-to-day needs and sustainability. The forestry was dominated by *S. fragilis* and *S. acmophylla* in Jahlma, which are likely managed for erosion control, protection from avalanches and subsistent needs. In forests, unique species like *S. kerelinii* (Khoksar) and *S. denticulata* (Hinsa) highlight natural ecological niches. North-facing slopes in the Himalaya retain cooler, moisture conditions due to reduced solar exposure, favoring shade-tolerant species like *S. kerelinii* and *S. acmophylla* (Rawat et al., 2023; Yang et al., 2020). On the other hand, the south-facing slopes, with higher insolation and aridity, favor drought-resistant, which exhibits plasticity in water-use efficiency (Edwards, 2023). The prevalence of *S. acmophylla* and *S. fragilis* in Jahlma and Hinsa agroforestry reflects traditional practices prioritizing willows for soil stabilization, fuelwood, fodder, and artisan

Table 2 Distribution of *Salix* species (Willows) in the North (N)- and the South (S)-facing slopes in cold desert of the Lahaul valley

Attributes	Khoksar		Jahlma		Hinsa		Kuthar	
Agroforestry	N	S	N	S	N	S	N	S
Salix acmophylla Bioss	–	–	✓	✓	–	–	✓	–
Salix daphnoides Vill	–	–	–	–	✓	–	–	–
Salix fragilis L	–	✓	✓	✓	✓	✓	✓	–
Forestry *Salix acmophylla* Bioss	–	–	✓	✓	–	–	–	–
Salix fragilis L	–	✓	✓	✓	✓	✓	–	–
Forests *Salix acmophylla* Bioss	–	–	✓	–	–	–	–	–
Salix denticulata Andress	–	–	–	–	✓	✓	–	–
Salix fragilis L		–	–	✓	✓	✓	✓	–
Salix kerelinii Turcz	✓	–	–	–	–	–	–	–

"–" represents absent

materials (Rawat & Everson, 2013). Their absence in Khoksar (except *S. fragilis* on south-facing slopes) may stem from higher-altitude constraints (3,200 m) limiting establishment (Rawat et al., 2006). *Salix kerelinii*, restricted to Khoksar's north-facing forests, is likely a cold desert specialist adapted to permafrost-edge conditions. Its limited range underscores vulnerability to climate warming and necessitates targeted conservation. *Salix denticulata* in the forest of Hinsa suggests a niche partitioning in riparian zones, a pattern observed in Himalayan headwaters (Akrim et al., 2023). Scaling *Salix* cultivation in Jahlma and Hinsa could enhance erosion control and livelihood resilience (Rawat et al., 2006). Monitoring *S. kerelinii* populations is critical, as warming may shrink its habitat. This analysis underscores the interplay of microclimate, land use, and species traits in shaping *Salix* distributions, offering actionable insights for cold desert ecosystem management. The distribution of *Salix* species across North-facing (N) and South-facing (S) slopes in agroforestry, forestry, and forest systems of the study sites is depicted in Table 2.

3.5 Floristic Composition and Ecological Suitability of **Salix** *Species and Associated Plants*

The floristic composition of *Salix* species and their associated plant communities across the North-facing (N) and South-facing (S) slopes in agroforestry, forestry, and forest systems of the Lahaul valley is depicted in Table 3. In general, the higher floristic composition occurred at the lower and middle altitudes of the Lahaul valley (Rawat et al., 2023). North-facing slopes are dominated by *Salix acmophylla*, *S. daphnoides*, and *S. fragilis*, co-occurring with broadleaf trees (*Acer*, *Prunus*) and nitrogen-fixers (*Robinia pseudoacacia*) in agroforestry. While, *S. fragilis* persists, but associated flora shifts to drought-tolerant shrubs (*Berberis jaeschkeana*, *Crataegus songarica*) and fruit trees (*Prunus avium*) on south-facing slopes. Rich in fruit/nut trees (*Juglans regia*, *Malus baccata*), reflecting human selection for subsistence in agroforestry. Non-native species (*Robinia pseudoacacia*) suggest intentional introduction for soil enrichment, fuelwood, and fodder purposes (Colling et al., 2025; Syngkli et al., 2025). *S. acmophylla* and *S. fragilis* associate with cold-adapted species (*Betula utilis*, *Hippophae rhamnoides*) in forestry of the north-facing slopes. While, *S. fragilis* pairs with xerophytic shrubs (*Cotoneaster obtusus*, *Juniperus macropoda*) on the south-facing slopes. Managed plantations prioritize erosion control (*Hippophae rhamnoides*) and timber (*Salix fragilis*). In forests, diverse *Salix* assemblage (*S. acmophylla*, *S. denticulata*, *S. fragilis*, *S. kerelinii*) coexists with conifers (*Abies pindrow*, *Picea smithiana*) and understory shrubs (*Rhododendron*, *Rubus*) on the north-facing slopes. While, reduced *Salix* diversity (*S. acmophylla*, *S. fragilis*) alongside drought-resilient species (*Cedrus deodara*, *Jasminum humile*) was found on the south-facing slopes. Natural assemblages include high-altitude specialists (*Pinus wallichiana*, *Viburnum continifolium*) and endemic *Salix kerelinii*. Higher moisture retention supports shade-tolerant, mesic species (e.g., *Abies pindrow*, *Rhododendron companulatum*) on the north-facing slopes. *Salix kerelinii* (endemic to Khoksar's forests) and *S. denticulata* are restricted to the north-facing slopes, indicating cold–humid niche specialization (Akrim et al., 2023). Aridity favors sclerophyllous shrubs (*Berberis*, *Cotoneaster*) and drought-adapted trees (*Celtis australis*, *Fraxinus xanthoxyloides*) on the south-facing slopes. *S. fragilis* dominates due to its plasticity in water-use efficiency.

The effects of microclimatic gradients and a niche partitioning on floristic composition were clearly visible on both the north- and south-facing slopes. The stark contrast between north- and south-facing slopes mirrors global patterns where aspect-driven microclimates dictate plant communities (Bennie et al., 2013). In the Lahaul valley, the north-facing slopes act as refugia for moisture-dependent *Salix* species (*S. kerelinii*) and boreal conifers (Abies), while the south-facing slopes favor xerophytes (*Crataegus*, *Juniperus*). This implies that the north-facing slopes provide a conducive environment for moisture-dependent *Salix* species by offering cooler and wetter conditions compared to other slopes. This niche partitioning minimizes competition and maximizes resource use in extreme environments (Akrim et al., 2023). While the effects of human-mediated floristic shifts can be observed in agroforestry and forestry management systems. The prevalence of *Prunus* and *Malus* alongside *Salix* underscores agro-ecological strategies to balance soil stabilization with *Salix* species and food security in agroforestry. However, non-native *Robinia pseudoacacia* may threaten native diversity through competitive exclusion, as observed in the Himalayan agroforestry systems (Colling et al., 2025; Rawat et al., 2023; Syngkli et al., 2025). In forestry and forests, focus on *Betula utilis* and *Hippophae* reflects utilitarian goals (timber, fodder), but limited *Salix* diversity here suggests prioritization of economic over ecological functions. In terms of conservation perspectives, *Salix kerelinii*, restricted to the north-facing forest slopes, is vulnerable to climate warming. Its survival hinges on preserving cold microhabitats and limiting

Table 3 Floristic composition of *Salix* species and associated plants in the north- and south-facing hill slopes in cold desert of the Lahaul valley

Agroforestry	North-facing slopes	South-facing slopes
Salix acmophylla Bioss., *Salix daphnoides* Vill., *Salix fragilis* L	<2700 m *Acer accuminatum L., Celtisaustralis* L., *Corylusjaquemontii* L., *Euonymousfimbriatus* Wall., *Fraxinus floribanda* Wall., *Malus baccata* (L.) Borkh. *Populus nigra* L., *Prunus cornuta* (Wall. ex Royle) Stued., *Robinia pseudoacacia* L., *Salix fragilis* L., *Fraxinus xanthoxyloides* (G.Don.) DC., *Prunus armeniaca* L., *Pyrus communis* L., *Pyrus malus* L., *Salix acmophylla* Bioss., *Berberis pseudumbellata* Parker, *Cotoneaster obtusus* Wall. ex Lindl., *Euonymus monbeigii* W.W. Sm., *Hippophae rhamnoides* L.ssp. turkestanica Rousi, *Juniperus communis* L.var. paxatilis Pallas, *Jasminum humile* L., *Lonicera hypoleuca* Decne., Parrotopsis jacquemontiana Render, *Prunus jacquemontii* Hook f. Fl. Brit.Ind., *Prunus sp., Prunus sp., Rhamnus prostrata* Jacq. ex Parker, *Ribes glaciale* Wall., *Ribes grossularia* auct. non L., *Rhamnus sp., Rosa webbiana* Wall. ex Royle, *Sorbus lanata* (D.Don) S.Schauer, *Sorbaria tomentosa* (Lindl.) Rehder, *Spiraea canescens* D.Don, *Viburnum continifolium* D. Don 2700–3000 m *Abies pindrow* Royle, *Juglans regia* L., *Malusbaccata* (L.) Borkh. *Populus nigra* L., *Prunus cornuta* (Wall. ex Royle) Stued., *Salix fragilis* L., *Crataegus songarica* C. Koch., *Fraxinus xanthoxyloides* (G.Don.) DC., *Prunus armeniaca* L., *Prunus prostrata* (Labill.), *Salix acmophylla* Bioss., *Berberis pseudumbellata* Parker, *Cotoneaster obtusus* Wall. ex Lindl., *Cotoneaster pruinosus* Klotz., *Hippophae rhamnoides* L.ssp. turkestanica Rousi, *Juniperus communis* L.var. paxatilis Pallas, *Lonicera hypoleuca* Decne., *Lonicera quinquelocularis* Hardwp., *Ribes grossularia* auct. non L., *Ribes orientale* Daesf., *Rosa webbiana* Wall. ex Royle, *Salix daphnoides* Vill., *Syringa emodi* Wall. ex Royle >3000 m –	<2700 m – 2700–3000 m *Juglans regia* L., *Juniperus polycarpos* C. Koch, *Malus baccata* (L.) Borkh., *Populus nigra* L., *Prunus cornuta* (Wall. ex royle) Stued., *Robinia pseudoacacia* L., *Salix fragilis* L., *Crataegus songarica* C. Koch., *Fraxinus xanthooxiloides* (G.Don.) DC., *Prunus armeniaca* L., *Prunus avium* L., *Prunus communis* Huds., *Prunus sp., Pyrus malus* L., *Salix acmophylla* Boiss., *Berberis jaeschkeana* Schneid., *Cotoneaster obtusus* Wall. ex Lindl., *Hippophae rhamnoides* L. ssp. turkestanica Rousi, *Juniperus communis* L. var. saxatilis Pallas, *Lonicera hypoleuca* Decne., *Ribes grossularia* auct.non L., *Rosa webbiana* Wall. ex Royle >3000 m *Populus nigra* L., *Salix fragilis* L., *Berberis jaeschkeana* Schneid. *Rosa webbiana* Wall. ex Royle, *Sorbus aucuparia* L

(continued)

Table 3 (continued)

Agroforestry	North-facing slopes	South-facing slopes
Forestry		
Salix acmophylla Bioss., *Salix fragilis* L	<2700 m – 2700–3000 m *Betula utilis* D.Don, *Juglans regia* L., *Malus baccata* (L.) Borkh., *Populus nigra* L., *Prunus cornuta* (Wall. ex Royle) Stued., *Salix fragilisI* L., *Pyrus mallus* L., *Salix acmophyla* Boiss., *Berberis pseudumbellata* Parker, *Hippophae rhamnoides* L. ssp. turkestanica Rousi, *Lonicera quinquelocularis* Hardwp., *Ribes grassularia* auct. non L., *Ribes orientale* Daesf., *Rosa webbiana* Wall. ex Royle, *Syringa emodi* Wall. ex Royle >3000 m –	<2700 m – 2700–3000 m *Juniperus polycarpos* C. Koch, *Malus baccata* (L.) Borkh., *Populus nigra* L., *Robinia peudoacacia* L., *Salix fragilis* L., *Fraxinus xanthoxyloides* (G.Don.) DC., *Prunus communis* Huds., *Pyrus malus* L., *Salix acmophylla* Boiss., *Berberis jaeschkeana* Schneid., *Hippophae rhamnoides* L. ssp. turkestanica Rousi, *Lonicera hypoleuca* Decne, *Ribes grossularia* auct.non L., *Rosa webbiana* Wall. ex Royle >3000 m *Salix fragilis* L., *Berberis jaeschkeana* Schneid., *Cotoneaster obtusus* Wall. ex Lindl., *Rosa webbiana* Wall. ex Royle

(continued)

Table 3 (continued)

Agroforestry	North-facing slopes	South-facing slopes
Forests		
Salix acmophylla (Bioss.), *Salix denticulata* Andress., *Salix fragilis* L., *Salix kerelinii* Turcz	<2700 m *Abies pindrow* Royle, *Betula utilis* D. Don, *Cedrus deodara* (Roxb. ex D.Don) G. Don, *Picea smithiana* (Wall.) Boiss., *Pinus wallichiana* A.B. Jackson, *Berberis pseudumbellata* Parker, *Cotoneaster pruinsesus* Kotz., *Indigofera hilerentha* Wall. ex Brand., *Juniperus communis* L.var. saxatilis Pallas, *Lonicera myrtillus* Hk.f. & Th., *Ribes grossularia* L., *Rosawebbiana* Wall. ex Royle, *Sorbaria tomentosa* (Lindl.) Rehder, *Spiraea canescens* D.Don, *Viburnum continifolium* D.Don 2700–3000 m *Abies pindrow* Royle, *Betula utilis* D. Don, *Picea smithiana* (Wall.) Boiss., *Pinus wallichiana* A.B. Jackson, *Salix fragilis* L., *Crataegus songarica* C. Koch. *Salix acmophylla* Boiss., *Berberis pseudumbellata* Parker, *Cotoneaster obtusus* Wall. ex Lindl., *Hippophae rhamnoides* L.ssp. turkestanica Rousi, *Juniperus communis* L.var. saxatilis Pallas, *Lonicera govaniana* Wall. ex DC., *Lonicera myrtillus* Hk.f. & Th., *Ribes grossularia* L., *Ribes orientale* Daesf., *Rosa webbiana* Wall. ex Royle, *Salix denticulata* Anderss., *Rubus irritants* Focke, *Viburnum continifolium* D.Don >3000 m *Betula utilis* D. Don, *Berberis jaeschkeana* Schneid., *Cotoneaster obtusus* Wall. ex Lindl., *Juniperus communis* L.var. saxatilis Pallas, *Lonicera quinquelocularis* Hardwp., *Rhododendron companulatum* L., *Rosawebbiana* Wall. ex Royle, *Salix karelinii* Turcz	<2700 m *Cedrus deodara* Roxb. ex D. Don, *Celtis australis* L., *Euonymous fimbriatus* Wall., *Fraxinus floribanda* Wall ex Roxb., *Populus nigra* L., *Robinia pseudoacacia* L., *Salix fragilis* L., *Crataegus songarica* C. Koch., *Berberis pseudumbellata* Parker, *Cotoneaster pruinosus* Kotz., *Euonymus monbeigii* W.W.Sm., *Indigofera hilerentha* Wall. ex Brand, *Jasminum humile* L., *Juniperus communis* L.var. saxatilis Pallas, *Lonicera hypoleuca* Decne, *Parrotopsis jaquemontiana* Render, *Rhamnus prostrata* Jacq. ex Parker, *Rosa webbiana* Wall. ex Royle, *Salix spp., Sorbus lanata* (D.Don) S.Schauer, *Spiraea canescens* D.Don 2700–3000 m *Juglans regia* L., *Juniperus polycarpos* C. Koch, *Malus baccata* (L.) Borkh., *Pinus wallichiana* A.B. Jackson, *Populus nigra* L., *Salix fragilis* L., *Fraxinus xanthoxyloides* (D. Don.)DC., *Berberis seudumbellata* Parker, *Cotoneaster obtusus* Wall. ex Lindl., *Hippophae rhamnoides* L.ssp. turkestanica Rousi, *Juniperus communis* L.var. saxatilis Pallas, *Lonicera hypoleuca* Decne., *Ribes grossularia* L., *Ribes orientale* Desf., *Rosa webbiana* Wall. ex Royle, *Salix spp.,* >3000 m *Berberis jaeschkeana* Schneid., *Cotoneaster microphylla* Wall. ex Lindl., *Juniperus indica* Bertol., *Juniperus communis* L.var. saxatilis Pallas, *Rosa webbiana* Wall. ex Royle, …

"–" indicates absent

anthropogenic disturbance. Invasive species *Robinia pseudoacacia* in agroforestry risks altering soil nitrogen cycles and displacing native flora (Colling et al., 2025; Masoodi et al., 2021; Syngkli et al., 2025). While, in the perspective of climate adaptation strategies, prioritize protection of *Abies* and *Rhododendron* to maintain moisture regimes critical for *Salix* survival on the north-facing slopes. While on the south-facing slopes need to promote *S. fragilis* and drought-adapted shrubs to combat desertification (Rawat & Everson, 2012). This study, therefore, highlights the interplay of geophysical gradients, human activity, and species traits in shaping the unique floristic composition of the Lahaul valley, offering actionable insights for sustainable management in the cold desert ecosystems.

Apart from Salix, seabuckthorn and poplar species highlight their ecological adaptability to cold deserts while planting and managing the species like Juniperus, Betula, Abies. Salix, and seabuckthorn are native pioneers known for erosion control, drought tolerance, fuelwood, fodder, and soil enrichment (Rawat et al., 2023), making them ideal for stabilizing slopes and water channels. Poplar, while fast-growing, may face survival challenges in arid conditions unless irrigated, potentially explaining its limited mention in later years. Budgetary cycles, staffing limitations, and harsh winters (which reduce seedling survival rates) could disrupt continuity, which is the main institutional challenge in the cold deserts. The slow decrease in participatory forestry (e.g., Joint Forest Management), which is usually followed by implementation delays, suggests a potential shift away from top-down departmental programs and toward community-led initiatives (Agrawal & Chhatre, 2006). Funding needs to ensure long-term restoration, particularly for erosion-prone zones in the Lahaul valley. Native species prioritization is needed to expand Salix, poplar, and seabuckthorn planting along with junipers and birch, given their ecological resilience, while there is a need to reassess suitability of other ecologically suitable plant species in the cold desert environment.

3.6 Causes, Effects, Environmental Challenges, and Remedial Measures

The harsh climatic conditions, avalanches/snowfall, soil erosion, irrigation water shortages, and eco-impacts from inorganic inputs affect the environmental conditions of the Lahaul valley (Table 4). Soil erosion and avalanches are commonly linked to the undulating topography, fragile geology, and extreme weather conditions of the region. This suggests that altitude-dependent vulnerabilities, where steeper slopes and prolonged freezing at higher elevations exacerbate erosion and hydrological disruptions. It was observed that concrete-lined *kuhls* (traditional irrigation channels) reduce seepage losses but may disrupt local hydrology if poorly designed. Participatory planning is critical to balance efficiency with ecological integrity in the region (Vishvakarma et al., 2005).

Table 4 Cause, effects, environmental challenges, village-specific prevalence, and remedial measures in the cold desert region of the Lahaul valley

Environmental problems	Villages				Remedial measures
	Kuthar	Hinsa	Khoksar	Jahlma	
1. Harsh climatic condition	√	√	√	√	• Resolving food security issue through establishing balance between traditional and cash crops • Scientific land-use planning and diversification • Widespread plantation of locally adaptable trees (e.g., Salix, Poplar, Seabuckthorn, and Juniper)
2. Frequent avalanches and snowfall	√	√	√	√	• Construction of check walls and snow fences • Gabion walls and contour trenches • Plantation and protection of downslopes with bushes and grasses
3. Slope destabilization, stoniness	-	√	√	√	• Construction of retreating walls with iron mesh to stabilize terraces • Use of gravels, pebbles in making terraces while removing from agricultural fields
4. Soil erosion	√	√	√	√	• Minimizing deforestation, plantation of bushes such as Salix, Poplar, Junipers, and Seabuckthorn in landslides prone areas • Prohibition of cultivation in the sloping areas >30° • Plantation of the bushes and grasses
5. Declining soil nutrients	√	√	√	√	• Increase vegetation cover in forests • Promotion of organic fertilizer • Reducing topsoil erosion using indigenous techniques through runoff • Mulching and appropriate agriculture techniques

(continued)

Table 4 (continued)

Environmental problems	Villages				Remedial measures
	Kuthar	Hinsa	Khoksar	Jahlma	
6. Drying up of natural springs	-	√	√	√	• Plantation of locally adaptable trees like Salix, Poplar, Juniper, and Seabuckthorn • Increase in vegetation cover • Trenches and contours
7. Irrigation—shortage of water through seepage and frequent damage to earthen irrigation channels	√	√	√	√	• Concreting *kuhls* to minimize water loss through seepage • Use of plastic pipes • Plantation of vegetation along the *kuhls* and around the villages
8. Other eco-impacts—excessive use of inorganic fertilizer and pesticides	√	√	√	√	• Applying proper and exact doses of fertilizer • Promoting organic farming • Use of appropriate agricultural techniques • Timely and regular irrigation

"–" indicates absent, "√" indicates present

Vegetation-based climate change mitigation is the plantation of cold hardy species like willow, poplar, and seabuckthorn; this needs to be prioritized for slope stabilization, spring revival, and erosion control (Doležal et al., 2022). Seabuckthorn and *Salix* plantations are proven for slope stabilization in cold deserts (Rawat & Everson, 2013; Rawat et al., 2006; Sharma et al., 2019), but their success hinges on community adherence and climate resilience. For instance, the deep root systems of seabuckthorn mitigate landslides, while water-retention capacity of *Salix* aids spring recharge in the area. In agricultural crops, organic fertilizers, precise agro-chemical use, and cultivation below 30° slopes aim to restore soil health and reduce runoff. The decline in soil nutrients underscores overcultivation and reliance on inorganic fertilizers; a trend exacerbated by market-driven shifts to cash crops (e.g., potatoes, peas). Promoting organic amendments and agroforestry could restore soil organic carbon, as demonstrated in cold deserts like Ladakh (Behera et al., 2014; Doležal et al., 2022).

Avalanches and spring drying could be exacerbated by rising temperatures and unpredictable precipitation, suggesting feasible but permanent solutions—like check walls—inadequate. Adaptive strategies must integrate real-time monitoring and flexible land-use zoning, as seen in Swiss Alpine regions (Stöckli et al., 2011). The success of measures like terracing with gravel or organic farming depends on labor availability and incentives for farmers. For example, relatively smaller household base ($n = 17$) at Kuthar may struggle with labor-intensive interventions compared to Hinsa (n = 52). Therefore, there is a need to prioritize community-led adaptive management to align measures according to local needs and ecological significance. In addition, integration of gender-sensitive policies to ensure equitable participation in land restoration in the cold deserts is a must under a sustainable approach.

4 Conclusions

Community-led FLR is a critical approach to sustainable land management, integrating agricultural productivity with environmental conservation. This plays a crucial role in enhancing ecosystem resilience, biodiversity conservation, and sustainable livelihoods. By integrating traditional ecological knowledge with a scientific approach, FLR could support the local communities by way of successfully rehabilitating agroforestry, forestry, and forest landscapes despite extreme climatic conditions. FLR enhances biodiversity, improves soil health, provides subsistence needs (e.g., fuelwood, fodder), carbon sequestration, and contributes to climate resilience and environmental sustainability. Community-led initiatives have demonstrated significant potential in restoring degraded landscapes and improving ecosystem services. The success of such efforts depends on continuous policy support, financial investment, and capacity-building programs. In addition, it needs training on sustainable forestry practices, agroforestry techniques, and climate-smart agriculture for FLR and ultimately the overall socio-economic development of the local communities. Strengthening participatory governance and fostering multi-stakeholder collaboration are the necessary ways for scaling up such restoration efforts for resilient ecosystems and sustainable landscapes in the cold deserts of the Himalayan region in particular and World Mountains in general.

Acknowledgements Author thanks the G. B. Pant National Institute of Himalayan Environment (NIHE), Kosi-Katarmal, Almora, Uttarakhand, for providing necessary facilities and the Federal Technical and Vocational Training Institute (TVTI), Addis Ababa, Ethiopia while finalizing the manuscript. The author also acknowledges the Council of Scientific and Industrial Research (CSIR), Govt. of India, New Delhi, for awarding the Senior Research Fellowship. The help and support provided during the study by the various sources and people are also gratefully acknowledged.

References

Agrawal, A., & Chhatre, A. (2006). Explaining success on the commons: Community forest governance in the Indian Himalaya. *World Development, 34*(1), 149–166. https://doi.org/10.1016/j.worlddev.2005.07.013

Akrim, F., Mahmood, T., Belant, J. L., Nadeem, M. S., Qasim, S., Dhendup, T., Fatima, H., Bukhari, S. A., Aslam, A., Younis, H., Rafique, A., Subhani, Z. A., Hashmi, S. A., & Munawar, N. (2023). Niche partitioning by sympatric civets in the Himalayan foothills of Pakistan. *PeerJ, 11*, Article e14741. https://doi.org/10.7717/peerj.14741

Aswal, B. S., & Mehrotra, B. N. (1994). *Flora of Lahual-Spiti. A Cold Desert in North-West Himalaya* (pp. 10–15). Dehradun, India: Bishan Singh Mahendra Pal Sigh.

Anonymous. (1993). *Action plan on cold desert: An integrated approach for sustainable development* (pp. 1–40). Solan, India: UHF.

Behera, M. D., Matin, S., & Roy, P. S. (2014). Biodiversity of Kargil cold desert in the Ladakh Himalaya. In S. I. Nakano, T. Yahara & T. Nakashizuka (Eds.), *Integrative observations and assessments. Ecological research monographs* (pp. 253–274). https://doi.org/10.1007/978-4-431-54783-9

Bennie, J., Hodgson, J. A., Lawson, C. R., Holloway, C. T. R., Roy, D. B., Brereton, T., Thomas, C. D., & Wilson, R. J. (2013). *Range expansion through fragmented landscapes under a variable climate.* https://doi.org/10.1111/ele.12129

Colling, G., Glaser, M., Buholzer, S., Bürger, J., Milan Chytrý, M., et al. (2025). Ninety years of alien plant species accumulation across regional and local scales in central European fields. *Agriculture, Ecosystems and Environment, 383*, Article 109483. https://doi.org/10.1016/j.agee.2025.109483

Doležal, J., Altman, J., Jandová, V., Milan Chytrý, M., et al. (2022). Climate warming and extended droughts drive establishment and growth dynamics in temperate grassland plants. *Agricultural and Forest Meteorology, 15*, Article 108762. https://doi.org/10.1016/j.agrformet.2021.108762

Edwards, P. E. (2023). Ecological traits of Saliceae and the species replacing them on the active floodplain. *River Research and Applications.* https://doi.org/10.1002/rra.4229

G.B. Pant Institute of Himalayan Environment and Development. (1995–1996). *Growth problems of tree species in the Lahaul valley in highland of North-West Himalaya. Annual Report (1995–1996).* Retrieved September 08, 2025. https://www.gbpihed.gov.in/PDF/Annual%20Reports/Annual_Report_2018_19_English.pdf

Kala, C. P. (2002). *Medicinal plants of Indian Trans-Himalaya.* Dehradun, India: Bishen Singh Mahendra Pal Singh.

Kuniyal, J. C., Vishvakarma, S. C. R., & Singh, G. S. (2004). Changing crop biodiversity and resource use efficiency of traditional versus introduced crops in cold desert of North-Western Indian Himalaya: A case of Lahaul valley. *Biodiversity and Conservation, 13*, 1271–1304.

Mbow, C., van Noordwijk, M., Luedeling, E., Neufeldt, H., Minang, P. A., & Godwin Kowero, G. (2014). Agroforestry solutions to address food security and climate change challenges in Africa. *Current Opinion in Environmental Sustainability, 6*, 61–67. https://doi.org/10.1016/j.cosust.2013.10.014

Masoodi, T. H., Shah, I. A., Dar, M. U. D., Sofi, P. A., & Mugloo, J. A. (2021). Evaluation of probability distribution functions applied to tree diameter in a mixed uneven Kiker (*Robinia pseudoacacia*) stand of Kashmir Himalaya, India. *Current Science, 121*(4), 573–577. https://doi.org/10.18520/cs/v121/i4/573-577

Negi, V. S., Joshi, B. C., Pathak, R., Rawal, R. S., & Sekar, K. C. (2018). Assessment of fuelwood diversity and consumption patterns in cold desert part of Indian Himalaya: Implication for conservation and quality of life. *Journal of Cleaner Production, 196*, 23–31.

Oinam, S. S., Rawat, Y. S., Khwairakpam, G. S., Khoiyangbam, R. S., Kuniyal, J. C., & Vishvakarma, S. C. R. (2005). Land use and land cover changes in Jahlma watershed of the Lahaul valley, cold desert region of the northwestern Himalaya, India. *Journal of Mountain Science, 2*(2), 129–136.

Rawat, Y. S., Oinam, S. S., Vishvakarma, S. C. R., Kuniyal, C. P., & Kuniyal, J. C. (2006). Willow (*Salix fragilis* Linn.): A multipurpose tree species under pest attack in the cold desert of Lahaul valley, North-Western Himalaya, India. *Ambio, 35*, 43–48.

Rawat, Y. S., Vishvakarma, S. C. R., & Todaria, N. P. (2009). Fuelwood consumption pattern of tribal communities in cold desert of the Lahaul valley, North-Western Himalaya, India. *Biomass and Bioenergy, 33*, 1547–1557.

Rawat, Y. S., & Vishvakarma, S. C. R. (2011). Ecological analysis of indigenous agroforestry systems in Kullu and Lahaul valleys, Himachal Pradesh, India. *Proceedings of National Academy of Science, India (Sect. B), 81*, 160–170.

Rawat, Y. S., & Everson, C. S. (2012). Ecological status and uses of juniper species in the cold desert environment of the Lahaul Valley, North-Western Himalaya, India. *Journal of Mountain Science, 9*, 676–686.

Rawat, Y. S., & Everson, C. S. (2013). Availability and use of willow species in representative cold desert areas of northwestern Himalaya, India. *Journal of Mountain Science, 10*(3), 472–481.

Rawat, Y. S., Negi, V. S., Moussa, I. M., Zaman, W., & Elansary, H. O. (2023). Diversity, distribution and vegetation assessment of woody plant species in the cold desert environment, North-Western Himalaya, India. *Sustainability, 15*, 10429. https://doi.org/10.3390/su151310429

Sharma, A., Singh, V., Sharma, A., & Negi, N. (2019). Sea buckthorn a new approach in ecological restoration of Himalayan Ecosystem: A review. *International Journal of Chemical Studies, 7*(1), 1219–1226.

Saxena, K. G., Liang, L., & Xue, X. (2011). *Global change, biodiversity and livelihood in cold desert region of Asia* (vol. 322). Dehradun, India: Bishen Singh Mahendra Pal Singh. ISBN 978-81-211-0780-8.

Sekar, K. C., & Srivastava, S. K. (2010). A supplement to the flora of Lahaul-Spiti. *Journal of Non-Timber Forest Products, 17*, 233–258.

Sekar, K. C., Pandey, A., Joshi, B. C., Bhatt, I. D., Bhojak, P., Dey, D., Thapliyal, N., Bisht, K., Bisht, M., Negi, V. S., et al. (2022). Floristic diversity in cold desert regions of Uttarakhand Himalaya, India. *Phytotaxa*, 537.

Srivastava, S. K., & Singh, D. K. (2005). *Glimpses of plant wealth of Uttaranchal*. Dehradun, India: Bishen Singh Mahendra Pal Singh.

Srivastava, S. K. (2010). Floristic diversity and conservation strategies in cold desert of western Himalaya, India. *Botanica Orientalis Journal of Plant Science, 7*, 18–25.

Shukla, A. N., & Srivastava, S. K. (2020). Flora of Ladakh: An annotated inventory of flowering plants. *Biodiversity of the Himalaya: Jammu and Kashmir State* (pp. 673–730). Springer.

Srivastava, S. K., & Shukla, A. N. (2015). *Flora of cold desert-Western Himalaya, India* (vol. 2). Kolkata, India: Botanical Survey of India.

Stöckli, V., Wipf, S., Nilsson, C., & Rixen, C. (2011). Using historical plant surveys to track biodiversity on mountain summits. *Plant Ecology & Diversity, 4*, 415–425.

Syngkli, R. B., & Rai, P. K. (2025). Expanding horizon of invasive alien plants under the interacting effects of global climate change: Multifaceted impacts and management prospects. *Climate Change Ecology, 9*, 100092. https://doi.org/10.1016/j.ecochg.2025.100092

Vishvakarma, S. C. R., Rawat, Y. S., Oinam, S. S., & Kuniyal, J. C. (2005). Irrigation (*Kuhls*): A community management system in cold desert of the Lahaul valley, northwestern Himalaya, India. In P. Pradhan & U. Gautam (Eds.), *Proceedings: Third International Seminar on Farmer Managed Irrigation Systems and Governance Alternatives* (pp. 425–444). Kathmandu, Nepal.

Yang, J., El-Kassaby, Y. A., & Guan, W. (2020). The effect of slope aspect on vegetation attributes in a mountainous dry valley, Southwest China. *Scientific Reports, 10*, 16465. https://doi.org/10.1038/s41598-020-73496-0

Agroforestry for Land Management, Climate Change Mitigation, and Adaptation in Cold Desert Ecosystems

Rushal Dogra, Praveen Kumar, Poonam, and Eepsa

1 Introduction

The global population has reached 8.14 billion, flourishing across diverse landscapes for its survival and sustenance (World Bank, 2024). With growing population, the need for space, food and shelter is also escalating alarmingly on the earth (Tal, 2025). This in turn is affecting a wide array of ecosystems in the world, due to which there is a growing tendency to protect these ecosystems sustainably. Among the most dominant ecosystems in the world, cold desert ecosystem is known for its unique biodiversity, population dynamics, cultural significance, and sustaining the most vulnerable floral and faunal communities of the world (Sottosanti, 2023; Tiwari & Kapoor, 2016; UNESCO World Heritage Centre, 2015). These ecosystems are sensitive to the global changes such as Greenhouse Gases (GHGs), global warming, and temperature increase.

Cold desert ecosystem receives scanty rainfall that is primarily in the form of snowfall (Aswal & Mehrotra, 1994; Behera et al., 2014; Rawat et al., 2023; Tiwari & Kapoor, 2016). This ecosystem generally thrives in the temperate to alpine climate

R. Dogra
Pandit Deendayal Upadhyay College of Horticulture and Forestry Piprakothi, Dr. Rajendra Prasad Central Agricultural University, Pusa, Samastipur, Bihar, India

P. Kumar (✉)
Department of Silviculture and Agroforestry, Dr. Yashwant Singh Parmar University of Horticulture and Forestry, Nauni, Solan, HP, India
e-mail: pkumaruhf@gmail.com

Poonam
Faculty of Applied Sciences and Biotechnology, Shoolini University, Solan, Himachal Pradesh, India

Eepsa
Department of Environment Science, Dr. Yashwant Singh Parmar University of Horticulture and Forestry, Nauni, Solan, HP, India

Y. S. Rawat and C. P. Kala (eds.), *Cold Deserts*,
https://doi.org/10.1007/978-3-032-29119-6_10

with vegetation generally adapted to withstand tough condition such as dry and cold weather along with nutrient deficient conditions. This ecosystem experiences a severe temperature fluctuation worldwide, with annual temperature ranging from −20° or lower in the winter to around 40 °C in the summer season. In India, cold desert is located in the Ladakh (Leh and Kargil districts), Himachal Pradesh (Lahaul-Spiti, North of Kinnaur and some parts of Chamba district), Uttarakhand, and Northern side of Sikkim (Tiwari & Kapoor, 2016). The high altitude of these region is an important factor adding more uniqueness to this ecosystem in term of aesthetics, harsh weather, and diversity of landscapes (Tiwari & Kapoor, 2016). A part from climate change impacts, the region is vulnerable to adverse events such as avalanches, strong winds, and deep frosts. The water supply is limited and that mainly from the glacier fed streams (Sottosanti, 2023).

Throughout the world, cold desert ecosystems are found in regions experiencing temperature fluctuations. These are extensively located in parts of western North America, South-eastern part of South America, Parts of central Asia, Arctic, and Antarctic region. The most well-known regions for cold deserts are Gobi desert, Pentagonia, and high altitude regions of Himalayas (Rawat et al., 2023; Sottosanti, 2023; Tiwari & Kapoor, 2016; UNESCO World Heritage Centre, 2015). These landscapes are highly untouched and unexplored throughout the world due to natural constraints and checks. The diversity is unique but scare in terms of both floral and faunal populations. Shrubs and grasses are the most dominant vegetation that primarily grow in some part of the year. Trees and shrubs are generally scarce in this ecosystem although occurred naturally in the form of patches. The communities thriving in these ecosystems (showing extremes of weather parameters in terms of temperature on lower ends along with scarce resources) are generally small and dispersed. The main source of livelihood in these region is pastoralism, limited agriculture, nomadism, and/or partially nomadic lifestyles throughout the year (Tiwari & Kapoor, 2016). Communities living in cold deserts keep herds of sheep, goats, and yak being well suited for the climate and edaphic conditions (Tiwari & Kapoor, 2016). The communities living in these regions exhibit a remarkable resilience to harsh conditions, limited resources, tough terrain, remoteness from the main population, and sub-zero temperature in most part of the year. In the Himalayas, this region is also home to genetically least diverse big cat, i.e., snow leopard (often called as the ghost of the mountains). In cold deserts of the Himalayas, there is a close association and/or interaction between wildlife and human population shaping the ecosystem in a complimentary manner (Moorthy, 2025), e.g., Kibber a high altitude village in Himalayas has led extensive efforts for human-wildlife conflict mitigation (i.e., livestock insurance program, awareness campaigns and wildlife tourism) (Moorthy, 2025; UNESCO World Heritage Centre, 2015). These efforts have protected an important keystone species along with fostering a sustainable economy to the local community.

2 Overview of the Global Challenges

Globally, ecosystems are facing the challenges of shrinking land area, owing to multiple factors such as climate change, greenhouse gas emission, population increase, agricultural expansion, and damage due to development (Chen et al., 2024).

(a) **Climate Change and Land Degradation**

Climate (an average of weather condition for extended period of time) is an integral factor for ecosystem functioning. Climate decides the nature of ecosystem in terms of temperature, precipitation, and the diversity associated (Weiskopf et al., 2020). Climate change is defined as change in the state of climate along an extended period of time, i.e., 10 years or more (Climate Change, 2025). Climate change is a fundamental threat to human health and nature. As per WHO (2023), around 3.6 billion people of the world are highly susceptible to climate change, and due to the risk of climate change on human health, total expenditure is estimated around 2–5 billion US $ annually up to 2030.

Land degradation is a change in the productivity, physical characteristics, and biological attributes leading to negative effects on its capability (FAO, 1999; UNDRR & ISC, 2025). Land resource containing water, minerals, and living organism act as a self-sustaining unit by interacting with each other (Das et al., 2022). Land as a resource is fundamental for laying the foundation for life sustenance (Das et al., 2022; Qi et al., 2025). However, throughput the world, there has been an increasing issue of depreciation in factors of land crucial for life sustenance (Chen et al., 2024). The effect is mainly escalated with human activities such as construction, mining, the use of land above its carrying capacity, and a prolonged loss of vegetation and wildlife. Further increasing water footprint of the agricultural activities is an add up in the issue of land degradation by lowering down the water availability of soil, leading to the water stress and scarcity in a long run (Dogra et al., 2025a). The effects are more pronounced along the altitude due to ecological factors, fragile ecosystem, desertification, low temperature, and resource scarcity. As in high altitude regions, the degradation of land is escalated by the limited growing season, less diversity, and prolonged winters (Rawat et al., 2023; Sottosanti, 2023; Tiwari & Kapoor, 2016).

(b) **The Need for Integrated Solutions for Future Generations**

There is a need for integrated solution for future generation (Brandt et al., 2024; Castle et al., 2022). This can be understood with the lowering down of agricultural productivity, increased habitat degradation, climate change, instances of human-wildlife conflict, and increasing population of the world. The most important concern in today's world is food security, shelter, fodder, and fuelwood in the high altitude cold desert regions. This has to be addressed in an integrated manner by combining different components of farms (such as multipurpose tree species, livestock, agricultural crops, horticultural components) altogether. In this regards, farming systems such as agroforestry could fulfill the substantial demands and mitigate the adverse effects associated on the bio-resources. There is a need for integration of trees, crops, and/or animals altogether for meeting the demands of world and mitigation of adverse effects (FAO, 1999). The tree-based agroforestry systems are a viable solution in the modern world for meeting the ever-increasing demands of growing population (animal as well as livestock) due to its ability to provide divers outputs from same unit of land. In environmental terms, agroforestry systems tend to lower down the soil erosion (Jaros et al., 2025; Song et al., 2019), act as a biomass reservoir, sequester carbon, recharge ground water, and improves the biodiversity. Tree-based agroforestry systems act as a support to the farmers against crop failures, and adverse climatic conditions (England et al., 2020; Gupta et al., 2024; Pardon et al., 2017).

People often depend on tree-based agroforestry systems for the sustenance of livelihood and fighting against poverty. Tree-based agroforestry systems act as a reliable source for food, fodder and fuelwood (fulfilling SDG 1 i.e., fighting against poverty). Further they help in water footprint minimization (Dogra et al., 2025a). Tree-based agroforestry systems by providing food security, affordable clean energy, better air quality, shade, microclimate regulation, and reduce heat stress in a comprehensive way meets SDG 1, 2, 3, 7, 8, 12, 13, and 15 (Dogra et al., 2025b; Gupta et al., 2024).

2.1 The Role of Tree-Based Agroforestry Systems in Addressing Challenges

Cold desert ecosystems are known for their tough terrain, low temperature, and limited vegetation (Rawat et al., 2023; Tiwari and Kapoor, 2016). As a result, the livelihood sustenance is quite challenging due to xeric soil, extreme climatic conditions, and limited bio-resources (Sharma et al., 2024). So, to provide a sustainable resource base for these fragile ecosystems, we require systems that offer multiple outputs throughout the year. Agroforestry is a viable solution for affording such multiple outputs round the year (tangible and intangible) under different climatic and edaphic conditions (Dogra et al., 2025b; England et al., 2020; Gupta et al., 2024; Pardon et al., 2017).

(a) **Introduction to Agroforestry as a Viable Solution**

Agroforestry is a well-known practice aimed for food security and livelihood sustenance (England et al., 2020; Gupta et al., 2024; Pardon et al., 2017). Agroforestry, an integration of trees with agricultural crops or livestock, has emerged as a sustainable land use practice offering significant environmental and economic benefits (Sharma et al., 2023a; Singh et al., 2024). In the context of a growing emphasis on the green growth development, agroforestry systems provide livelihoods, improve soil health, and mitigate climate-related risks. It allowed farmers to harvest a variety of products, including timber, fruits, and non-timber forest goods. On the other hand, agroforestry reduces dependence on single crops, thus stabilizing income and enhancing resilience to market fluctuations and environmental challenges. Additionally, agroforestry improves soil fertility through organic matter, reducing the need for chemical fertilizers, which in turn cuts costs and boosts crop productivity (Bhushan et al., 2025; Dogra et al., 2025b). The incorporation of trees into land use systems also contributes to water retention, erosion control, and acts as windbreaks, further protecting crops and promoting long-term yield stability. Agroforestry presents opportunities for carbon sequestration, enabling farmers to engage in carbon trading and access payments for ecosystem services, making the practice economically more attractive (Gupta et al., 2024; Pardon et al., 2017). Ultimately, this chapter positions agroforestry as a promising strategy that supports both economic profitability and environmental sustainability, while contributing to global goals such as poverty reduction and climate change adaptation.

(b) **Multidisciplinary Approach of Agroforestry System**

Agroforestry system adopts a multidisciplinary approach by combining principles and practices that form diverse subjects such as agronomy, forestry, soil science, animal husbandry, ecology, ergonomics, economics, and social sciences for development as productive and sustainable land use systems (Nair, 1993; Panthera et al., 2021). Forestry and agronomy govern the selection of tree, crop integration in order to optimize the output in harmony with the environment (Fig. 1). Under the similar aspect in cold desert region, the multidisciplinary approach of agroforestry adapts to address harsh climatic condition such as low temperature, limited precipitation, strong winds (planning for shelterbelts and windbreaks), and short growing season. This practice contributes to the judicious selection of hardy trees and shrubs such as *Salix* spp., *Hipophae* spp., and Junipers along with suitable crop such as Peas, Potatoes, and *Humulus lapulus, Fagopyrum* spp., *Hordeum himalayense, Saussurea costus*, and *Inula recemosa* (Aswal & Mehrotra, 1994; Behera et al., 2014; Rawat et al., 2023; Tiwari & Kapoor, 2016). However, soil science contributes to maintaining the fertility of soil, nutrient regime, biodiversity, fauna, and flora. Animal science plays a crucial role in governing the optimum interaction of livestock with farm under silvipature and other pasture integrating practices (Nair, 1993). Aspects such as focus on biodiversity conservation, microclimate, carbon sequestration, and climate change are handled under ecology (Khambalkar et al., 2025). One of the crucial and corner stone among these practices is finding the feasibility and viability of any farm practice. Further economics is integral for calculation of cost, benefits, and planning of any long-term venture. Overall, multidisciplinary approach of agroforestry ensures that this practice is economically viable, socially acceptable, and stable for the environment in cold desert ecosystems (Bhushan et al., 2025; Dogra et al., 2025b; Nair, 1993; Panthera et al., 2021; Sharma et al., 2024).

3 Definition and History of Agroforestry

Although agroforestry is practiced traditionally from centuries, but agroforestry got its global recognition as a scientific discipline in 1970 from International Centre for Research in Agroforestry (ICRAF) to address the issues such as degradation of

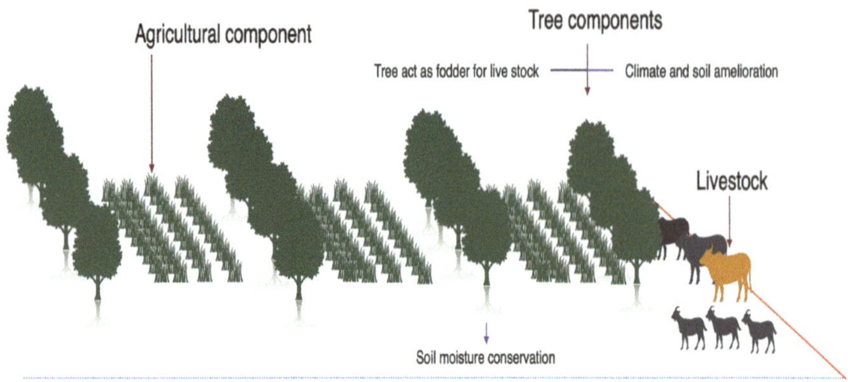

Fig. 1 Brief and simple concept of agroforestry system

land, deforestation, and food insecurity. As per Nair (1993), the most frequently used terms are agrisilviculture system, silvopastoral system, and agrosilvopastoral systems, under which the most frequently practiced are home garden, shelterbelts, windbreaks, and alley cropping. But the most dominant practice in India is retaining trees on bunds of the farms and along the terrace, in order to optimize that space and productivity.

3.1 Global Adoption of Agroforestry

(a) **Trends and Scenarios**

The global trend in agroforestry shows that it is increasingly recognized as a key strategy for sustainable agriculture, climate resilience, biodiversity conservation, and rural development (Singh et al., 2023, 2024). Adoption of agroforestry is growing across regions such as African continent, Asia, Latin America, and parts of Europe: mainly driven by agroforestry's ability to enhance soil health, improve water retention, boost ecosystem services, ecosystem functions, and provide diversified income stream for farmers (Terasaki et al., 2023; Tomar et al., 2025). The integration of modern technologies such as drones, satellite, remote sensing, data analytics, and mapping enables monitoring of tree and crop health, soil moisture, and carbon stocks helps farmer optimize management and meet environmental compliance while tapping into carbon markets (Kala, 2025; Montes et al., 2020). Further the innovations such as blockchain traceability strengthen supply chain for timber and non-timber products, enhancing value addition. Partnership like major first initiatives by global companies and governments supports large-scale tree planting and farmer training programs. Global events such as sixth World Congress on agroforestry bring together research and practitioner to share the latest research policy insight and best practices reflecting a broader commitment to scaling property for people planet and profitability in long run. Additionally, National agroforestry policy (2014) support India in easing out tree harvest from farm to dedicated markets by dealing with legal framework to promote agroforestry expansion, therefore contributing to climate change mitigation and fulfilling Sustainable Development Goals (Stopfer et al., 2023). The policy focuses on improving livelihood security, food, and fodder availability on farms. The policy is important due to its focus on tree outside forest to minimize the pressure for output from forest lands alongside focusing on strengthening the supply chain and market for tree-based and/or forest-based outputs (Dogra et al., 2023; National Agroforestry Policy, 2014).

4 Agroforestry and Climate Change Mitigation in Cold Deserts

Agroforestry in modern world is portrayed as a viable practice for livelihood sustenance and poverty alleviation owing to multiple output associated (Fig. 2). The term itself signifies the integration of tree and agriculture all together for efficient utilization of scare resource and better output. Agroforestry yields multiple benefits under the category of production, protection, amenity, and recreations. However, under the scenario of limited resources, we need to increase the output efficiency of cold desert agroforestry systems. As today, world is focusing on green and environmentally friendly food and energy production practices (Brandt et al., 2024; Castle et al., 2022; England et al., 2020; Nicholas & Scown, 2024; Terasaki et al., 2023).

Green energy and food are two of the major concerns in today's world that all are stressing upon. The gaps in the demand and supply of resources and environment

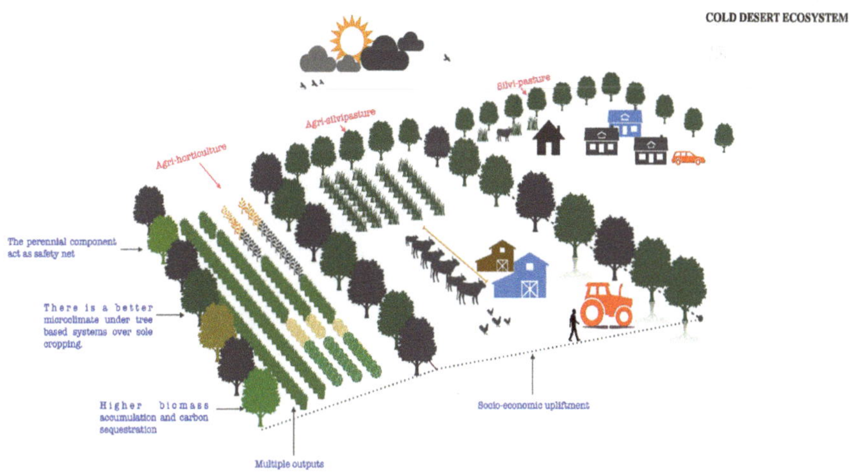

Fig. 2 Agroforestry and climate change mitigation in cold deserts

concerns could be supplemented while practicing the agroforestry systems support the food production and sustainability (Nicholas & Scown, 2024; Qi et al., 2025).

4.1 Climate Change and Environmental Benefits of Agroforestry Systems

(a) **Carbon Sequestration and Greenhouse Gas Reduction**

Agroforestry is a highly effective climate change mitigation approach because it integrates trees with crops and/or livestock in ways that simultaneously enhance carbon sequestration and reduce greenhouse gas emissions while maintaining agricultural productivity (Gupta et al., 2024). Trees in agroforestry systems capture and store carbon in above-ground biomass such as trunks, branches, and leaves, as well as in extensive root systems that contribute long-term carbon storage below ground. Continuous inputs of leaf litter and reduced soil disturbance increase soil organic carbon stocks, often making agroforestry soils significantly richer in carbon than those under conventional monocropping. In addition, agroforestry reduces greenhouse gas emissions by carbon sequestration, improving nutrient cycling and nitrogen use efficiency, which lowers dependence on synthetic fertilizers and reduces nitrous oxide emissions from soils (Getnet et al., 2023). The inclusion of nitrogen-fixing tree species in agroforestry systems further supports this process by providing biologically fixed nitrogen in a more stable form. In silvopastoral systems, trees improve microclimatic conditions and forage quality, reducing heat stress on animals and lowering methane emissions per unit of livestock product. Agroforestry also contributes to reduced emissions by supplying on farm fuelwood and timber, reducing pressure on natural forests, and decreasing energy use for irrigation and farm inputs (Brandt et al., 2024; Castle et al., 2022; England et al., 2020; Nicholas & Scown, 2024; Terasaki et al., 2023). Overall, agroforestry offers a multifunctional, long-term strategy to mitigate climate change impacts while enhancing soil health, biodiversity, farm resilience, and livelihood security (Ali & Wang, 2021; Mugabowindekwe, 2023; Nicholas & Scown, 2024; Singh et al., 2024). Tree-based agroforestry systems act as a carbon sink by sequestering carbon from atmosphere and retaining it as biomass (Singh et al., 2024) besides offering better air quality, shade, reducing heat stress, and helping in climate mitigation (Castle et al., 2022; England et al., 2020).

(b) **Energy Efficiency of Cold Desert Ecosystems**

How Tree-based Agroforestry Systems Reduce Energy Consumption and Promote Sustainable Energy Production

Tree-based agroforestry system reduces energy consumption mainly by modifying the local environment and farm management in such a manner that lower the need for external energy input (Hodges & Brandle, 1996; Huang et al., 2024). Tree provides shade and regulate microclimate, which reduce temperature extreme and lowers energy demand for irrigation, cooling, water pumping particularly in hot and dry regions. Windbreak and shelterbelt decrease wind speed reducing the loss of water from crops and soil and thereby lowering irrigation and pumping energy requirements in the long run (Hodges & Brandle, 1996). Tree-based agroforestry system improved soil structure and higher organic matter increases water and filtration and retention that reduce the energy needed for water management. Tree also supplies on farm resources that decrease reliance on energy intensive external inputs and its subsequent transportation additionally the better ecosystem and improved nutrient cycling and reduces dependency on synthetic fertilizers and field operations (Huang et al., 2024).

(c) **Agroforestry and Soil Management**

Trees reduce soil erosion through a combination of processes as physical protection, biological stabilization, and improvement of soil structures (Fig. 3). Trees intercept rainfall, break the impact of raindrop, and reduce splash erosion that detach and transport soil particle from one place to another. The plant parts such as leaf, twigs, and other deadwood debris make a protective layer above the soil that further slow surface water flow and encourage water infiltration and percolation rather than runoff (Jaros et al., 2025; Panthera et al., 2021; Song et al., 2019). The root system of trees binds all particle together, increasing soil particle to particle attachments and reinforce soil structure, making the soil more resistance for erosion and associated causes (Hairiah et al., 2020; Jaros et al., 2025; Panthera et al., 2021). Tree roots also create channel in the soil that enhance porosity. Trees reduce the speed of surface runoff that causes sheet, rill, and gully or ravine erosion. Trees provide reinforcement that stabilize slope, prevent landslide, and bank collapse. Additionally, the continuous input of organic matter from woody component improves soil structure and water holding capacity over time leading to healthier and more resilient soils that are far less susceptible to erosion (Jaros et al., 2025). Studies have stated that most of the physical chemical and biological properties are better under tree canopy as compared to open land, demonstrating the positive effect of trees on soil and its properties associated (Bhushan et al., 2025; Tomar et al., 2025).

5 Agroforestry and Sustainable Development Goals (SDGs)

Agroforestry supports multiple SDGs, such as climate action, life on land, and zero hunger. Trees grown on farmers' fields offer multiple benefits and fulfills several Sustainable Development Goals (SDGs), particularly SDG 1, 2, 3, 7, 8 by improving food security through reliable fodder sources, fuelwood, economic growth and sustenance, and SDG 12, 13, 15 (Climate Action) by fixing biomass, carbon sequestrations, and ameliorating adverse climatic phenomena in agroecosystems (Dogra et al., 2025a; Nicholas & Scown, 2024). SDGs are interlinked with one another, i.e., fulfilling one goals is directly supporting fulfilling of the other (Dogra et al., 2025b). Agroforestry is about the poverty reduction, food security, health, affordable clean energy, and economic growth perspectives (Brandt et al., 2024; Terasaki

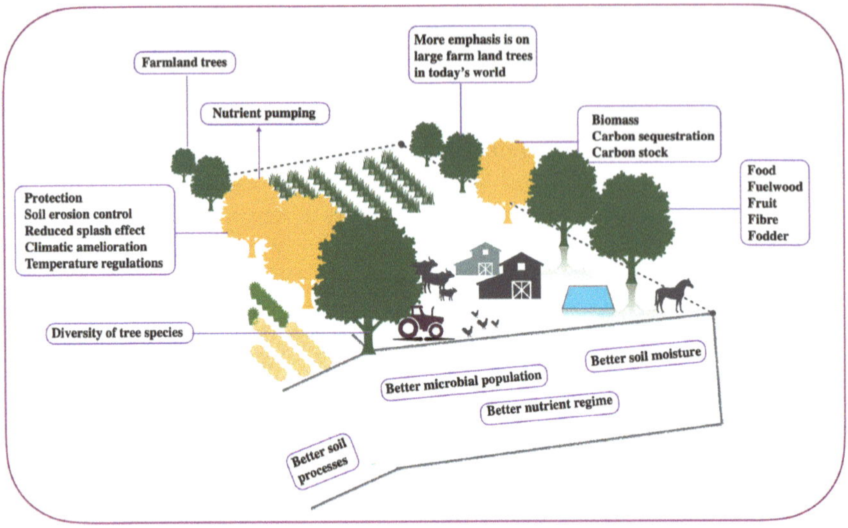

Fig. 3 Demonstration of agroforestry helping in restoring land degradation

et al., 2023). Agroforestry has immense potential for providing maximum benefits under land scarce conditions (Dogra et al., 2025a; Mbow et al., 2014).

5.1 Rural Upliftment and Economic Sustainability

(a) **Economic Benefits for Smallholder Farmers**

Agroforestry has potential to provide significant economic benefits for small holder farmers by diverse outputs, income sources, reducing production risks, and improving long-term sustainability (Arvola et al., 2020; Bettles et al., 2021). In harsh conditions of cold desert, this attribute of agroforestry offers multiple benefits in terms of food, fodder, medicinal plants, and non-timber forest products (Aswal & Mehrotra, 1994; Behera et al., 2014; Bhardwaj et al., 2024; Dogra et al., 2025a, 2025b; Safi et al., 2024). The diversification in cold desert ecosystem reduces dependence on a single crop and provides income at different times of the year thereby improving households' socioeconomic stability. Trees also enhance the resilience to climate shock such as snow, floods, and draughts thereby lowering the risk of total crop failure and protecting farmer's farms sustainability. Agroforestry system contributes to the reduction in the total cost by increased productivity (Bhardwaj et al., 2024; Bhushan et al., 2025; Tomar et al., 2025). Improved soil structure and water retention increase crop yield over time, while the other attributes of tree-based systems such as shades and protection against wind also increase the productivity. On a long-term basis, tree-based agroforestry systems enhance asset building and market opportunities for small holder farmers (National Agroforestry Policy, 2014). Tree acts as an insurance or saving as they can be harvested or sold during the due course of time. Agroforestry products often have high end market value and can give access to farmers in modern markets such as climate smart agriculture and carbon credit schemes (Awazi et al., 2025). Further government is pressing on retention of trees on farm, so that farmers get income stability, and build long-term economic resilience (National Agroforestry Policy, 2014).

6 Emerging Trends and Challenges in Agroforestry

Different kinds of land use practices have been done by tribals in cold desert regions of the world and in India specific. The tribes in Lahaul and Spiti region of Himachal Pradesh practice Agri-horticulture, Silvi-pasture, Horti-pasture, and Horticulture (Aswal & Mehrotra, 1994; Behera et al., 2014; Sharma et al., 2023b, 2024). There is a stark demonstration of these land use system in terms of resilience, sustenance, diversity, soil conservation, soil protection, water conservation, and promotion of ecological stability. Currently, pertaining to the modern world, there's a need to focus on aspects such as an integration of energy security alongside meeting the requirement for food, fodder, and fuelwood of the world population alongside biodiversity protection.

(a) **The Role of Research in Shaping Future Practices in Cold Desert Ecosystems**

Agroforestry adoption is constrained by a range of technical, social, and financial barriers that interact and reinforce one another. Particularly for small holder farmers from a technical perspective, agroforestry systems are often knowledge intensive and require careful planning regarding tree crop and/or animal combinations, spacing, species selection of trees along with crop, and long run management (Alemagi et al., 2014; Kpoviwanou et al., 2024). Many farmers lack access to adequate training and extension services, making it difficult to implement agroforestry systems effectively (Alemagi et al., 2014). The same situation is accelerated by the region, where it is being planned particularly focusing on cold desert regions where resources are limited, and lacks providing technical knowledge by the government. Additionally, cold desert region supplemented by slow growth of trees simply means that the benefit of well managed system will be delivered in future, which leads to discouragement in the adoption of agroforestry (Sharma & Sharma, 2022; Tiwari & Kapoor, 2016).

Further the social acceptance also plays a significance role in governing the adoption of agroforestry practices (Arvola et al., 2020; Bettles et al., 2021). As in cold desert region, land is scarce and important asset where insecure land tenure and unclear tree ownership interface with the short-term gains from agricultural outputs (Tiwari & Kapoor, 2016). The cultural practices and traditional farming generally favor annual crop and monocropping leading to the resistance of people to retain more trees on their farmland. The demand for the labor at the time of establishment and maintenance of these complex systems discourages farmer from the adoption.

The limited awareness of farmers regarding the importance of tree-based outputs and less awareness about the long-term benefits associated leads to less adoption of this practice (Arvola et al., 2020; Bettles et al., 2021; Tiwari & Kapoor, 2016). Economic feasibility and/or financial barriers are one of the most significant facts that controls the adoption of agroforestry practices in any part of the region and that too gets applied to the cold desert. Establishment of agroforestry in any farm sitting often required a more capital investment for seed, seedling protection, tools, and labor; that is not a luxury every farmer can afford. Trees are long-term enterprise and give return after a long period of time which increases the sense of uncertainty among farmers about their market demands and benefits associated. In comparison to the credit availability for monocropping, there are very limited availability of funds for starting of any agroforestry project on farm, as it is observed as a high-risk investment because the market demand for any tree species might change in the due course of time. Limited access of farmers for the market of tree-based produce aggravates the situation. The absence of supportive policies, insurance scheme, and/or limited financial incentives makes agroforestry a less attractive compared to conventional farming system in cold desert regions (Tiwari & Kapoor, 2016).

(b) **Policy and Governance Implications**

Adoption of agroforestry in cold desert region is very closely related to the policy and governance, as unsupportive institutional framework often intensifies the technical, social, and financial barriers faced by farmers in such an extreme tough weather. From technical aspect, there's a need for government investment in the agriculture research, extension services, and capacity building for sharing knowledge, training, and quality planting material for agroforestry practices (National Agroforestry Policy, 2014). Lack of proper coordination between forestry and agricultural institution often results in fragmented policies. In some cases, the restriction for tree harvesting and transport also discourages farmers from integrating some specific trees, even when the practices are environmentally sustainable. In Indian scenario, national agroforestry policy 2014 is the benchmark achievement for the promotion of agroforestry at national scale. This policy has addressed the key issues of Indian market by suggesting for establishment of specific market for the sellout of agroforestry produce (National Agroforestry Policy, 2014). National agroforestry policy 2014 has also focused on providing institutional backup, flexibility in transportation of tree outputs from farm sitting. Agroforestry shall be focused on the governance, and policy shall be framed for strengthening the market for agroforestry sector. As weak market governance, including poor infrastructure, price regulation, and limited support for value chain development reduces the profitability of agroforestry products. Furthermore, insufficient recognition of the ecosystem services provided by agroforestry such as biodiversity conservation, carbon sequestration, soil, and water protection means that farmers are rarely aware and capable of harnessing this benefit. Overall, we shall focus on a holistic policy development for agroforestry based on future requirement and solving the governance issue regarding the adoption and promotion of agroforestry particularly with reference to the cold desert region of the world (Arvola et al., 2020; Bettles et al., 2021; Sharma et al., 2024; Tiwari & Kapoor, 2016).

7 Future Prospects

(a) Opportunities for Scaling Agroforestry in Cold Desert Ecosystems

Integration of agroforestry with agrivoltaics can be a viable model for energy and food security along with meeting SDGs. Therefore, we need to study the potential possible synergistic effects of energy and food production together in cold desert agroforestry systems (Fig. 4). As there is a very limited literature available in the direction, we shall access the literature of agriculture, tree-based agroforestry systems, and make extensive efforts to understand the potential effects as well as processes governing the system as a unit. Further we shall interlink different process driven approaches of agroforestry with agrivoltaics. We shall also incorporate different SDGs that agroforestry and green energy source has potential to full fill along with environmental protection. We must make every effort to cover co-benefits of agrivoltaics and agroforestry in real world (Rosati et al., 2023). There shall be focus on studying the effect of varying light intensity on integration of agrivoltaics in tree-based system. In this manner, we address the current concerns for environment, food, and energy security, by providing a unified mean of addressing multiple concerns. As with increasing population and land scarcity, we must focus on integrating energy production with other land use systems in near future.

(b) Collaboration between Governments, NGOs, and Local Communities

Addressing agroforestry challenges for a group for wider community requires collaboration between government organizations, non-governmental organizations, and local communities (Arvola et al., 2020; Bettles et al., 2021). Government creates policies, provides extension services, ensures secure land tenure, and offers financial

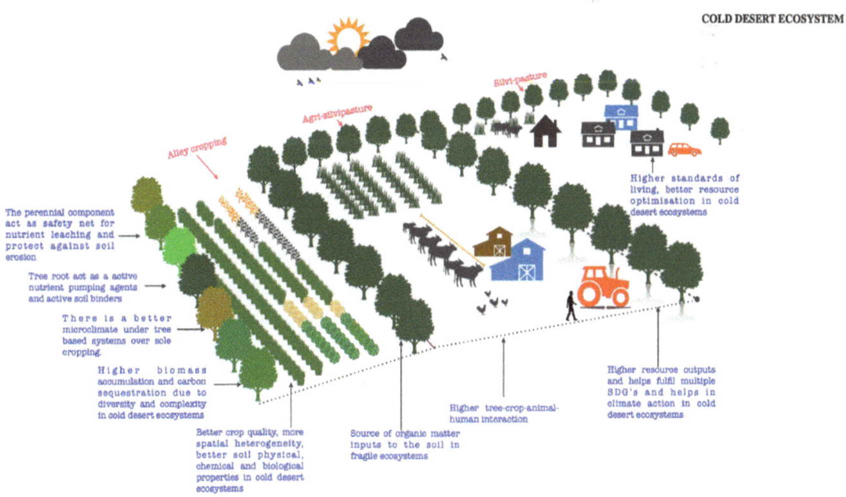

Fig. 4 Opportunities of scaling agroforestry in cold desert ecosystems

gains to reduce risk and cost for farmers. The efforts of the government are supplemented by non-governmental organizations by assisting in technical training, raising awareness, advocating for inclusive policies, and helping marginalized group access resources. Local community is central for implementation, contribution indigenous knowledge, labor, insight into locally appropriate practices farm producer organization, farmer cooperatives, and participatory committees enhance communication, resource sharing, and joint problem solving, ensuring that agroforestry systems are socially acceptable, economically viable and technically feasible, and also, the calibration from different domains can also strengthen market access for agroforestry products and establish long-term support mechanism, creating a sustainable environment for widespread adoption (Alemagi et al., 2014; Bettles et al., 2021; Kpoviwanou et al., 2024). Although agroforestry affords multiple benefits, but their adoption is generally limited by technical, social, financial, and government aspects prevalent in the region. Therefore, addressing these issues require integrated strategies that combines financial incentives and active collaborations between governing bodies of the state, NGOs, and local communities. Diverse approach of agroforestry ensures as a viable and attractive livelihood strategies for farmers in the cold desert region (Aswal & Mehrotra, 1994; Behera et al., 2014; Rawat et al., 2023).

8 Conclusion

Agroforestry is one of the major and most focused practice in modern world for increasing the green cover and maximizing food production. In cold desert ecosystems, agroforestry (land use practice) promises long-term sustenance under limited resources. The land with high capability class is limited and that too shared for multiple outputs such as agriculture, animal rearing, and living chores in these fragile ecosystems. During winter period, there is scarcity of fodder resources and fuelwood and is met up with the tree raised on farm lands. Further fruit-based agroforestry systems add more dimensions in the concept by providing avenues in terms of monetary gains and socioeconomic upliftment. The integration of modern concepts such as agrivoltaics in agroforestry can add new domain in the field by providing energy security and reducing expense on the power consumption. There is a need to explore the unexplored avenues in cold desert ecosystem, where diversity and vulnerability run parallel to each other. As cold desert ecosystems are diverse but very sensitive to small changes due to their vulnerability to temperature, limited diversity, short growing season, and long winters. Therefore, we shall focus on improving the traditional systems with modern interventions without disturbing the base and building blocks.

References

Alemagi, D., Duguma, L., Minang, P. A., Nkeumoe, F., Feudjio, M., & Tchoundjeu, Z. (2014). Intensification of cocoa agroforestry systems as a REDD+ strategy in Cameroon: Hurdles, motivations, and challenges. *International Journal of Agricultural Sustainability, 13*(3), 187–203. https://doi.org/10.1080/14735903.2014.940705

Ali, A., Wang, L. Q. (2021). Big-sized trees and forest functioning: Current knowledge and future perspectives. *Ecological Indicators, 127*, 107760. https://doi.org/10.1016/j.ecolind.2021.107760

Arvola, A., Brockhaus, M., Kallio, M., Pham, T. T., Chi, D. T. L., Long, H. T., Nawir, A. A., Phimmavong, S., Mwamakimbullah, R., & Jacovelli, P. (2020). What drives smallholder tree growing? Enabling conditions in a changing policy environment. *Forest Policy and Economics, 116*, 102173. https://doi.org/10.1016/j.forpol.2020.102173

Aswal, B. S., & Mehrotra, B. N. (1994). *Flora of Lahual-Spiti. A cold desert in north-west Himalaya* (pp. 10–15). Bishan Singh Mahendra Pal Sigh.

Awazi, N. P., Alemagi, D., & Ambebe, T. F. (2025). Promoting the carbon market in agroforestry systems: The role of global, national and sectoral initiatives. *Discover Forests, 1*(1). https://doi.org/10.1007/s44415-025-00013-4

Behera, M. D., Matin, S., & Roy, P. S. (2014). Biodiversity of Kargil cold desert in the Ladakh Himalaya. In S.-I. Nakano, T. Yahara, & T. Nakashizuka (Eds.), *Integrative observations and assessments* (pp. 253–274). https://doi.org/10.1007/978-4-431-54783-9

Bettles, J., Battisti, D. S., Cook-Patton, S. C., Kroeger, T., Spector, J. T., Wolff, N. H., & Masuda, Y. J. (2021). Agroforestry and non-state actors: A review. *Forest Policy and Economics, 130*, 102538. https://doi.org/10.1016/j.forpol.2021.102538

Bhardwaj, D. R., Navale, M. R., Dogra, R., Verma, T., Sharma, P., & Keprate, A. (2024). Biomass tables and allometric equations for predicting the fodder and fuelwood production of prominent tree resources in agricultural landscapes of the mid-hills of the North-Western Himalayas. *Agroforestry Systems, 98*, 3129–3144. https://doi.org/10.1007/s10457-024-01078-4

Bhushan, P., Bhardwaj, D. R., Sharma, P., Dogra, R., Pathania, S., & Kumar, D. (2025). Assessment of pea (*Pisum sativum* L.) and cauliflower (*Brassica oleracea* L. var. botrytis) productivity in mulberry (*Morus alba* L.) based agroforestry system in mid-hills of the North-Western Himalayas: Effects of tree canopy, fertiliser and bio-stimulants. *Agroforestry Systems*, 99–198. https://doi.org/10.1007/s10457-025-01276-8

Brandt, M., Ominski, D. G., Reiner, F., Kariryaa, A., Guthula, V. B., Ciais, P., et al. (2024). Severe decline in large farmland trees in India over the past decade. *Nature Sustainability, 7*, 860–868. https://doi.org/10.1038/s41893-024-01356-0

Castle, S. E., Miller, D. C., Merten, N., Ordonez, P. J., & Baylis, K. (2022). Evidence for the impacts of agroforestry on ecosystem services and human Well-being in high-income countries: A systematic map. *Environmental Evidence, 11*, 10. https://doi.org/10.1186/s13750-022-00260-4

Chen, Y., Fuller, R. A., Lee, T. M., & Hua, F. (2024). Disproportionate low-elevation forest loss in over 65% of the world's mountains calls for targeted conservation. *One Earth, 7*(10), 1833–1845. https://doi.org/10.1016/j.oneear.2024.08.006

Climate Change. (2025, March 18). World meteorological organization. https://wmo.int/topics/climate-change

Das, B. S., Wani, S. P., Benbi, D. K., Muddu, S., Bhattacharyya, T., Mandal, B., Santra, P., Chakraborty, D., Bhattacharyya, R., Basak, N., & Reddy, N. N. (2022). Soil health and its relationship with food security and human health to meet the sustainable development goals in India. *Soil Security, 8*, 100071. https://doi.org/10.1016/j.soisec.2022.100071

Dogra, R., Gupta, H., Prajapati, A., & Eepsa. (2023). Agroforestry practices and potential advantages. *Imminent Farming*, 81–98.

Dogra, R., Bhardwaj, D. R., Sharma, P., Pooja, N., Thakur, S., Keprate, A., Sharma, V., & Chauhan, S. (2025a). Agroforestry water footprint: Pathways to environmentally conscious farming. In *Elsevier eBooks* (pp. 183–200). https://doi.org/10.1016/B978-0-443-30054-7.00004-5

Dogra, R., Pant, K. S., Bishist, R., et al. (2025b). Impact of multipurpose farm trees in subtemperate agroecosystems: Benefits for soil, biomass, fodder, and climate resilience. *Agroforestry Systems, 99*, 116. https://doi.org/10.1007/s10457-025-01217-5

England, J. R., O'Grady, A. P., Fleming, A., Marais, Z., & Mendham, D. (2020). Trees on farms to support natural capital: An evidence-based review for grazed dairy systems. *Science of the Total Environment, 704*, 135345. https://doi.org/10.1016/j.scitotenv.2019.135345

FAO. (1999). Poverty alleviation and food security in Asia. In *Land degradation*, chapter 3. Retrieved January 21, 2025, from Food and Agricultural Organization of the United Nations (FAO).

Getnet, D., Mekonnen, Z., & Anjulo, A. (2023). The potential of traditional agroforestry practices as nature-based carbon sinks in Ethiopia. *Nature-Based Solutions, 4*, 100079. https://doi.org/10.1016/j.nbsj.2023.100079

Gupta, H., Sanodiya, P., Singh, A., & Dogra, R. (2024). Land-use systems for biomass, carbon storage, and carbon credit: Implications for climate change mitigation in subtropical pockets of Vindhyan region, India. *Biofuels Bioproducts and Biorefining, 18*(4), 1047–1064. https://doi.org/10.1002/bbb.2664

Hairiah, K., Widianto, W., Suprayogo, D., & Van Noordwijk, M. (2020). Tree roots anchoring and binding soil: Reducing landslide risk in Indonesian agroforestry. *Land, 9*(8), 256. https://doi.org/10.3390/land9080256

Hodges, L., & Brandle, J. R. (1996). Windbreaks: An important component in a plasticulture system. *HortTechnology, 6*(3), 177–181. https://doi.org/10.21273/horttech.6.3.177

Huang, J., Echeverri, D. P., & Zhang, Z. (2024). Planting trees is a cost-effective way to reduce residential electricity consumption and abate atmospheric CO_2. *Applied Energy, 373*, 123842. https://doi.org/10.1016/j.apenergy.2024.123842

Jaroš, J., Šamonil, P., Daněk, P., Adam, D., Tikhomirov, D., Ediriweera, S., Christl, M., & Egli, M. (2025). Trees slow down erosion and allow soil progression in an extremely high-rainfall old-growth mixed dipterocarp forest of Southwest Sri Lanka. *Catena, 254*, 108911. https://doi.org/10.1016/j.catena.2025.108911

Kala, C. P. (2025). Agroforestry in a changing climate: Challenges, opportunities and solutions. *Ecological Frontiers, 45*(2), 269–276. https://doi.org/10.1016/j.ecofro.2024.11.003

Khambalkar, P. A., Agrawal, S., Dhaliwal, S., Yadav, S. S. S., Sadawarti, M. J., Singh, A., Yadav, I. R., Yadav, K., Shivansh, Prasad, D., Singh, A., & Afreen, N. (2025). Sustainable nutrient management balancing soil health and food security for future generations. *Applied Food Research, 5*(2), 101087. https://doi.org/10.1016/j.afres.2025.101087

Kpoviwanou, M. R. J. H., Sourou, B. N. K., & Ouinsavi, C. A. A. N. (2024). Challenges in adoption and wide use of agroforestry technologies in Africa and pathways for improvement: A systematic review. *Trees Forests and People, 17*, 100642. https://doi.org/10.1016/j.tfp.2024.100642

Mbow, C., van Noordwijk, M., Prabhu, R., & Simons, T. (2014). Knowledge gaps and research needs concerning agroforestry's contribution to sustainable development goals in Africa. *Current Opinion in Environmental Sustainability, 6*, 162–170.

Montes, O., Uribe, M., Castro, R., Villanueva, C., Pérez, M., & Lara, A. (2020). Policy forum: Proposal of a Mexican precision agroforestry policy. *Forest Policy and Economics, 119*, 102292. https://doi.org/10.1016/j.forpol.2020.102292

Moorthy, M. (2025, November 12). How Kibber turned snow leopard conflict into coexistence. Frontline. Retrieved November 24, 2025, from https://frontline.thehindu.com/environment/spiti-snow-leopard-conservation-success-story/article70157703.ece?utm_source=engagement&utm_medium=email&utm_campaign=spotlightsubs

Mugabowindekwe, M., Brandt, M., Chave, J., Reiner, F., Skole, D. L., Kariryaa, A., et al. (2023). Nationwide mapping of tree-level aboveground carbon stocks in Rwanda. *Nature Climate Change, 13*(1), 91–97. https://doi.org/10.1038/s41558-022-01544-w

Nair, P. K. R. (1993). *An introduction to agroforestry*. Kluwer Academic Publishers.

National Agroforestry Policy. (2014). Department of Agriculture and Cooperation. Ministry of Agriculture and Farmer Welfare. Government of India. 2014: 1–13. Retrieved January 29, 2026, from https://www.nitiforstates.gov.in/public-assets/Policy/policy-repo/agriculture-and-allied-services/PNC511A000509.pdf

Nicholas, K. A., & Scown, M. (2024). Rural land systems both support and hinder the sustainable development goals across Europe. *Communications Earth & Environment, 5*, 692. https://doi.org/10.1038/s43247-024-01736-6

Pantera, A., Mosquera-Losada, M. R., Herzog, F., & Herder, M. D. (2021). Agroforestry and the environment. *Agroforestry Systems, 95*(5), 767–774. https://doi.org/10.1007/s10457-021-00640-8

Pardon, P., Reubens, B., Reheul, D., Mertens, J., De Frenne, P., Coussement, T., Janssens, P., & Verheyen, K. (2017). Trees increase soil organic carbon and nutrient availability in temperate agroforestry systems. *Agriculture Ecosystems & Environment, 247*, 98–111. https://doi.org/10.1016/j.agee.2017.06.018

Qi, C., Peng, H., Xu, M., et al. (2025). Comprehensive assessment of land degradation in the Sanjiangyuan region based on multi-source indicators. *Scientific Reports, 15*, 41392. https://doi.org/10.1038/s41598-025-25277-w

Rawat, Y. S., Negi, V. S., Moussa, I. M., Zaman, W., & Elansary, H. O. (2023). Diversity, distribution and vegetation assessment of woody plant species in the cold desert environment, North-Western Himalaya, India. *Sustainability, 15*(13), 10429. https://doi.org/10.3390/su151310429

Rosati, A., Proctor, K., Dazaea, A., Graham, M., Ates, S., Kirschten, H. M., & Higgins, C. W. (2023). Agroforestry versus agrivoltaic: Spectral composition of transmitted radiation and implications for understory crops. *Agroforestry Systems, 98*(8), 2697–2710. https://doi.org/10.1007/s10457-023-00914-3

Safi, M., Bishist, R., Bhardwaj, N. D., Thakur, N. C., Prakash, P., Gautam, K. L., Dogra, R., Sharma, H., Kishore, K., & Kengoo, N. (2024). Influence of different nutrient sources on growth, yield and nutritive value of fodder maize. *Bangladesh Journal of Botany, 53*(4), 923–929. https://doi.org/10.3329/bjb.v53i4.77836

Sharma, R. K., & Sharma, D. (2022). Drying willow (*Salix fragilis* L.) population under agroforestry system in cold desert region of trans-Himalaya: A possible consequence of repeated vegetative propagation. *International Journal of Ecology and Environmental Sciences*. https://www.nieindia.org/Journal/index.php/ijees/article/view/2421?utm_source=chatgpt.com

Sharma, A., Sharma, J. C., Shukla, Y. R., Verma, M. L., Singh, U., Spehia, R. S., Sharma, D., Gautam, K. L., Dogra, R., Tecimen, H. B., Kumar, M., & Kumar, A. (2023a). Nutrient management influences root characteristics and nitrogen use efficiency in the vegetable-based agroecosystem in the Northwestern Himalayas. *Sustainability, 15*(13), 10593. https://doi.org/10.3390/su151310593

Sharma, H., Pant, K., Bishist, R., Gautam, K. L., Ludarmani, N., Dogra, R., Kumar, M., & Kumar, A. (2023b). Estimation of biomass and carbon storage potential in agroforestry systems of northwestern Himalayas, India. *Catena, 225*, 107009. https://doi.org/10.1016/j.catena.2023.107009

Sharma, A., Shashni, S., & Rathore, S. (2024). Traditional practices of agroforestry systems in the cold desert area of Himachal Himalayas. In A. Borthakur & P. Singh (Eds.), *Addressing the climate crisis in the Indian Himalayas*. Springer. https://doi.org/10.1007/978-3-031-50097-8_6

Singh, G., Chauhan, V., Thakur, C., Verma, M., Bishist, R., Prakash, P., Gautam, K. L., Sharma, H., Dogra, R., Kishore, K., & Kumar, M. (2023). Studies on natural regeneration, floristic composition, biomass, carbon density and soil properties along an altitudinal gradient in the North-Western Himalayas, India. *Forest Ecology and Management, 548*, 121391. https://doi.org/10.1016/j.foreco.2023.121391

Singh, R., Biradar, C., Behera, M., Prakash, A., Das, P., Mohanta, et al. (2024). Optimising carbon fixation through agroforestry: Estimation of aboveground biomass using multi-sensor data synergy and machine learning. *Ecological Informatics, 79*, 102408. https://doi.org/10.1016/j.ecoinf.2023.102408

Song, Z., Seitz, S., Li, J., Goebes, P., Schmidt, K., Kühn, P., Shi, X., & Scholten, T. (2019). Tree diversity reduced soil erosion by affecting tree canopy and biological soil crust development in a subtropical forest experiment. *Forest Ecology and Management, 444*, 69–77. https://doi.org/10.1016/j.foreco.2019.04.015

Sottosanti, K. (2023). *Cold desert*. Encyclopedia Britannica. https://www.britannica.com/science/cold-desert

Stopfer, L., Kaulen, A., & Purfürst, T. (2023). Potential of blockchain technology in wood supply chains. *Computers and Electronics in Agriculture, 216*, 108496. https://doi.org/10.1016/j.compag.2023.108496

Tal, A. (2025). The environmental impacts of overpopulation. *Encyclopedia, 5*(2), 45. https://doi.org/10.3390/encyclopedia5020045

Terasaki, H. D. E., Yeo, S., Almaraz, M., Beillouin, D., Cardinael, R., Garcia, E., et al. (2023). Priority science can accelerate agroforestry as a natural climate solution. *Nature Climate Change, 13*, 1179–1190. https://doi.org/10.1038/s41558-023-01810-5

Tewari, V. P., & Kapoor, K. S. (2016). Western Himalayan cold deserts: Biodiversity, eco-restoration, ecological concerns and securities. *Annals of Arid Zone, 52*(3–4). https://doi.org/10.56093/aaz.v52i3-4.63386

Tomar, S., Bhardwaj, D. R., Kumar, J., Sharma, P., & Dogra, R. (2025). Soil carbon and nitrogen dynamics in agroforestry land use systems of North-Western Himalayas: Implications for environmental sustainability. *Agroforestry Systems, 99*(5). https://doi.org/10.1007/s10457-025-01193-w

UNDRR and ISC. (2025). *UNDRR–ISC Hazard information profiles—2025 update: EN0301 land degradation United Nations Office for disaster risk reduction*. International Science Council. https://www.undrr.org/terms/hips/EN0301

UNESCO World Heritage Centre. (2015). Cold desert cultural landscape of India. Retrieved January 24, 2026, from https://whc.unesco.org/en/tentativelists/6055/

Weiskopf, S. R., Rubenstein, M. A., Crozier, L. G., Gaichas, S., Griffis, R., Halofsky, J. E., Hyde, K. J., Morelli, T. L., Morisette, J. T., Muñoz, R. C., Pershing, A. J., Peterson, D. L., Poudel, R., Staudinger, M. D., Sutton-Grier, A. E., Thompson, L., Vose, J., Weltzin, J. F., & Whyte, K. P. (2020). Climate change effects on biodiversity, ecosystems, ecosystem services, and natural resource management in the United States. *The Science of the Total Environment, 733*, 137782. https://doi.org/10.1016/j.scitotenv.2020.137782

World Bank Open Data. (2024). World Bank open data. Retrieved November 23, 2025, from https://data.worldbank.org/

World Health Organization: WHO. (2023). Climate change. Retrieved November 24, 2025, from https://www.who.int/news-room/fact-sheets/detail/climate-change-and-health

Diversity, Traditional Knowledge, and Vulnerability of Medicinal Plants to Climate Change in Cold Desert Areas of Indian Himalaya

Chandra Prakash Kala

1 Introduction

The cold deserts in Indian Himalaya are home to a unique set of medicinal plants adapted to some of the harshest conditions on earth, including low temperatures, low humidity, little rainfall, high level of ultraviolet radiations, short growing seasons, and thin rocky soils (Kala, 2003, 2011; Maharjan et al., 2022). In comparison to the Himalaya, the trans-Himalayan cold desert areas host fewer species but still contribute significantly to the region's medicinal plants wealth (Kala, 2026). These medicinal plants are central to traditional healing systems such as Amchi (Sowa-Rigpa) and Ayurveda (Kala, 2005a). Local communities rely on these plants for curing a wide range of ailments, and traditional knowledge about their use is deeply rooted in the culture and daily life of the inhabitants of these regions. Many medicinal plant species in Indian cold deserts are endemic or have highly localized distributions, making them rare and ecologically significant (Kala, 2000a, 2005b). The diversity and distribution of medicinal plants are also influenced by elevations, harsh weather, short growing seasons, and poor soil conditions (Kala, 2004, 2005a; Mishra et al., 2023). Adaptation of plant species to such harsh climatic conditions and environmental stressors results in unique phytochemical profiles and higher concentrations of secondary metabolites (viz., alkaloids, flavonoids, and essential oils), making these plants valuable for medicinal and pharmaceutical uses (Verma and Tewari, 2016; Gairola et al., 2010).

A large number of medicinal plants growing in Indian cold desert areas are used as raw materials in modern pharmaceuticals; hence, these plant species have high commercial demand due to their unique medicinal compounds (Kala, 2003; Gairola et al., 2010). The rising demand for commercial purpose, coupled with unsustainable and unscientific harvesting, has led to the endangerment of many medicinal plant

C. P. Kala (✉)
Indian Institute of Forest Management, Bhopal, Madhya Pradesh, India
e-mail: cpkala@yahoo.co.uk

Y. S. Rawat and C. P. Kala (eds.), *Cold Deserts*,
https://doi.org/10.1007/978-3-032-29119-6_11

species (Kala, 2011; Sharma and Kala, 2018). In addition, many medicinal plants in cold deserts are threatened due to loss of habitats, habitat degradation, anthropogenic climate change, increased tourism, and the spread of invasive species (Kala, 2005b; Joshi et al., 2016; Paul et al., 2021). Since many cold desert medicinal plants have become rare and endangered, they are centered to conservation efforts because their availability is critical for sustaining traditional medicinal systems prevalent across the cold deserts of Indian Himalaya (Kala & Manjrekar, 1999; Kala, 2000b, 2005b). These medicinal plant species remain vital for healthcare and livelihood not only for the inhabitants of remote cold desert areas but also to the pharmaceutical industries and the consumers for health care depend on them (Kala, 2003).

The medicinal plants in cold deserts have special adaptations for harsh and high-altitude environments, which is within their limit of adaptive capacity. Their survival, distribution, and abundance can be greatly affected by rising temperatures, altered precipitation patterns, and increased frequency of extreme weather events (Kala, 2013, 2026; Maharjan et al., 2022). Overharvesting, habitat loss, and other anthropogenic pressures combined with ongoing climate change may increase the risk of local extinction for many cold desert medicinal plant species. Most of these plants often have restricted distribution ranges and specialized habitats, making them particularly vulnerable to environmental change and habitat fragmentation (Kala, 2005b). Nonetheless, the impact of climate change on cold desert medicinal plants, including their population dynamics, is yet to be understood as there is a lack of comprehensive data on how climate change specifically affects cold desert medicinal plants. The decline in plant populations or changes in their medicinal value can have a negative impact on community health and economic stability.

In view of the high significance of medicinal plants of Indian cold deserts, it is realized to study such a unique and invaluable natural resource, which supports not only the health care but also the livelihood to the local people, and equally significant for maintaining ecological balance of fragile and sensitive cold desert ecosystems. The present study, therefore, aims to document diverse uses of medicinal plants, their distribution along the altitudinal gradients and habitats in Indian cold deserts, especially in Spiti and Ladakh region of India. Given the threats these cold deserts medicinal plants face from overexploitation, habitat degradation, and environmental changes, at present, this study is considered quite essential for future course of actions and sustainable development.

2 Methodology

2.1 Study Area

2.1.1 Spiti

Spiti, a cold desert area, is located in the Indian state of Himachal Pradesh between 31°42′-32°58’ N latitude and 77°21′-78°35′ E longitude. It is bounded by Tibet (China) on the northeast and by the Great Himalayan ranges on the south. Administrative boundaries of Spiti include Kinnaur in the southeast, Kullu in the west, and Ladakh in the north with average elevation of 4270 m. Spiti is divisible into 4 ecological units namely Sham (the lower region), Pin (lies by the Pin river), Bhar (the middle tract), and Tud (the high territories) (Kala, 2000a). The entire terrain is dry with an annual rainfall of about 17 cm (Kapadia, 1996). More than 800 vascular plant species have been recorded from this area (Aswal & Mehrotra, 1994), of which many plant species are used as medicine. To conserve the biodiversity of this area, Pin Valley National Park and Kibber Wildlife Sanctuary have been notified by the State Government. A total of 378 plant species have been recorded within Pin Valley National Park (Manjrekar, 1997). The inhabitants of Spiti are mainly the followers of Buddhism, who have been using plants for health care since time immemorial. The local herbal practitioners are called *amchis*, and due to the remoteness of the area, a majority of the inhabitants are still dependent on the “amchis medical system” for curing diseases (Kala and Manjrekar, 1999). This practice of herbal use in medicine is also known as the ‘Tibetan system of medicine’.

2.1.2 Ladakh

Ladakh is located in the rain shadow of the main Himalayan massif between $32^0 15'$ to 36^0 N and $75^0 15'$ to 80^0 15′ E. It is bounded by eastern range of Karakoram Mountains on the north and western extreme of the main Himalaya on the south. The altitude, in general, ranges from 3000 m (lower Indus and Nubra valleys) to 7680 m (Saser Kangri). This cold desert mountain landscape is broken by a host of small to large rivers, notably the Indus, and its tributaries like Zanskar, Markha, Shyok, Nubra, and Suru. Ladakh is extremely cold during winter when temperature goes down at places to minus 30 to 40 degrees celsius. The vegetation of Ladakh is dry alpine steppe and depicts typical high-altitude desert characteristics (Kala & Mathur, 2002). A total of 611 plant species have been reported from different parts of Ladakh, of which 540 are dicots, 65 monocots, and 2 the gymnosperms (Kachroo et al., 1977). Annual and perennial herbs followed by few dwarf shrubs dominate the flora of Ladakh. The remarkable feature of Ladakh is a continuous and successive change in the floristic pattern during growth period. Plants like *Draba, Christolea,* and *Arabis* are the first to bloom, followed by species of *Corydalis, Primula, Saxifraga,* etc. *Saussurea, Anaphalis, Leontopodium, Jurinea,* and *Waldheimia* bloom during the late summer (Aug-Sep). To protect the biodiversity of Ladakh, three protected areas viz., Hemis National Park (4100 km^2), Karakoram Wildlife Sanctuary (5000 km^2), and Changthang Wildlife Sanctuary (4000 km^2), have been notified so far. The people of eastern Ladakh (Leh) are predominantly Buddhist while people of western Ladakh (Kargil) have a high proportion of Muslims. Ladakhis have their own social customs like Tibetans right from childbirth to death. In Buddhist inhabited areas, Tibetan or amchi medical system prevails for the treatment of diseases, whereas in the Muslim inhabited areas such as Drass, Kargil, Sankhoo, and Mulbek Unani system is predominated.

2.2 *Survey of Medicinal Plants*

A comprehensive literature survey was carried out to gather information on indigenous uses of medicinal plants growing in the cold deserts of India. Besides, the ethnobotanical surveys were conducted across Spiti and Ladakh during the snow free period. A total of 40 *amchis* in Spiti and 43 amchis in Ladakh were approached for interview and questionnaire surveys. In Spiti, a total of 19 localities were visited from Pin valley to Kibber, Tabo, Kaza, Tarbak, and Pilshur for understanding the uses of medicinal plants by *amchis*, and also the distribution of medicinal plants across the various habitats. In Ladakh, *amchis* resided in different valleys were approached for the collection of information on use of medicinal plants. Besides, the field surveys were conducted in Nubra, Kargil, Drass, Changthang, Hemis, and Leh valleys of Ladakh for understanding the distribution of medicinal plants across various habitats. Unstructured and semistructured questionnaire surveys were conducted and information was collected on the medicinal plants in use, plant parts used, period of flowering and fruiting, and area of their collection. Field visits were made with a*mchis* for identification and verification of plant species in use, and also for determination of their occurrence in the specific localities, habitats, and elevations. The survey areas were stratified into five zones based on the range of elevations, such as <3500 m, ≥ 3500 m to <4000 m, ≥ 4000 to <4500 m, ≥ 4500 to <5000 m and ≥ 5000 m. The occurrence of prominent medicinal plant species across the elevations was recorded and analyzed. The nomenclature as given in Aswal and Mehrotra (1994) was followed.

3 Results and Discussion

3.1 *Diversity and Distribution of Medicinal Plants in Indian Cold Deserts*

The literature review and the primary surveys across Spiti and Ladakh region of Indian cold deserts have resulted in documentation of 337 medicinal plant species, which are used by the *amchis* for treatment of ailments. These medicinal plat species are distributed over 55 families. Of all the plant families, Asteraceae is the most dominant family, having 71 medicinal plant species, followed by Lamiaceae (25), Ranunculaceae (23), Brassicaceae (17), and Fabaceae (17). Of these medicinal plants, a total of 45 species are placed in various threat categories, which include 8 species in critically endangered, 7 endangered, and 9 vulnerable categories (Kala, 2003, 2006).

The distribution of medicinal plants in cold desert areas of India is influenced by geographical, geological, and climatological factors. In Ladakh, vegetation changes gradually from primarily alpine meadow (Carex, Potentilla, Nepeta) on the north side of the Himalayan crest to steppe vegetation (Caragana, Artemisia, Stachys, Ephedra, Stipa) to the north and east, with shrublands (Hippophae, Salix, Myricaria) along the river courses due to various geographical and climatological factors (Kachroo et al., 1977; Kala, 2003). During present study, it was observed that *Hippophae rhamnoides*, a rare medicinal plant, was the most dominant shrub species along the Indus, Shyok, and Nubra river valleys, followed by Myricaria. Apparently, juniper and birch have disappeared from many areas within historical times and only a few relict stands still survive in some of the more remote river gorges (Fox et al., 1994).

There are a few localities which possess a great diversity of medicinal plants. Pilshur and Tarbak in the Pin valley of Spiti region of Himachal Pradesh and Sapi Nala, Padum, Rangdum, Ralakunb, and Sinku-la in the Zanskar region of Ladakh harbor a rich diversity of medicinal plant species. These areas sustained good populations of rare and endangered medicinal plants as well which include *Dactylorhiza hatagirea, Bergenia stracheyi, Delphinium cashmerianum, Saussurea bracteata, Podophyllum hexandrum,* and *Picrorhiza kurrooa* due to their habitat diversity, relatively higher soil moisture and humidity (Kala, 2003, 2006). Such localities may have the greater species diversity because of their proximity to moister Himalayan region, which has both Himalayan and trans-Himalayan elements. A study conducted in cold deserts of India reports that Zanskar region in Ladakh endows with largest proportion of the medicinal plant (33%), followed by the Pin valley (31%) in Spiti and the Leh valley (26%) in Ladakh (Kala, 2003).

A study in Spiti cold desert of Himachal Pradesh reports that rare and endangered medicinal plants are quite localized in Spiti, and they are found in patches, except *Arnebia euchroma, Ferula jaeschkeana,* and *Ephedra gerardiana.* Although, the average land area under the *Arnebia euchroma* is highest (100,000 m^2), it is more or less uniformly distributed in dry scrub/scree slopes (Kala, 2000a).

The Pin Valley is known to harbor the highest number of rare and endangered medicinal plants (23 species), followed by the Zanskar Valley (21 species), the Leh

Valley's high passes and meadows (18 species), Nubra (12 species), and Malari (5 species) (Kala, 2003).

Elevation is an important geographical factor, which determines the distribution and upper altitudinal limits of medicinal plant species in the high-altitude cold desert areas. The highest number of medicinal plant species is reported to occur in the ≥3500 to <4000 m altitudes, followed by ≥4000 to <4500 m and < 3500 m altitudes. Above 5000 m elevation, the environmental conditions become so stressful even in the cold desert areas that only a few medicinal plants can grow in such elevations (Kala, 2003). The medicinal plant species predominated in the lower elevations (viz., <3500 m of Ladakh and Spiti are: *Artemisia sieversiana, Clematis tibetina, Aconitum violaceum, Geranium sibiricum, Dactylorhiza hatazirea, Artemisia dracunculus, Dracocephalum staminium, Lonicera spinosa, Physochlaina praealta, Echinops cornigerus, Taraxacum officinale,* and *Ephedra gerardiana* (Table 1).

A total of 117 species of medicinal plants are reported from the elevation ranges ≥3500 to <4000 m. *Artemisia desertorum, Clematis orientalis, Lychnis nutans, Circium wallichii, Actinocarya tibetica, Hyoscyamus niger, Ferula jaeschkeana, Gentianella paludosa, Tanacetum tibeticum, Gentianella moorcroftiana, Sagina saginoides, Thalictrum alpinum, Artemisia salsoloides, Cousinia thomsonii, Hyssopus officinale, Erodium tibetanum, Koelpinia linearis, Rheum moorcroftianum,* and *Euphorbia stracheyi* are the important medicinal plants growing between ≥3500 to <4000 m elevations in the cold deserts of Spiti and Ladakh.

The elevation ranges between ≥4000 and < 4500 m are the preferred range of altitudes for 105 medicinal plant species growing in the cold desert of India. The prominent medicinal plant species growing to this elevation ranges include *Chrysanthemum tibeticum, Lactuca serriola, Paraquilegia microphylla, Silene moorcroftiana, Swertia thomsonii, Thermopsis inflata, Thymus linearis, Lancea tibetica, Aconitum rotundifolium, Bergenia stracheyi, Christolea crassifolia, Oxyria digyna, Senecio tibeticus,* and *Corydalis meifolia.*

The high elevation ranges between ≥4500 and < 5000 m possess the limited number of medicinal plants in comparison to lower elevations ranges. The important medicinal plants inhabited to this elevation range are *Rhodiola imbricata, Thlaspi montanum, Artemisia stracheyi, Echinops niveus, Corydalis tibetica, Rhodiola heterodonta, Acantholimon lycopodioides, Waldheimia tomentosa, Gentiana olivieri, Sedum ewersii, Saxifraga flagillaris, Chrysanthemum griffithi, Delphinium brunonianum, Delphinium cashmerianum,* and *Leontopodium nanum.* The lowest number of medicinal plant species are inhabited above >5000 m elevations, which include *Delphinium viscosum, Gentiana nubigena, Gentiana squarrosa, Gentiana stracheyi, Arenaria bryophylla, Adonis chrysocyathus,* and *Cremanthodium arnicoides* (Table 1).

The length of flowering and fruiting period for the majority of medicinal plant species is relatively enlarged (e.g., June to September) at the lower elevations (<3500 m), whereas at the high elevations (>5000 m), the flowering and fruiting for majority of medicinal plants is squeezed to July and August only. Among the various habitats of medicinal plants, most of the medicinal plant species prefer to grow in the moist areas.

Table 1 Prominent medicinal plant species in the Indian cold deserts, their habitats, plant part used, traditional uses, locality of occurrence, and period for flowering and fruiting

Medicinal plant species	Habitat type	Part used	Uses	Locality	Flowering and fruiting
Medicinal plant species growing below <3500 m elevation					
Artemisia sieversiana Willd.	Rocky slopes	Leaf, flower, root	Intestinal worms, bronchitis	Leh, Changthang valley	July-Sep
Clematis tibetina Kuntze	Cultivated areas	Root, stem	Itching	Leh valley	July-Sep
Aconitum violaceum Jac. Ex Stapf	Moist areas	Root	Cough	Nubra valley	June-Aug
Geranium sibiricum Linn.	Sandy & wastelands	Leaf	Diarrhea	Nubra valley	July-Sep
Dactylorhiza hatazirea (D.Don) Soo.	Moist places	Tuber	Kidney complaints, aphrodisiac	Nubra valley	June-Aug
Artemisia dracunculus Linn.	Wastelands	Leaf	Toothache, menstrual cycle	Nubra valley	June-Aug
Dracocephalum staminium Kar. & Kir.	Slopes	Whole	Cough, headache	Changthang, Nubra valley	July-Sep
Lonicera spinosa (Jac. Ex Decne.) Walp.	Rocky slopes	Flower	Asthma, headache	Leh valley	July-Sep
Physochlaina praealta (Decne.) Miers	Stony slopes, roadsides	Seeds, flower	Toothache	Leh valley	July-Sep
Echinops cornigerus DC.	Village areas	Leaf, stem, seed, flower	Tonic, septic wounds, food poisoning	Leh valley	July-Sep
Taraxacum officinale Wigg. Boiss.	Marshy	Root, stem	Tonic, mouth blisters	Leh, Nubra valley	June-Aug
Ephedra gerardiana wall. ex Stapf.	Stony slopes, gravelly soil	Stem	Asthma, fever, bronchitis, rheumatism	Leh, Nubra valley	June-Sep
Medicinal plant species growing ≥3500 m to <4000 m elevation					
Artemisia desertorum Spreng	Wastelands	Leaf	Intestinal pain	Zanskar valley	June-Sep
Clematis orientalis L.	Dry slopes	Flower, leaf	Fever	Leh, Kargil, Nubra	July-Sep
Lychnis nutans Benth.	Moist places	Root	Dysentery	Zanskar valley	June-Sep
Circium wallichii DC.	River valley	Leaf	Stomach pain	Leh, Changthang valley	July-Sep

(continued)

Table 1 (continued)

Medicinal plant species	Habitat type	Part used	Uses	Locality	Flowering and fruiting
Actinocarya tibetica Benth.	Undulating areas	Root	Fever	Leh, Nubra valley	July-Sep
Hyoscyamus niger L.	Village site	Fruit	Toothache	Leh, Nubra, Hemis	July-Sep
Ferula jaeschkeana Vatke	Dry slopes	Root, seeds	Chest pain	Drass	July-Sep
Gentianella paludosa (Hk.) H. Smith	Moist slopes	Root	Tonic	Zanskar valley	July-Sep
Tanacetum tibeticum Hk. f. and Thoms. ex cl.	Sandy places, stony slopes	Leaf., flower	Jaundice, intestinal worms	Leh valley	July-Sep
Gentianella moorcroftiana (wall. ex. G. Don) airy Shaw	Moist slopes	Leaf., flower	Fever, headache, jaundice, acidity	Zanskar	July-Sep
Sagina saginoides (L.) karst.	Moist area	Whole	Headache	Zanskar	July-Aug
Thalictrum alpinum Linn.	Plateau	Root, leaf	Fever	Changthang	July-Aug
Artemisia salsoloides Willd.	Dry slopes	Shoot	Intestinal complaints	Zanskar valley	June-Aug
Cousinia thomsonii Cass.	Wastelands	Leaf., flower	Fever, arthritis	Leh, Nubra, Hemis	July-Sep
Hyssopus officinale L.	Dry slopes	Leaf	Fever, stomachache	Zanskar	July-Sep
Erodium tibetanum Edgew and Hk. f.	Alpine slopes	Shoot	Indigestion	Nubra	June–July
Koelpinia linearis pall.	Alpine pasture	Shoot	Rheumatism	Drass	July-Aug
Rheum moorcroftianum Meissn	Marshy	Root	Internal injury	Hemis, Markha	July-Aug
Euphorbia stracheyi Boiss.	Sandy places	Latex	Boils	Leh valley	July-Sep
Medicinal plant species growing below ≥4000 m to <4500 m elevation					
Chrysanthemum tibeticum Hk. F. and T.	Rocky slopes	Shoot	Septic wounds	Zanskar valley	June-Aug
Lactuca serriola L.	Alpine slopes	Shoot	Jaundice, malaria	Leh, Nubra, Markha	July-Sep
Paraquilegia microphylla (Royle) Drumm. Et hutch.	Moist areas	Flower	Acidity	Zanskar	July-Sep

(continued)

Table 1 (continued)

Medicinal plant species	Habitat type	Part used	Uses	Locality	Flowering and fruiting
Silene moorcroftiana wall. ex Benth.	Marshy	Leaf., flower	Fever	Zanskar	June-Sep
Swertia thomsonii cl. ex Hk. f. & Th.	Moist area	Whole	Fever, headache	Hunder, Zanskar	July-Aug
Thermopsis inflata Camb.	Scree	Fruit	Swelling	Zanskar	July-Sep
Thymus linearis Benth.	Rocky	Whole	Tonic	Rumbek, Zanskar	June-Aug
Lancea tibetica Hk. f. Th.	Marshy	Seeds	Cough, blood vomiting	Zanskar	July-Aug
Aconitum rotundifolium Kar. and Kir.	Dry scrub	Root	Cough, blood purification	Zanskar	July-Aug
Bergenia stracheyi (Hk. f. & T.) Engl.	Rocky	Root, flower	Blisters	Zanskar	July-Sep
Christolea crassifolia Camb.	Stony slopes	Shoot	Liver complaints	Nubra	July-Sep
Oxyria digyna (L.) hill	Sandy	Flower	Diarrhea	Leh, Nubra, Khardungla, Markha	July-Sep
Senecio tibeticus Hk. f.	Plateau	Leaf	Diuretic	Changthang	June-Aug
Corydalis meifolia wall.	Marshy	Shoot, root	Leprosy, headache, stomachache	Zanskar	July-Aug
Medicinal plant species growing ≥4500 m to <5000 m elevation					
Rhodiola imbricata Edgew.	Moist Rocky	Shoot	Cough	Zanskar	July-Sep
Thlaspi montanum L.	Rocky	Leaf	Cuts & wounds	Zanskar valley	June-Aug
Artemisia stracheyi Hk. f. and T. Ex comp.	Plateau	Shoot	Toothache	Changthang	June-Aug
Echinops niveus wall. ex DC.	Plateau	Shoot	Septic wounds	Changthang	July-Aug
Corydalis tibetica Hk. f. and Th.	Plateau	Whole	Fever	Changthang	July-Aug
Rhodiola heterodonta (Hk. f. and T.) Boriss.	Rocky	Shoot	Cough, lung infection	Zanskar	July-Sep
Acantholimon lycopodioides (Girard) Boiss.	Rocky slopes	Whole	Cardiac disorder	Leh valley	July-Aug
Waldheimia tomentosa (Decne.) regel.	Stony slopes	Leaf., flower	Septic wounds	Leh valley	July-Sep

(continued)

Table 1 (continued)

Medicinal plant species	Habitat type	Part used	Uses	Locality	Flowering and fruiting
Gentiana olivieri Griseb.	Moist slopes	Shoot, root	Fever	Zanskar	July-Aug
Sedum ewersii Ledeb.	Moist bouldery	Shoot	Appetizer, toothache	Zanskar	July-Sep
Saxifraga flagillaris Willd. Ex Sternb.	Glacial moraines	Leaf., flower, stem	Fever, jaundice, hepatitis	Khardung la	July-Sep
Chrysanthemum griffithi cl.	Plateau	Flower	Menstruation	Changthang	July-Aug
Delphinium brunonianum Royle	Moist rocky slopes	Seeds	Insecticide	Changthang	July-Sep
Delphinium cashmerianum Royle	Stony slopes	Seeds, shoot, flower	Insecticide, colic, throat-ache	Leh valley	July-Sep
Leontopodium nanum (Hk. f.) hand.	High passes	Whole	Septic wounds	Leh valley	July-Aug
Medicinal plant species growing ≥5000 m elevation					
Delphinium viscosum Hk. f. Th.	Moist stony area	Flower, Root	Rheumatism, edema	Khardungla	July-Aug
Gentiana nubigena Edgew.	Moist slopes	Flower	Cough	Khardungla	July-Aug
Gentiana squarrosa Ledeb.	Alpine slopes	Root	Diuretic	Zanskar valley	July-Aug
Gentiana stracheyi (cl.) Kitam.	Moist places	Root	Tonic	Zanskar valley	July-Aug
Arenaria bryophylla Fern.	Stony slopes	Shoot	Kidney disorder	Leh, Changthang valley	Aug-Sep
Adonis chrysocyathus Hk. f. and T.	Plateau, passes	Shoot	Skin diseases	Changthang	July-Aug
Cremanthodium arnicoides DC.	Moist areas	Shoot	Fever, diarrhea	Khardungla	July-Aug

3.2 *Medicinal Uses of Plants and the Role of* Amchis

The inhabitants of Ladakh and Lahual-Spiti have developed their own healthcare system based on the available natural resources, which include medicinal plants, animal organs, and a few minerals. This system of therapy is known as Tibetan medical system. It is also known as *Amchi* medical system because *Amchis* are the practitioners of this system. In Ladakh, the *amchis* are known to look after 60% of public health, whereas in Lahaul-Spiti it is reported up to 80% (Kala, 2005a).

Monasteries also play an important role in the development of Tibetan medical system through establishing Lord Buddha as a Medicine Buddha, who is considered as the Supreme Healer in this system of medicine (Kala, 2000b).

The different parts of plant species are gathered by *amchis* for the preparation of medicine to prescribe for curing ailments prevalent in cold deserts of Ladakh and Spiti. The leaves of plant species in Amchi Medical System or Sowa Rigpa are used quite frequently, as the leaves of 103 species of medicinal plants are reported to use for preparing medicine by *amchis*, followed by the use of flowers, roots, shoots, seeds, and fruits. *Amchis* also use lichens, in addition to flowering plants or angiosperms for curing ailments. Some of the medicinal plants are used frequently due to their use for common ailments such as cold, cough, fever, headache, and cut and wounds. *Arnebia euchroma* is one such species belongs to family Boraginaceae and has relatively good population between 3500 and 4200 m elevation, and it is used to cure number of ailments, including earache, eye diseases, cuts, wounds, apart from to strengthen the hair. This species is also reported from upper areas of the Kinnaur district of Himachal Pradesh, mainly Pooh sub division, which falls in the cold desert areas of India (Verma & Tewari, 2016). The stem of *Ephedra gerardiana*, another important species in cold desert, is crushed to powder after sun drying, which is mixed with rhizomes of *Zingiber officinale*, and the mixture is prescribed to consume with hot water twice a day for treatment of cold and cough. The roots of *Saussurea costus*, a species belongs to family Asteraceae and locally called as Kuth, are consumed after crushing to powder for treatment of fever, cold, and cough. The flowers of *Saussurea bracteata* after drying are consumed twice a day with hot water for curing of cold and cough (Kala, 2003; Ballabh & Chaurasia, 2007). The extract of *Artemisia dracunculus* is used to reduce fever and relieve toothache, along with the treatment of gastrointestinal problems in the Nubra valley of Ladakh, Kibber Wildlife Sanctuary, and the Lahual valley in Himachal Pradesh (Singh & Chauhan, 2005; Devi et al., 2013).

Ferula jaeschkeana, a perennial herb from the Apiaceae family, distributed in Ladakh, and Lahaul and Spiti, grows on dry, sunny slopes between $\geq$3500 m to $<$4000 m elevations, is traditionally used in folk medicine for a range of disorders, including chest pain, tumors, ulcers, chronic conditions, asthma, whooping cough, flatulent colic, pneumonia, and bronchitis. In Ayurveda, it is considered helpful for treating flatulent colic, pneumonia, and respiratory ailments (Yaqoob & Nawchoo, 2017). *Hippophae rhamnoides* is one of the most important species in cold deserts of India, which is inhabited along riversides in sandy and gravel ground at elevations from 2700 to 5200 m (Acharya et al., 2010). Its fruits are used for number of purposes, including treatment of tumors (Verma & Tewari, 2016). It has high amount of Vitamin C. The use of plant species for curing different ailments and growing across the elevations and habitats in Ladakh and Spiti region of India are described in Table 1.

3.3 Climate Change Impacts on Medicinal Plants

The present study has explicitly expressed that the occurrence and distribution of medicinal plant species in cold deserts of Spiti and Ladakh are influenced by elevations, the presence of soil moisture, and microclimatic and habitat conditions. Temperature controls plant metabolic rates, which are essential for plant growth, survival, and reproduction, and it decreases with elevation. Temperature is a fundamental factor shaping plant species' physiology, distribution, diversity, and ecosystem roles, with its influence becoming increasingly significant under climate change scenarios (Hemp, 2006; Kala & Silori, 2013). There are studies elsewhere which claim that mean annual temperature is the major determinant of plant species distribution in Himalayas, as >54% species responded to temperature for their distribution, followed by soil characteristics and degree of slope (Maharjan et al., 2022). It is also considered that in water deficit ecoregions, precipitation plays a significant role in phenological development in combination with temperature (Gerst et al., 2017).

Climate change in response to rising temperature and changing precipitation patterns in cold desert areas of Indian Himalayas can force medicinal plants to shift their distribution to higher altitudes (Kala, 2013). This upward shifts of species leads to habitat fragmentation and a reduction in suitable habitats, especially for species that are already restricted to specific microhabitats or altitudinal zones. Some high-altitude cold-adapted species are at the risk of extinction as they may eventually run out of suitable higher ground to colonize (Gairola et al., 2010).

Altered temperature and precipitation regimes are changing the timing of plant life cycle events (phenology), such as earlier flowering in high-altitude medicinal plants (Ratha et al., 2012). These changes can disrupt plant reproduction and reduce seed germination rates, threatening the long-term survival of certain species as depicted in Table **2**. Climate-induced stress (e.g., temperature extremes, drought) can alter the production of secondary metabolites, the compounds responsible for the medicinal properties of these plants (Gairola et al., 2010). Changes in secondary metabolites may affect the efficacy and potency of traditional medicines derived from these plants, posing a challenge to *amchi* medical system that rely heavily on consistent medicinal plants quality. The loss or decline of medicinal plant resources threatens the livelihoods of local communities who depend on traditional medicine for healthcare and income. Erosion of traditional knowledge and ethno-medical practices is also observed as plant availability and reliability decrease (Ratha et al., 2012).

The cold desert region in India supports some unique and valuable but threatened medicinal plant species. For instance, *Ferula jaeschkeana*, is considered a threatened medicinal plant due to habitat loss and overexploitation in its native Himalayan range. The plant is included in several traditional herbal formulations and is valued for its broad therapeutic applications, but its wild populations are declining, highlighting the need for conservation efforts (Kala, 2005b, 2026). Like *Ferula jaeschkeana*, there are many medicinal plants which have become rare and endangered due to various reasons, including loss of habitats, over exploitation,

Table 2 Effects of climate change on medicinal plants and potential consequences on cold desert ecosystems

Climate change effects	Impact on medicinal plants	Potential consequences for cold desert regions	Source
Warming, rising temperature	Upward shift of species	Habitat loss, possible local extinction of species	Kala (2013).
Altered precipitation	Change in phenology	Reduced reproduction, lower germination	Ratha et al. (2012); Gerst et al. (2017); Ojo et al. (2024).
Increased stress due to altered environmental factors	Altered secondary metabolites	Change medicinal efficacy	Gairola et al. (2010).
Invasive species	Competition with native medicinal plant species	Loss of medicinal plants and overall biodiversity	Zargar et al. (2025).
Socio-economic pressures	Overharvesting, loss of traditional indigenous knowledge	Threat to traditional medicinal system	Kala and Silori (2013).

and global climate change (Kala, 2005b). There is a major threat to those species which are uprooted and their underground parts such as rhizomes, tubers, bulbs, and roots are used for medicine. *Dactylorhiza hatagirea* (Angbolakpa, Hathajari, Salampanja), *Picrorhiza kurrooa* (Honglen, Katuki), *Saussurea gnaphaloides* (Pangtsi), *Podophyllum hexandrum* (Olmoselo, tandik), *Lancea tibetica* (Payogtsa), *Rheum webbianum* (Tukshu, tatra), *Rheum tibeticum* (Tukshu), *Aconitum heterophyllum* (Bonna karpo, Atis), *Aconitum violaceum*, etc. are such species which have become endangered throughout the Himalayan and trans-Himalayan region due to over-collection. It is realized that expanding the coverage of protected areas can help in safeguarding the medicinal plants diversity along with their ecosystem in cold desert areas of Indian Himalaya. Though Ladakh boasts of distinct assemblage of high-altitude flora and fauna, it still remains the least protected ecosystem with only one national park, viz., Hemis National Park covering Rumbak and Markha valleys.

4 Conclusion

The cold deserts in Indian Himalaya stands as a global hotspot for unique medicinal plants diversity, supporting traditional medicinal systems, local economies, and healthcare. The distribution and abundance of medicinal plants are determined by climatological, geographical, and geological factors. Elevation is one of the important factors, which determines the distribution and upper altitudinal limits of medicinal plant species in the high-altitude cold desert areas. The elevation ranges from ≥3500 m to <4000 m endows with highest number of medicinal plant species, followed by ≥4000 to <4500 m and < 3500 m altitudes. Many species of medicinal plants, especially those have commercial values, low population size, and narrow range of distribution, are in danger due to overharvesting and habitat destruction. At present, the climate change is also known to lead the shift in distribution of medicinal plants to higher elevations due to alterations in temperature and precipitation. The adverse impacts of overharvesting, habitat degradation coupled with climate change not only threaten the diversity of medicinal plants in Indian cold deserts, but also weaken the traditional healthcare systems and local livelihoods dependent on these invaluable plant species.

References

Acharya, S., Stobdan, T., & Singh, S. B. (2010). Seabuckthorn (*Hippophae* sp. L.): New crop opportunity for biodiversity conservation in cold arid trans-Himalayas. *Journal of Soil and Water Conservation, 9*(3), 201–204.

Aswal, B. S., & Mehrotra, B. N. (1994). *Flora of Lahaul-Spiti (a Cold Desert in north west Himalaya)*. Bishen Singh Mahendra Pal Singh.

Ballabh, B., & Chaurasia, O. P. (2007). Traditional medicinal plants of cold desert Ladakh—Used in treatment of cold, cough and fever. *Journal of Ethnopharmacology, 112*(2), 341–349. https://doi.org/10.1016/j.jep.2007.03.020

Devi, U., Seth, M. K., Sharma, P., & Rana, J. C. (2013). Study on ethnomedicinal plants of Kibber wildlife sanctuary: A cold desert in trans Himalaya, India. *Journal of Medicinal Plants Research, 7*, 3400–3419.

Fox, J. L., Nurbu, C., Bhatt, S., & Chandola, A. (1994). Wildlife conservation and land-use changes in the trans-Himalayan region of Ladakh, India. *Mountain Research and Development, 14*, 39–60.

Gairola, S., Shariff, N. M., Bhatt, A., & Kala, C. P. (2010). Influence of climate change on production of secondary chemicals in high altitude medicinal plants: Issues needs immediate attention. *Journal of Medicinal Plants Research, 4*(18), 1825–1829.

Gerst, K. L., Rossington, N. L., & Mazer, S. J. (2017). Phenological responsiveness to climate differs among four species of Quercus in North America. *Journal of Ecology, 105*(6), 1610–1622. https://doi.org/10.1111/1365-2745.12774

Hemp, A. (2006). Continuum or zonation? Altitudinal gradients in the forest vegetation of Mt Kilimanjaro. *Plant Ecology, 184*, 27–42.

Joshi, R. K., Satyal, P., & Setzer, W. N. (2016). Himalayan aromatic medicinal plants: A review of their ethnopharmacology, volatile phytochemistry, and biological activities. *Medicine, 3*(1), 6. https://doi.org/10.3390/medicines3010006

Kachroo, P., Sapru, B. L., & Dhar, U. (1977). *Flora of Ladakh: An ecological and taxonomic appraisal*. Dehradun, India: Bishen Singh Mahendra Pal Singh.

Kala, C. P. (2000a). Status and conservation of rare and endangered medicinal plants in the Indian trans-Himalaya. *Biological Conservation, 93*(3), 371–379.

Kala, C. P. (2000b). *A study of traditional health care and medicinal plants of Ladakh*. Wildlife Institute of India.

Kala, C. P. (2003). *Medicinal plants of Indian trans-Himalaya: Focus on Tibetan use of medicinal resources*. Dehradun, India: Bishen Singh Mahendra Pal Singh.
Kala, C. P. (2004). Assessment of species rarity. *Current Science, 86*(8), 1058–1059.
Kala, C. P. (2005a). Health traditions of Buddhist community and role of amchis in trans-Himalayan region of India. *Current Science, 89*(8), 1331–1338.
Kala, C. P. (2005b). Indigenous uses, population density, and conservation of threatened medicinal plants in protected areas of the Indian Himalayas. *Conservation Biology, 19*(2), 368–378.
Kala, C. P. (2006). Medicinal plants of the high altitude cold desert in India: Diversity, distribution and traditional uses. *The International Journal of Biodiversity Science and Management, 2*(1), 43–56.
Kala, C. P. (2011). Floral diversity and distribution in the high altitude cold desert of Ladakh, India. *Journal of Sustainable Forestry, 30*, 360–369. https://doi.org/10.1080/10549811.2011.534036
Kala, C. P. (2013). Climate change and challenges of biodiversity conservation. In C. P. Kala & C. S. Silori (Eds.), *Biodiversity, communities and climate change* (pp. 259–269). New Delhi: The Energy and Resource Institute (TERI).
Kala, C. P. (2026). High altitude medicinal plants: Traditional uses and distribution patterns. In S. Chandra & H. Lata (Eds.), *High altitude medicinal plants: Botany, conservation and cultivation* (pp. 59–68). Springer. https://doi.org/10.1007/978-3-031-99406-7_4
Kala, C. P., & Manjrekar, N. (1999). Ethno-medicobotany of Indian trans-Himalaya: A case study from Spiti. *Journal of Economic and Taxonomic Botany, 23*(1), 177–183.
Kala, C. P., & Mathur, V. B. (2002). Patterns of plant species distribution in the trans-Himalayan region of Ladakh India. *Journal of Vegetation Science, 13*(6), 751–754.
Kala, C. P., & Silori, C. S. (2013). *Biodiversity, Communities and Climate Change*. New Delhi: TERI Publications.
Kapadia, H. (1996). *SPITI: Adventures in the trans-Himalaya*. Indus Publishing Company.
Maharjan, S. K., Sterck, F. J., Raes, N., & Poorter, L. (2022). Temperature and soils predict the distribution of plant species along the Himalayan elevational gradient. *Journal of Tropical Ecology, 38*(2), 58–70.
Manjrekar, N. (1997). *Feeding ecology of ibex (Capra ibex sibirica) in Pin Valley National Park, Himachal Pradesh*. Ph.D. Thesis, Unpublished, Saurastra University, Rajkot, India.
Mishra, A. P., Kumar, A., & Yadav, S. N. (2023). Ecology and conservation of threatened medicinal plants in the trans-Himalayan region of Nanda Devi biosphere reserve, Western Himalaya. *Trees, Forests and People, 14*, 100451. https://doi.org/10.1016/j.tfp.2023.100451
Ojo, T. A., Kirkman, K., & Tedder, M. (2024). Effects of warming and rainfall variation on grass phenology and regenerative responses in Mesic grassland. *South African Journal of Botany, 174*, 107–115. https://doi.org/10.1016/j.sajb.2024.08.050
Paul, S., Kanwal, K. S., Sharma, S., Bhatt, I. D., & Sundriyal, R. C. (2021). Diversity, distribution, indigenous uses and conservation of medicinal and aromatic plants in Changthang landscape of Ladakh, trans Himalaya. *Journal of Non-Timber Forest Products, 28*(4), 138–146. https://doi.org/10.54207/bsmps2000-2022-BNX122
Ratha, K. K., Sruajani, M. S., Arya, J. C., & Joshi, G. C. (2012). Impact of climate change on diversity of Himalayan medicinal plant: A threat to Ayurvedic system of medicine. *International Journal of Research in Ayurveda & Pharmacy, 3*(3), 327–331.
Sharma, N., & Kala, C. P. (2018). Harvesting and management of medicinal and aromatic plants in the Himalaya. *Journal of Applied Research on Medicinal and Aromatic Plants, 8*, 1–9. https://doi.org/10.1016/j.jarmap.2017.09.003
Singh, V., & Chauhan, N. S. (2005). Traditional practices of herbal medicines in the Lahaul valleys, Himachal Himalayas. *Indian Journal of Traditional Knowledge, 4*, 208–220.
Verma, R.K., Tiwari, V.P. (2016). Some important medicinal plants of cold desert regions of district Kinnaur of Himachal Pradesh state in India: Their uses and chemical ingredients. Journal of Plant Chemistry and Ecophysiology *1*(2): 1009. https://austinpublishinggroup.com/jpceonline/fulltext/jpce-v1-id1009.php (Accessed on: 21 May, 2025).

Yaqoob, U., & Nawchoo, I. A. (2017). Impact of habitat variability and altitude on growth dynamics and reproductive allocation in ferula jaeschkeana Vatke. *Journal of King Saud University-Science, 29*(1), 19–27. https://doi.org/10.1016/j.jksus.2015.10.002

Zargar, S. A., Malik, R. A., Khuroo, A. A., Ganie, A. H., & Reshi, Z. A. (2025). Alien flora of the Himalayan highlands: Naturalised and invasive plants in the trans-Himalayan region of Ladakh India. *Biological Invasions, 27*(4), 1–19. https://doi.org/10.1007/s10530-025-03562-z

Ethnobotany and Conservation Challenges in the Cold Desert Ecosystems of Uttarakhand, India

Vikram S. Negi, Rishav Rawal, Priyanka Rana, Yashwant Rana, K. C. Sekar, I. D. Bhatt, and R. K. Maikhuri

1 Introduction

The Trans-Himalayan zone, located in the rain shadow of the main Himalayan range, is often described as a "High Altitude Cold Desert" (Saxena et al., 2011). These cold desert areas are characterized by extremely low temperatures, annual rainfall of about 500–600 mm, and sparse vegetation dominated by annual and perennial herbs with few stunted shrubs (Chowdhery and Rao, 1990; Saxena et al., 2011). Therefore, cold desert regions have rarely been at the forefront of biodiversity and sustainability research, largely due to low species richness, harsh environment, and socio-economic marginality (Rawat and Everson, 2012; Negi et al., 2018a; Rawat et al., 2023). The high-altitude cold desert in the Indian Himalaya, including Lahaul-Spiti, Ladakh, parts of Sikkim, and Uttarakhand, is known as the Indian trans-Himalaya. These regions are home to traditional indigenous communities with a rich cultural heritage and extensive knowledge of medicinal plants (Kala, 2006; Negi et al., 2018b). The high-altitude cold desert of the Indian Himalaya also remained isolated due to poor

V. S. Negi (✉) · Y. Rana
Department of Botany and Microbiology, Hemvati Nandan Bahuguna Garhwal University (HNBGU), B.G.R. Campus, Pauri 246001, Uttarakhand, India
e-mail: vikramsnegii@gmail.com

R. Rawal
Department of History, Kumaun University, DSB Campus, Nainital, Uttarakhand, India

P. Rana · R. K. Maikhuri
Department of Environmental Science, HNBGU, Srinagar Garhwal 246174, Uttarakhand, India

K. C. Sekar
Garhwal Regional Center, G.B. Pant National Institute of Himalayan Environment, Srinagar Garhwal, Uttarakhand, India

I. D. Bhatt
G.B. Pant National Institute of Himalayan Environment, Kosi-Katarmal, Almora 263643, Uttarakhand, India

Y. S. Rawat and C. P. Kala (eds.), *Cold Deserts*,
https://doi.org/10.1007/978-3-032-29119-6_12

accessibility. The geographical and cultural isolation of these regions, along with the unique diversity of aromatic, medicinal, and rare plants, has led to the development of a distinct healthcare system to survive in harsh environmental conditions (Kala, 2006; Saxena et al., 2011). The trans-Himalaya sustains about 337 species of medicinal plants (Kala and Mathur, 2002; Kala 2006), which are low compared to other areas of the Himalaya due to the distinct geography and harsh environmental conditions (Kala, 2003). However, the diversity in cold deserts is unique, and with a rich diversity of rare and threatened species of medicinal plants (Pandey et al., 2018; Sekar et al., 2022). The traditional system of medicine in cold desert regions is known as *Amchi* system of medicine; *amchi* being the name given to a local practitioner of this system (Kala, 2005a, 2006), and in Uttarakhand, they are known as *Vaidyas* (Negi et al., 2010; Phondani et al., 2010, 2014).

Around 50,000 pharmaceutical companies in India source most of their medicinal raw materials from wild resources (Kala, 2006). This has placed a significant pressure on medicinal plants in alpine ecosystems, leading to a decline in their population density; many medicinal plants are becoming rare and endangered (Hamilton, 2004; Vashistha et al., 2006; Kala, 2005b; Negi et al., 2019). More than 80% of plant species used in the herbal industries are extracted from wild, particularly from alpine regions of the Himalaya (Kala, 2005b; Singh et al., 2021). About 70% of medicinal plants in the Indian Himalaya face destructive harvesting (Dhar et al., 2000; Negi et al., 2018b, 2019; Singh et al., 2021), resulting in habitat loss and raising concerns about the sustainability of both their populations and the medicinal plant-based sector. Further, climate change has impacted the availability, phenology, and gradual loss of regeneration potential and diversity of medicinal plants in the alpine ecosystems (Negi et al., 2017, 2023; Maikhuri et al., 2018). The continuous exploitation of medicinal plants from the wild, along with significant habitat loss over the past decades, has led to a decline in the populations of several high-value medicinal plants (Kala, 2003; Kumar et al., 2015; Goraya and Ved, 2017; Negi et al., 2019; Singh et al., 2021). The depletion and loss of medicinal plant diversity can directly affect human health and livelihoods, particularly when ecosystem services are no longer sufficient to meet societal needs. Therefore, this study was undertaken in a relatively underexplored, tribal-dominated region to document ethnobotanical knowledge of medicinal plants and to support the preservation of unique traditional herbal practices.

2 Methodology

2.1 Study Area

The study was conducted in the cold desert of Byans, Darma, Johar valley, located in the Kailash Sacred Landscape (KSL), under Pithoragarh district, and Niti valley in Chamoli district of Uttarakhand, bordering India, Nepal, and China. The region is inhabited by the "Bhotiyas," an ethnic community of Mongoloid origin, who traditionally engaged in trans-border trade with China (Tibet) and Nepal until the closure of trade routes in 1962 following the Sino-Indian conflict. These communities inhabit the valleys during the summer months (May to October) and migrate to lower-altitude winter settlements (below 1200 m asl) from November to April. Niti valley is inhabited by Rongpa, who are also known as the Marchha and Tolcha people. The Johar valley is inhabited by Johari, while the Darma and Byans valleys are inhabited by the Rung community. The Rung community has truly transboundary connections as these people inhabit both the KSL-India and KSL-Nepal, and share their cultural heritage (Negi et al., 2018b). Traditionally, the "Rung" culture spreads across the watersheds of the Dhauliganga and Kali rivers of KSL-India and some parts of Nepal. Bhotiya tribes of Uttarakhand have survived successfully in relatively inhospitable terrains either by aligning their activities with climatic events or by inventing unique practices. These communities have adapted to a diversified livelihood system that integrates agriculture, pastoralism, and trade in minor forest products (Maikhuri et al., 1998; Nautiyal et al., 2002; Farooquee et al., 2004; Garbyal et al., 2005; Negi et al., 2018b). They possess a rich repository of ethnobotanical knowledge on medicinal plants, reflecting the deep symbiotic relationship between biodiversity and cultural identity. They not only use medicinal plant for their healthcare, but also earn their livelihoods through the trade and commerce of plant resources.

2.2 Methods

A rapid rural appraisal survey was conducted in all the selected villages of the Johar, Darma, Byans, and Niti valleys of Uttarakhand. Indigenous people, actively engaged in traditional healthcare practice, were interviewed, followed by the filling of semi-structured questionnaires to document their unique traditional knowledge on the medicinal value of different plant species, following past studies (Maikhuri et al., 1998; Kala, 2006; Negi et al., 2010). More than 30% of villagers engaged in traditional healthcare in the valleys were interviewed and consulted for documenting the traditional and herbal knowledge system.

After establishing a prior informed consent orally, the ethno-medicinal data were gathered from the local people, especially from women and elderly people, through interviews and semi-structured questionnaires. Information on local names, modes of use, and collection practices of each medicinal plant species was documented (Farooquee et al., 2004; Phondani et al., 2014). The ethnobotanical data were validated with focused group discussions and village-level meetings were held with Vaidyas (traditional healers), medicinal plant cultivators, and local inhabitants. Cross-verification was ensured by interviewing at least five traditional healers for each reported species and its associated herbal formulations, past studies (Kala, 2006; Phondani et al., 2014).

In a few cases, the plant specimens were collected, and the subsequent interviewing of the informants for names and uses of medicinal plants were conducted. Field visits to alpine pasture sites were conducted jointly with traditional healers and other elderly knowledgeable community members to identify specific medicinal plants. More than 30% of villagers engaged in traditional healthcare in the valleys were consulted to gather perceptions on the availability of medicinal plants and phenological changes in medicinal plants due to various factors, including climate change, and their subsequent impacts on the traditional healthcare system (Negi et al., 2017; Maikhuri et al., 2018). Additional information on community perceptions of climate change impacts were obtained through personal interviews, focused group discussions, and village-level meetings, and these observations were further verified through field surveys.

3 Results

3.1 Traditional Practices and Herbal Knowledge

The Bhotiya tribe of Uttarakhand practices Kuncha, a form of vertical transhumance in which families and livestock migrate seasonally between fixed summer and winter settlements. With the arrival of spring in April and May, these communities migrate to their native settlements in Niti, Darma, Johar, and Byans. As winter approaches in October and November, they return to their winter settlements. In the present study, a total of 81 plant species belonging to 36 families have been documented in the region commonly used by the inhabitants for the treatment of various ailments in the traditional system of healthcare (Table 1). Asteraceae (12) was the most dominant family, followed by Polygonaceae (8) and Liliaceae (7) in the treatment of ailments in cold deserts. Among the documented species, 4 were trees, 8 shrubs, and 69 herbs (Fig. 1). These plants are used for the treatment of common diseases, i.e., fever, cough, cuts and wounds, jaundice, stomachache, rheumatism, skin diseases, and digestive disorders. Frequency analysis showed that cold and cough were treated with the maximum number of species (20), followed by cuts and wounds (18), and rheumatism (17) (Fig. 2). A few traditional crops cultivated in the villages of cold desert areas are also used for the treatment of ailments (Table 2).

The methods of utilizing medicinal plants vary depending on the medicinal value of the plant and the specific nature of the disease being treated. Different plant parts were reported for their medicinal use. Roots and rhizomes were the most frequently utilized plant parts, accounting for 39 ailments. Leaves were used in 11 ailments, while seeds, fruits, and flowers together were reported in 7 ailments. The whole plant was employed in 6 ailments, followed by bark in 5 and stems in 3 ailments (Fig. 3). A few MPs were prepared as formulations by combining them with other plants or edible products. For example, *Dactylorhiza hatagirea*, *Polygonatum verticillatum*, and *Picrorhiza kurrooa* are used in diabetes, while *Prinsepia utilis*, *Nardostachys grandiflora*, and *Aconitum balfourii* are commonly used for rheumatism. The root powders of *N. grandiflora* and *P. kurrooa* mixed with warm water are used for jaundice and fever. *Pinus wallichiana* and *Betula utilis* are used in bone fractures; the

Table 1 List of important medicinal plants used in the traditional system of herbal medicine in the cold desert ecosystems, Uttarakhand

S. No.	Plant name	Family	Traditional use of medicinal plants
1	*Abies pindrow* (Royle ex D.Don) Royle	Pinaceae	Paste of bark is used for the treatment of cough and cold, and the paste of the leaf is applied on cuts & wounds.
2	*Aconitum lethale* Griff.	Ranunculaceae	Tuber used in fever, rheumatism, and arthritis, and as an antidote for scorpion bites.
3	*Aconitum heterophyllum* wall. *Ex* Royle	Ranunculaceae	Paste of root is used in the treatment of fever, diabetes, stomach problems, cold & cough
4	*Aconitum violaceum* Jacquem. *Ex* Stapf	Ranunculaceae	Root of the plant used to treat fever and scorpion bites
5	*Angelica glauca* Edgew.	Apiaceae	Roots are used to treat indigestion and gastric complaints
6	*Allium stracheyi* baker	Liliaceae	Decoction of leaves is used to cure cold, cough, and jaundice
7	*Allium humile* Kunth	Liliaceae	Decoction of leaves is used for cold and cough
8	*Anaphalis nepalensis* (Spreng.) hand.-Mazz.	Asteraceae	Flowers are used for arthritis and bone fractures
9	*Crucihimalaya tibetica* (Hook.f. & Thomson) Al-Shehbaz, D.A.German & M.Koch	Brassicaceae	Leaves are used to cure wounds
10	*Eremogone bryophylla* (Fernald) Pusalkar & D.K.Singh	Caryophyllaceae	Shots are used in kidney disorders
11	*Arnebia benthamii* (wall. *Ex* G.Don) I.M.Johnst.	Boraginaceae	Roots are used to prepare a hair tonic to reduce baldness.
12	*Arnebia euchroma* (Royle *ex* Benth.) I.M.Johnst.	Boraginaceae	Roots are used as a hair tonic
13	*Artemisia maritima* L.	Asteraceae	The whole plant is used to treat cuts and gastric complications
14	*Artemisia gmelinii* Weber ex Steckm.	Asteraceae	Leaves are used to cure cough and cold
15	*Artemisia nilagirica* (C.B.Clarke) Pamp.	Asteraceae	The juice of leaves is used to treat skin disease
16	*Artemisia salsoloides* Willd.	Asteraceae	Shoot is used in intestinal complaints
17	*Astragalus rhizanthus* Benth.	Fabaceae	Roots are used in skin diseases
18	*Berberis pseudumbellata* R.Parker	Berberidaceae	Root powder is used for liver disorders and diabetes

(continued)

Table 1 (continued)

S. No.	Plant name	Family	Traditional use of medicinal plants
19	*Berberis jaeschkeana* C.K. Schneid.	Berberidaceae	Roots are used in eye diseases
20	*Bergenia stracheyi* (Hook.f. & Thomson) Engl.	Saxifragaceae	Rhizome is used for cuts, wounds, kidney stones, and liver diseases
21	*Betula utilis* D. Don	Betulaceae	Bark is used for plaster during fractures, and also used in wounds and jaundice
22	*Bistorta affinis* (D. Don) Greene	Polygonaceae	It is traditionally used to treat digestive issues and dysentery
23	*Cannabis sativa* L.	Cannabaceae	Seeds are used to cure arthritis and digestive problems
24	*Campanula pallida* wall.	Campanulaceae	The shoot of the plant is used to cure wounds
25	*Carum carvi* L.	Apiaceae	Seeds increase appetite and are used in the treatment of digestive problems and cough
26	*Cicer microphyllum* Royle *ex* Benth.	Fabaceae	Seeds are useful in jaundice
27	*Coleus barbatus* (Andrews) Benth. *Ex* G.Don	Lamiaceae	The roots are used in the treatment of worms and cough
28	*Corydalis govaniana* wall.	Papaveraceae	Root paste is applied to cuts and wounds and joint pain
29	*Cedrus deodara* (Roxb. Ex D.Don) G.Don	Pinaceae	Bark is used to cure rheumatism, back pain; oil is used to cure wounds and cuts
30	*Cypripedium himalaicum* Rolfe	Orchidaceae	The whole plant is used to treat cough
31	*Dactylorhiza hatagirea* (D. Don) Soo	Orchidaceae	Tubers are useful for cough, diabetes, wounds, and cuts
32	*Ephedra gerardiana* wall. *Ex* Klotzsch & Garcke	Ephedraceae	Root is used to cure asthma, rheumatism, and headaches
33	*Ephedra intermedia* Schrenk & C.A.Mey.	Ephedraceae	Juice extracted from shoots is used to cure joint pain
34	*Erigeron multiradiatus* (Lindl. *Ex* DC.) Benth. & Hook.f.	Asteraceae	The whole plant is used to cure stomachache
35	*Fritillaria cirrhosa* D.Don	Liliaceae	Tubers are used to treat asthma, fever, and tuberculosis
36	*Geranium wallichianum* D.Don *ex* sweet.	Geraniaceae	Decoction of leaves is used for cuts and wounds
37	*Heracleum candicans* wall. *Ex* DC.	Apiaceae	Fruits are used as aphrodisiacs and to treat intestinal parasites

(continued)

Table 1 (continued)

S. No.	Plant name	Family	Traditional use of medicinal plants
38	*Dolomiaea costus* (Falc.) Kasana & A.K.Pandey	Asteraceae	It is used to treat diarrhea, skin disease, and stomachaches
39	*Hippophae salicifolia* D.Don	Elaeagnaceae	The juice of fruits is used to cure stomachache and cough
40	*Hippophae tibetana* Schltdl.	Elaeagnaceae	It is used to treat coughs, jaundice, and rheumatism
41	*Impatiens balsamina* L.	Balsaminaceae	Leaves are used in rheumatism
42	*Iris kemaonensis* wall. *Ex* D.Don	Iridaceae	The roots are used to cure various liver complaints and constipation
43	*Juniperus communis* L.	Cupressaceae	Leaf paste is applied to the skin for disease
44	*Dolomiaea macrocephala* DC. *Ex* Royle	Asteraceae	Roots are considered to be stimulants and given in fever
45	*Lilium oxypetalum* (D. Don) baker	Liliaceae	Its paste is applied to swellings and muscular pain
46	*Malaxis muscifera* (Lindl.) Kuntze	Orchidaceae	The tonic of the plant is believed to promote overall health and vitality. It is used to treat dysentery and diarrhea
47	*Meconopsis aculeata* Royle	Papaveraceae	The whole plant is used as a tonic and also in the treatment of fever and inflammation
48	*Morina longifolia* wall. *Ex* DC	Caprifoliaceae	Roots are used in boils, and flowers in stomach disorders
49	*Nardostachys jatamansi* (D.Don) DC.	Caprifoliaceae	It is traditionally used for heart disease, rheumatism, and arthritis
50	*Nepeta eriostachya* Benth.	Lamiaceae	Bark soaked in mustard oil is used to cure hair fall
51	*Origanum vulgare* L.	Lamiaceae	Tea is prepared using dry leaves. Leaf paste is applied for skin diseases
52	*Oxyria digyna* (L.) hill	Polygonaceae	Traditional uses also include treating wounds, jaundice, and as a stomach tonic
53	*Oxytropis lapponica* (Wahlenb.) J.Gay	Fabaceae	The shoot of the plant used for joint pain
54	*Paeonia emodi* wall. Ex Royle	Paeoniaceae	The infusion of roots is given for cough, fever, diabetes, and diarrhea. Roots are used in the treatment of cuts & wounds
55	*Paris polyphylla* Sm	Melanthiaceae	Paste of dried rhizomes used for the treatment of boils and insect bites

(continued)

Table 1 (continued)

S. No.	Plant name	Family	Traditional use of medicinal plants
56	*Pedicularis gracilis* wall ex Benth.	Orobanchaceae	It is traditionally used to reduce muscle spasms, shoulder pain, and back pain
57	*Polygonatum cirrhifolium* (wall.) Royle	Asparagaceae	The root of the plant is used to treat leucorrhoea, diabetes, and urinary complaints
58	*Polygonatum verticillatum* (L.) all.	Asparagaceae	Rhizome is edible and is used as a tonic, diabetes and is also used to cure skin diseases
59	*Podophyllum hexandrum* Royle	Berberidaceae	Tubers are used in the treatment of septic wounds, fever, leucorrhoea, diabetes, and constipation
60	*Pterocyclus angelicoides* (DC.) Klotzsch	Apiaceae	Tubers are used in the treatment of stomach disorders, typhoid, and rheumatism
61	*Picrorhiza kurroa* Royle *ex* Benth.	Plantaginaceae	Roots are used to cure stomachache, cold & cough, fever, cardiac problems, diabetes, jaundice, and indigestion
62	*Sibbaldia cuneata* Edgew	Rosaceae	Paste of the root used as an ointment around the infected teeth
63	*Prinsepia utilis* Royle	Rosaceae	Seed oil is used to treat arthritis and rheumatism
64	*Rheum webbianum* Royle	Polygonaceae	Roots are used in the treatment of abdominal disorders, constipation, and wounds
65	*Rheum australe* D.Don	Polygonaceae	Rhizomes are considered very effective in the treatment of fractures of limbs and wounds in animals
66	*Rheum moorcroftianum* Royle	Polygonaceae	Roots are used in the treatment of cold & cough and asthma
67	*Rumex nepalensis* Spreng.	Polygonaceae	Young leaves are edible, and the decoction of leaves is used to cure dysentery, rheumatism, and skin disease
68	*Rhododendron campanulatum* D. Don	Ericaceae	Traditionally used to treat skin diseases, throat pain, and body aches
69	*Rhododendron anthopogon* D. Don	Ericaceae	The leaves are used to treat colds, coughs, and skin diseases
70	*Rhodiola heterodonta* (Hook. f. & Thomson) Boriss.	Crassulaceae	Shoots are used to cure cough, fever, and lung infection

(continued)

Table 1 (continued)

S. No.	Plant name	Family	Traditional use of medicinal plants
71	*Saussurea laniceps* hand.-Mazz.	Asteraceae	The infusion is used to cure stomach disorders
72	*Saussurea obvallata* (DC.) Sch.Bip.	Asteraceae	Roots are used to treat jaundice, cough, cuts & wounds
73	*Silene kumaonensis* F.NWilliams	Caryophyllaceae	It is used to cure hair dandruff
74	*Ligusticopsis wallichiana* (DC.) Pimenov & Kljuykov	Apiaceae	It is used to cure rheumatism, digestive issues, and treat cuts and wounds
75	*Selinum vaginatum* (Edgew.) C.B.Clarke	Apiaceae	The decoction of roots used in dysmenorrhea
76	*Swertia petiolata* D. Don	Gentianaceae	The whole plant is used in the treatment of stomachache, cough, and fever
77	*Taraxacum officinale* F.HWigg.	Asteraceae	Roots are used to treat liver ailments and jaundice
78	*Taxus wallichiana* Zucc	Taxaceae	Bark is used to cure rheumatism and fever and as a plaster for fractured bones
79	*Thalictrum foliolosum* DC.	Ranunculaceae	Roots are useful in the treatment of fever, jaundice, and abdominal pain
80	*Thymus linearis* Benth.	Lamiaceae	The whole plant is used to prepare a local beverage and to cure skin diseases
81	*Trigonella emodi* Benth	Fabaceae	Seeds are used to relieve indigestion and sores

resin of *Pinus* is mildly heated and plastered on the fracture part, which is immediately covered by the bark of *B. utilis*. A local formulation prepared from *Picrorhiza kurrooa* and *A. heterophylla* is traditionally used to treat fever and other ailments during childbirth by women.

3.2 Threatened Medicinal Plants in Cold Deserts

Indigenous people of the region obtained a substantial part of their income from the sale of medicinal plants. Species like *Picrorhiza kurrooa, Swertia chirayta, Arnebia benthamii, Angelica glauca, Allium stracheyi,* and *Paris polyphylla* are the most traded species in the region. However, in a changing scenario of market-driven economic changes, the uninterrupted harvesting of selected species has caused extensive damage to certain habitats and populations, which require adequate management

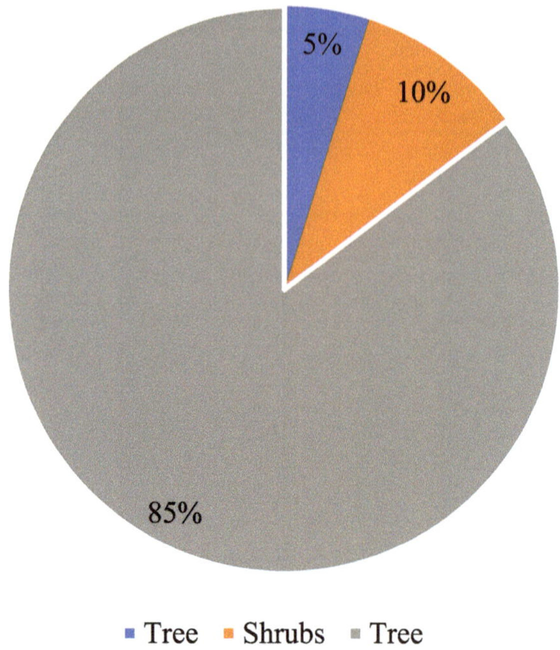

Fig. 1 Life forms of plants used in the herbal system of medicine

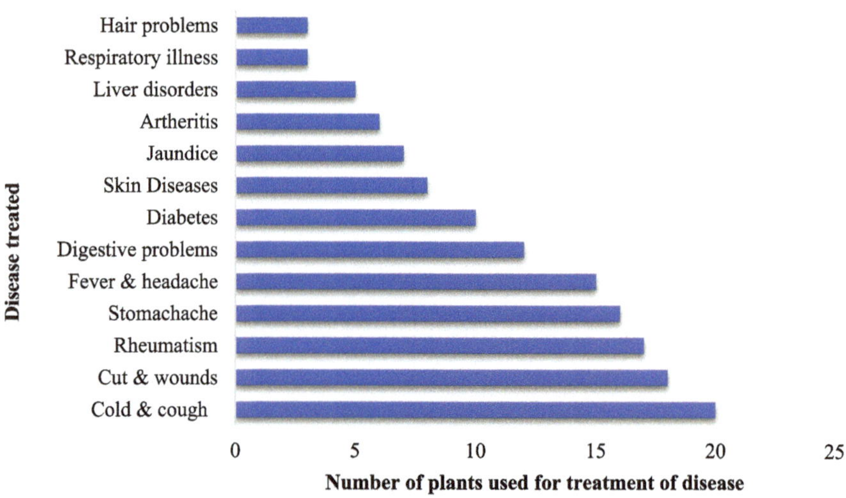

Fig. 2 Number of plants used for the treatment of ailments in herbal medicine

Table 2 Traditional knowledge associated with lesser-known crops cultivated in the Cold Desert villages, KSL, India

Crop species	Vernacular name	Ethnobotanical knowledge
Fagopyrum esculentum Moench	Polygonaceae	The grains are used to cure fever and all kinds of abdominal ailments. Juice of the root is used for urinary disorders
Fagopyrum tataricum (L.) Gaertn	Polygonaceae	Leaves are used in the treatment of fever and headache. Decoction of grains is used to cure colic pain
Hordeum vulgare L	Poaceae	The powder grains are to improve the digestive system
Phaseolus vulgaris L	Fabaceae	Traditionally used as a folk remedy for kidney stones
Macrotyloma uniflorum (lam.) Verdc	Fabaceae	Grain soup is administered as a cure for kidney stones
Amaranthus cruentus L	Amaranthaceae	The husk of the grains is used to cure skin diseases

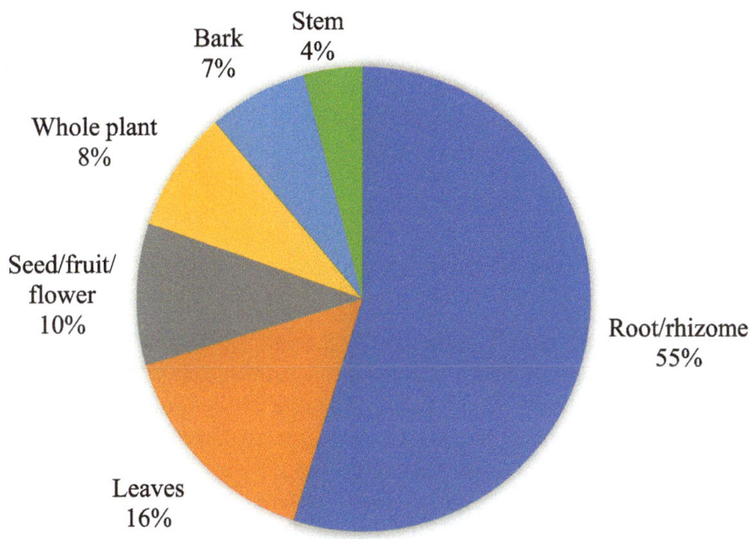

Fig. 3 Percentage of plant parts for the treatment of ailments in herbal medicine

interventions. The availability and population density of some high-value medicinal plants are decreasing rapidly due to overharvesting and environmental changes. Among the high-value medicinal plants in the cold desert region of Uttarakhand are those under various categories of threats assessed by the IUCN, including *Aconitum heterophyllum, Angelica glauca, Picrorhiza kurrooa, Ephedra gerardiana, Saussurea obvallata,* and *Fritillaria roylei* under the "Endangered" category. *Aconitum violaceum, Polygonatum cirrhifolium, Paris polyphylla, Polygonatum verticillatum, Podophyllum hexandrum, Rheum webbianum, Allium stracheyi, Malaxis muscifera, Pleurospermum angelicoides, Rheum webbianum,* and *Aconitum balfourii* are under "Vulnerable." *Nardostachys jatamansi, S. costus, Arnebia benthamii, Lilium oxypetalum,* and *Dactylorhiza hatagirea* are under "Critically Endangered" due to high threat and decreasing global population (Table 3).

3.3 Climate Change and Medicinal Plants

It is well known that high-elevation ecosystems are widely recognized as being highly sensitive and vulnerable to the impacts of climate change. A questionnaire-based survey conducted with local inhabitants and traditional herbal healers revealed a noticeable decline in the availability of medicinal plants over the past two decades. The primary factors contributing to this decline include habitat degradation, grazing pressure, over-exploitation, and climate change. Species that were once abundant near village peripheries are now largely restricted to distant alpine meadows. Furthermore, both local inhabitants and herbal healers reported observable shifts in the phenology of these plants. In comparison to their normal phenological phases, most medicinal plants now exhibit earlier flowering and fruiting by 15–25 days, while a few species display delayed phenological events (Table 3).

3.4 Indigenous Biodiversity Conservation Practices

In the cold desert valleys of Uttarakhand, especially in the KSL, several sacred pastures and landscapes have been preserved to ensure sustainable resource utilization and follow the religious belief system. The role of socio-cultural practices and traditional knowledge in biodiversity conservation is demonstrated by the Rung community in the KSL, India. A variety of biodiversity conservation initiatives were observed in the valleys, each precisely tailored to address specific aspects of biodiversity preservation and climate change adaptation. For instance, the *Abies spectabilis* forest patch in Baling village, extending uphill to the settlement near Byakshi Bugyal, is conserved with the belief that this forest serves as a protective barrier against glacier-related hazards in the area. Similarly, a patch of *Betula utilis* is strictly protected under the worship of the deity Larungdev in the Darma valley. A designated 15–30 days allocated for the collection of medicinal plants in Bon and Dugtu village in the Darma. A sacred patch of *Betula utilis* forest in the Milam valleys is conserved for religious beliefs. *Pinus wallichiana* forests near Gunji have been conserved; even the collection of fallen bark and wood is not permitted from that forest. Pandav Kutti is a sacred alpine patch situated near Kutti village in Byans,

Table 3 List of threatened MPs and people's perception on the phenology of MPs used in THCS, KSL, India

Plant species	CAMP (Conservation Assessment and Management Plan) (2003)	IUCN (2020)	General flowering and fruiting time	Change in flowering and fruiting time
Aconitum lethale Griff.	VU	VU	August–October	August–September
Aconitum violaceum Jacquem. *Ex* Stapf	VU	VU	September–October	August–October
Aconitum heterophyllum wall. *Ex* Royle	CE	EN	June–July	June–august
Allium stracheyi baker	VU	–	June–July	June–august
Arnabia benthamii (wall. *Ex* G.Don) I.M.Johnst.	CR	–	June–July	June–august
Angelica glauca Edgew.	EN	EN	June–august	July–September
Betula utilis D. Don	LC	LC	March–June	Mid-march-June
Dactylorhiza hatagirea (D. Don) Soo	NT	CR	June–September	June–October
Ephedra intermedia Schrenk & C.A.Mey.	CR	LC	July–august	Mid-June-august
Fritillaria cirrhosa D.Don	EN	EN	June–august	June–October
Hippophae salicifolia D.Don	NT	–	July–October	September–October
Lilium oxypetalum (D. Don) baker		CR	June–July	June–august
Malaxis muscifera (Lindl.) Kuntze	EN	VU	June–July	June–august
Nardostachys jatamansi (D.Don) DC	CR	CR	August–October	July–October
Paris polyphylla Sm.	VU	–	August–October	August-mid-October

(continued)

Table 3 (continued)

Plant species	CAMP (Conservation Assessment and Management Plan) (2003)	IUCN (2020)	General flowering and fruiting time	Change in flowering and fruiting time
Picrorhiza kurroa Royle *ex* Benth.	CR	CR	June–July	May–June
Podophyllum hexandrum Royle	EN	–	April–October	April–September
Polygonatum verticillatum (L.) all.	VU	–	June–September	April–September
Pterocyclus angelicoides (DC.) Klotzsch	VU	–	June–July	June–august
Rheum webbianum Royle	VU	–	June–august	June–September
Dolomiaea costus (Falc.) Kasana & A.K.Pandey	EN	CR	June–July	June–august
Saussurea laniceps hand.-Mazz	VU	–	August–September	September–October
Saussurea obvallata (DC.) Sch.Bip.	EN	–	July–august	July–September
Taxus wallichiana Zucc	EN	CR	August–September	September–October

THCS—Traditional Health Care System; KSL—Kailash Sacred Landscape; EN—Endangered; VU—Vulnerable; CR—Critical; (Source: Gowthami et al., 2021; Mehta et al., 2020, 2021) where harvesting of resources is completely restricted. Hya-Roshe Bugyal (alpine meadow) near Napalchu village (3400 m) and Putuk-tu Bugyal near Kutti village (4000 m) are protected through deeply rooted socio-cultural practices and dedicated as offerings to local deities. The extraction of resources from sacred forests and alpine areas in the villages of all three valleys involves seeking permission from the village authority, and after offerings to the local deities.

4 Discussion

Traditional knowledge is collectively owned and forms an integral part of the cultural and spiritual identity of the community in which it is practiced and preserved (Burton et al., 2015; WIPO, 2020; Fakih et al., 2022). Ethnobotany, the study of human interactions with plants, has important applications in food security, climate change, biodiversity conservation, and human health (Pei et al., 2020; Albuquerque et al., 2020; Chashike et al., 2025). Ethnobotanical studies can integrate local and scientific knowledge, contributing significantly to the advancement of traditional herbal knowledge and food security (Gaoue et al., 2017). The present study documents the knowledge of medicinal herbs of the Bhotiya community, developed over centuries through exposure to diverse plant species during their seasonal migrations between summer and winter settlements, as also reported in earlier studies (Maikhuri et al., 1998; Nautiyal et al., 2002; Farooquee et al., 2004; Garbyal et al., 2005; Phondani et al., 2010; Pandey et al., 2018; Negi et al., 2018b). The geographic isolation of communities in the cold desert areas has facilitated the strengthening of unique indigenous traditional and herbal knowledge systems.

Traditional herbal healers (*Amchis* or *Vaidyas*) usually collect medicinal plants following some conservation practices and religious belief systems. They possess detailed knowledge about the correct timing of collection, the specific plant parts to be used, and doge preparation methods, the composition of the drug, as elsewhere across the world (Gaoue et al., 2017; Albuquerque et al., 2020; Chashike et al., 2025). In the majority of cases, decoctions prepared from leaves, and tubers/roots are used for the treatment of various ailments; these decoctions are usually prepared by crushing the plant materials using a mortar and pestle in combination with other plants as reported in past studies (Maikhuri et al., 1998; Kala, 2005b, 2006; Phondani et al., 2014; Negi et al., 2018a, b, c). The paste of certain plants is applied to the skin to set dislocated or fractured bones and to relieve muscular pain. Depending on the nature of the disease, specific dietary restrictions are often prescribed during the course of treatment for various ailments to enhance their effectiveness.

These communities are often effective managers of their bio-resources, and in many cases, indigenous methods of resource utilization are considered sustainable, i.e., rotational practice of resource harvesting, conservation through development of sacred sites, the cultivation of important medicinal plants, as also reported in past studies (Nautiyal et al., 2002; Phondani et al., 2010; Negi et al., 2018b, 2023). Community-driven conservation practices, embedded in socio-cultural and religious traditions, play a vital role in preserving ecologically sensitive landscapes, maintaining ecosystem services, and promoting sustainable resource use. These practices demonstrate how indigenous belief systems and traditional knowledge operate as effective informal institutions for environmental stewardship in fragile mountain ecosystems. However, excessive and illegal harvesting of medicinal plants also persists to obtain higher incomes in the short term, with little concern for sustainability in many parts of the landscapes. Many species are under severe threat due to the collection of underground parts such as rhizomes, tubers, bulbs, and roots for medicinal use in the region, and other parts of the world (Olsen and Larsen, 1997, 2003; Pauls and Franz, 2013).

Among the high-value medicinal plants documented in the cold desert region, *Arnebia benthamii, Aconitum heterophyllum, A. violaceum, Angelica glauca, Ephedra gerardiana, Fritillaria roylei, Dactylorhiza hatagirea, Malaxis muscifera, Picrorhiza kurrooa, Polygonatum cirrhifolium, Polygonatum verticillatum, Podophyllum hexandrum, Rheum webbianum* are under various categories of threats according to IUCN (Mehta et al., 2020, 2021; Gowthami et al., 2021). Every year, about 960 species of medicinal plants are traded in the country, of which 178 species are extracted in surplus of 100 Metric tons (Goraya and Ved, 2017). This increasing commercialization has led to the overharvesting of several economically important medicinal plants from different parts of the Himalaya (Vashistha et al., 2006; Kala and Ratajc, 2012; Chauhan et al., 2013; Kuniyal et al., 2015; Negi et al., 2018c, 2019). The respondents reported that the populations of species such as *Aconitum balfourii, A. violaceum, Angelica glauca, Arnebia benthamii, Dactylorhiza hatagirea, Fritillaria roylei, Malaxis muscifera, Picrorhiza kurrooa, Podophyllum hexandrum, Rheum webbianum,* and *Sausurrea obvallata* have declined in the wild in recent decades due to high use value in the traditional system of healthcare and unregulated extraction. Further, *Angelica glauca, Dactylorhiza hatagirea, Podophyllum hexandrum, Picrorhiza kurrooa, Saussurea costus, Nardostachys jatamansi*, and *Polygonatum verticillatum* are in high demand in both national and international markets (Ghimire et al., 2005; Chauhan et al., 2013; Kuniyal et al., 2014; Thakur et al., 2016), indicating an immense threat to these species and a need for conservation measures. The high-altitude areas are more vulnerable to the impacts of climate change in the Himalayan region (IPCC 2007, 2022; Shrestha et al., 2012; Negi et al., 2022). Climate change in the region is posing new threats to biodiversity elements, especially in high-altitude alpine ecosystems.

High-altitude forests are particularly susceptible to climate change, with plant species responding through reduced diversity, shifts in functional groups, and migration to new habitats (Parmesan and Yohe, 2003; IPCC, 2007). Respondents in the present study perceived that the climate is changing and is impacting various ecosystems, i.e., agro-ecosystems, forest and alpine, and animal husbandry. It is cited that climate change is affecting medicinal plants in high-altitude regions, with herbal healers reporting advances in flowering and fruiting that disrupt traditional harvesting practices, as also reported in past studies (Maikhuri et al., 2018; Negi et al., 2017, 2021). Declining population and the upward range shift of a few species further highlight these impacts, consistent with similar observations from other regions (Maikhuri et al., 2018; Gaira et al., 2014; Sharma and Shrestha, 2016; Applequist et al., 2020). The high level of endemism in alpine species and their inability to shift upward means that these species are more vulnerable and are on the verge of extinction (Parmesan and Yohe, 2003; Jump et al., 2012). Thus, local observations of climate change are an important indication to policy planners for developing adaptation and management strategies (Savo et al., 2016; Reyes-García et al., 2019).

Conservation and cultivation of medicinal plants is important, keeping in view the fact that 21% fall under the Endangered category according to the International Union for Conservation of Nature (Schippmann et al., 2006; Chen et al., 2016). Cultivation of medicinal plants is widely viewed not only as an option for meeting current and future demands for large-volume production of plant-based drugs and herbal remedies, but also as a strategy for relieving harvest pressure on wild populations (Dhar et al., 2000; Maikhuri et al., 2018; Negi et al., 2018b, 2019; Singh et al., 2021). If properly planned, high-altitude medicinal plants can become unique selling points for these areas of the state. Sustainable collection practices followed by indigenous communities play an important role in conserving natural resources while meeting their daily needs. Rural people often use rotational harvesting for collecting medicinal plants, a method that helps protect wild plant populations (Negi et al., 2023). Such practices ensure a continuous and sustainable supply of medicinal plants while providing additional benefits to nearby villagers. Indigenous communities also make efforts to conserve species most vulnerable to destructive harvesting, helping maintain their viable populations in natural habitats. The possible approaches that can be adopted for the conservation of medicinal plants include: (i) encouraging farmers to cultivate medicinal plants on private lands rather than exploiting wild populations (Kuniyal et al., 2015; Maikhuri et al., 2018); (ii) establishing herbal gardens in schools to serve as germplasm repositories; (iii) promoting the plantation of medicinal plants in community forests; and (iv) adoption of modern technologies such as in vitro propagation for large-scale production of planting material and secondary metabolites. In the cold desert villages of Uttarakhand, the inhabitants of Gunji in Byans, Milam in Johar, and Bonn and Dugtu in Darma valley of KSL, and Lata, Tolma, Malari, and Gamsali in Niti valley cultivate medicinal plants, only on a small scale though, which needs promotion. Strengthening these efforts through the provision of planting material, training, and market opportunities could significantly promote the cultivation of medicinal plants in the region.

5 Conclusion

The cold desert ecosystems of Uttarakhand support a rich repository of medicinal plants and associated traditional knowledge that continues to play a key role in the healthcare, culture, and livelihoods of indigenous communities. However, overharvesting, habitat degradation, grazing pressure, and climate change are threatening the availability and distribution of many high-value species, several of which are already categorized as endangered, vulnerable, or critically endangered. The observed shifts in phenology further highlight the sensitivity of these alpine ecosystems to climatic variability. At the same time, the persistence of community-based conservation practices, such as the protection of sacred groves and regulated harvesting systems, shows the potential of indigenous knowledge in ensuring sustainable use and long-term conservation. Integrating these traditional systems with scientific approaches, including cultivation practices, in vitro propagation, and market-based incentives, can provide a balanced strategy to address both biodiversity conservation and livelihood security. Strengthening local capacity, promoting community participation, and developing supportive policy frameworks are essential steps for safeguarding medicinal plant diversity in the fragile high-altitude landscapes. Based on field observations in the landscapes, available literature, and findings of the present study, our recommendations are as follows:

- Provide quality planting material, technical training, and financial incentives to farmers for on-farm cultivation of threatened and traded species.
- Support *Van Panchayats* (community forests) and sacred groves as in situ conservation sites.
- Recognize and strengthen traditional institutions that regulate seasonal harvesting and protect sacred landscapes.
- Apply in vitro propagation and tissue culture techniques for large-scale multiplication of threatened species, and develop nurseries and germplasm banks in collaboration with research institutions.
- Develop sustainable value chains and market linkages for locally cultivated medicinal plants.
- Facilitate fair trade and ensure benefit-sharing mechanisms with local communities.
- Formulate policies that integrate traditional knowledge with scientific conservation approaches.

Acknowledgments We are grateful to the Director of the G.B. Pant National Institute of Himalayan Environment (NIHE), Kosi-Katarmal, Almora, and to the Hon'ble Vice Chancellor of HNB Garhwal University, Srinagar Garhwal, for providing facilities and encouragement. The authors also acknowledge the support of NMSHE Task Force 3, Forest Resource and Plant Biodiversity—Phase II, DST, New Delhi, Government of India, and the KSLCDI project funded by ICIMOD.

References

Albuquerque, U. P., Ferreira Júnior, W. S., Santoro, F. R., Torres-Avilez, W., & Medeiros, P. M. (2020). The role of ethnobotany and ethnopharmacology in inter-and transdisciplinary research. *Journal of Ethnopharmacology, 262*, 113230.

Applequist, W. L., Brinckmann, J. A., Cunningham, A. B., Hart, R. E., Heinrich, M., Katerere, D. R., & Van Andel, T. (2020). Scientists' warning on climate change and medicinal plants. *Planta Medica, 86*(1), 10–18.

Burton, A., Smith, M., & Falkenberg, T. (2015). Building WHO's global strategy for traditional medicine. *European Journal of Integrative Medicine, 7*(1), 13–15.
Chashike, A., Shibru, S., Gebre, T., & Uncha, A. (2025). Ethnobotanical study of traditional medicinal plants and associated indigenous knowledge in Melokoza District, South Ethiopia. *Trees, Forests and People, 20*, 100849.
Chauhan, R. S., Nautiyal, B. P., & Nautiyal, M. C. (2013). Trade of threatened Himalayan medicinal and aromatic plants–socioeconomy, management and conservation issues in Garhwal Himalaya India. *Global Journal of Medical Research, 13*(2), 9–17.
Chen, S. L., Yu, H., Luo, H. M., Wu, Q., Li, C. F., & Steinmetz, A. (2016). Conservation and sustainable use of medicinal plants: problems, progress, and prospects. *Chinese Medicine, 11*(1), 37.
Chowdhery, H. J., & Rao, R. R. (1990). Plant life in Himalayan cold deserts: Some adaptive strategies. *Bulletin of the Botanical Survey of India, 32*, 43–56.
Dhar, U., Rawal, R. S., & Upreti, J. (2000). Setting priorities for conservation of medicinal plants–a case study in the Indian Himalaya. *Biological Conservation, 95*, 57–65.
Fakih, M., Perbawati, C., & Monalisa. (2022). Relevance of WHO traditional medicine strategy (2014-2023) with traditional health care policy in the perspective of national law and international law. *Asian Journal of Legal Studies, 1*(1), 25–34.
Farooquee, N. A., Majila, B. S., & Kala, C. P. (2004). Indigenous knowledge systems and sustainable management of natural resources in a high-altitude society in Kumaun Himalaya India. *Journal of Human Ecology, 16*(1), 33–42.
Gaira, K. S., Rawal, R. S., Rawat, B., & Bhatt, I. D. (2014). Impact of climate change on the flowering of *Rhododendron arboreum* in central Himalaya India. *Current Science, 106*(12), 1735–1738.
Gaoue, O. G., Coe, M. A., Bond, M., Hart, G., Seyler, B. C., & McMillen, H. (2017). Theories and major hypotheses in ethnobotany. *Economic Botany, 71*(3), 269–287.
Garbyal, S. S., Aggarwal, K. K., & Babu, C. R. (2005). Return of biodiversity in Darma Valley, Dharchula Himalayas, Uttaranchal, North India following fortuitous changes in traditional lifestyle of the local inhabitants. *Current Science, 88*(5), 722–725.
Ghimire, S. K., McKey, D., & Aumeeruddy, T. Y. (2005). Conservation of Himalayan medicinal plants: Harvesting patterns and ecology of two threatened species, *Nardostachys grandiflora* DC. And *Neopicrorhiza scrophulariiflora* (Pennell) Hong. *Biological Conservation, 124*, 463–475.
Goraya, G. S., & Ved, D. K. (2017). *Medicinal plants in India: an assessment of their demand and supply*. National Medicinal Plants Board, Ministry of AYUSH, Government of India, New Delhi & Indian Council of Forestry Research and Education, Dehradun.
Gowthami, R., Sharma, N., Pandey, R., & Agrawal, A. (2021). Status and consolidated list of threatened medicinal plants of India. *Genetic Resources and Crop Evolution, 68*(6), 2235–2263.
Hamilton, A. C. (2004). Medicinal plants, conservation and livelihoods. *Biodiversity and Conservation, 13*, 1477–1517.
IPCC. (2007). Climate change 2007: Impacts, adaptation, and vulnerability. In *Contribution of working group II to the fourth assessment report of the IPCC*. Cambridge University Press.
IPCC. (2022). Climate change 2022: Impacts, adaptation, and vulnerability. In H.-O. Pörtner, D. C. Roberts, M. Tignor, E. S. Poloczanska, K. Mintenbeck, A. Alegría, M. Craig, S. Langsdorf, S. Löschke, V. Möller, A. Okem, & B. Rama (Eds.), *Contribution of working group II to the sixth assessment report of the intergovernmental panel on climate change*. Cambridge University Press.
Jump, A. S., Huang, T. J., & Chou, C. H. (2012). Rapid altitudinal migration of mountain plants in Taiwan and its implications for high altitude biodiversity. *Ecography, 35*(3), 204–210.
Kala, C. P. (2003). *Medicinal plants of Indian trans-Himalaya: Focus on Tibetan use of medicinal resources*. Bishen Singh Mahendra Pal Singh.
Kala, C. P., & Ratajc, P. (2012). High altitude biodiversity of the Alps and the Himalayas: Ethnobotany, plant distribution and conservation perspective. *Biodiversity and Conservation, 21*(4), 1115–1126.

Kala, C. P. (2005a). Indigenous uses, population density, and conservation of threatened medicinal plants in protected areas of the Indian Himalayas. *Conservation Biology, 19*(2), 368–378.
Kala, C. P. (2005b). Health traditions of Buddhist community and role of amchis in trans-Himalayan region of India. *Current Science, 89*(8), 1331–1338.
Kala, C. P. (2006). Medicinal plants of the high-altitude cold desert in India: Diversity, distribution and traditional uses. *International Journal of Biodiversity Science & Management, 2*(1), 43–56.
Kala, C. P., & Mathur, V. B. (2002). Patterns of plant species distribution in the Trans-Himalayan region of Ladakh, India. *Journal of Vegetation Science, 13*(6), 751–754.
Kumar, A., Mitra, M., Adhikari, B. S., & Rawat, G. S. (2015). Depleting indigenous knowledge of medicinal plants in cold-arid region of Nanda Devi biosphere reserve Western Himalaya. *Medicinal & Aromatic Plants, 4*(195), 2167–2178.
Kuniyal, C. P., Bhatt, V. P., Butola, J. S., & Sundriyal, R. C. (2014). Promoting nursery enterprise in high altitude villages: A participatory approach for conservation and commercialization of Himalayan threatened medicinal plants. *Journal of Medicinal Plants Research, 8*(48), 1399–1407.
Kuniyal, C. P., Bisht, V. K., Negi, J. S., Bhatt, V. P., Bisht, D. S., Butola, J. S., et al. (2015). Progress and prospect in the integrated development of medicinal and aromatic plants (MAPs) sector in Uttarakhand, Western Himalaya. *Environment, Development and Sustainability, 17*(5), 1141–1162.
Maikhuri, R. K., Nautiyal, S., Rao, K. S., & Saxena, K. G. (1998). Role of medicinal plants in the traditional health care system: A case study from Nanda Devi biosphere reserve Himalaya. *Current Science, 75*(2), 152–157.
Maikhuri, R. K., Phondani, P. C., Dhyani, D., Rawat, L. S., Jha, N. K., & Kandari, L. S. (2018). Assessment of climate change impacts and its implications on medicinal plants-based traditional healthcare system in central Himalaya, India. *Iranian Journal of Science and Technology, Transactions A: Science, 42*(4), 1827–1835.
Mehta, P., Bisht, K., & Sekar, K. C. (2021). Diversity of threatened medicinal plants of Indian Himalayan region. *Plant Biosystems, 155*(6), 1121–1132.
Mehta, P., Sekar, K. C., Bhatt, D., Tewari, A., Bisht, K., Upadhyay, S., et al. (2020). Conservation and prioritization of threatened plants in Indian Himalayan region. *Biodiversity and Conservation, 29*(6), 1723–1745.
Nautiyal, S., Rao, K. S., Maikhuri, R. K., Semwal, R. L., & Saxena, K. G. (2000). Traditional knowledge related to medicinal and aromatic plants in tri'al societies in a part of Himalaya. *Journal of Medicinal and Aromatic Plant Sciences, 23/4A & 23/1A*, 528–441.
Negi, V. S., Joshi, B. C., Pathak, R., Rawal, R. S., & Sekar, K. C. (2018a). Assessment of fuelwood diversity and consumption patterns in cold desert part of Indian Himalaya: Implication for conservation and quality of life. *Journal of Cleaner Production, 196*, 23–31.
Negi, V. S., Kewlani, P., Pathak, R., Bhatt, D., Bhatt, I. D., Rawal, R. S., Sundriyal, R. C., & Nandi, S. K. (2018c). Criteria and indicators for promoting cultivation and conservation of medicinal and aromatic plants in Western Himalaya, India. *Ecological indicators, 93*, 434–446.
Negi, V. S., Maikhuri, R. K., Maletha, A., & Phondani, P. C. (2019). Ethnobotanical knowledge and population density of threatened medicinal plants of Nanda Devi biosphere reserve, Western Himalaya, India. *Iranian Journal of Science and Technology, Transactions A: Science, 43*(1), 63–73.
Negi, V. S., Maikhuri, R. K., Pharswan, D., Thakur, S., & Dhyani, P. P. (2017). Climate change impact in the Western Himalaya: people's perception and adaptive strategies. *Journal of Mountain Science, 14*(2), 403–416.
Negi, V. S., Maikhuri, R. K., Phondani, P. C., & Rawat, L. S. (2010). An inventory of indigenous knowledge and cultivation practices of medicinal plants in Govind Pashu Vihar wildlife sanctuary, central Himalaya, India. *International Journal of Biodiversity Science, Ecosystem Services & Management, 6*(3–4), 96–105.
Negi, V. S., Pathak, R., Sekar, K. C., Rawal, R. S., Bhatt, I. D., Nandi, S. K., & Dhyani, P. P. (2018b). Traditional knowledge and biodiversity conservation: A case study from Byans Valley

in Kailash sacred landscape, India. *Journal of Environmental Planning and Management, 61*(10), 1722–1743.
Negi, V. S., Pathak, R., Thakur, S., Joshi, R. K., Bhatt, I. D., & Rawal, R. S. (2023). Scoping the need of mainstreaming indigenous knowledge for sustainable use of bioresources in the Indian Himalayan region. *Environmental Management, 72*(1), 135–146.
Negi, V. S., Thakur, S., Dhyani, R., Bhatt, I. D., & Rawal, R. S. (2021). Climate change observations of indigenous communities in the Indian Himalaya. *Weather, Climate, and Society, 13*(2), 245–257.
Negi, V. S., Tiwari, D. C., Singh, L., Thakur, S., & Bhatt, I. D. (2022). Review and synthesis of climate change studies in the Himalayan region. *Environment, Development and Sustainability, 24*(9), 10471–10502.
Olsen, C. S., & Helles, F. (1997). Making the poorest poorer: Policies, laws and trade in medicinal plants in Nepal. *Journal of World Forest Resource Management, 8*, 137–158.
Olsen, C. S., & Larsen, H. O. (2003). Alpine medicinal plant trade and Himalayan mountain livelihood strategies. *The Geographical Journal, 169*, 243–254.
Pandey, A., Sekar, K. C., Joshi, B., & Rawal, R. S. (2018). Threat assessment of high-value medicinal plants of cold desert areas in Johar Valley, Kailash sacred landscape India. *Plant Biosystems, 153*(1), 39–47.
Parmesan, C., & Yohe, G. (2003). A globally coherent fingerprint of climate change impacts across natural systems. *Nature, 421*(6918), 37–42.
Pauls, T., & Franz, M. (2013). Trading in the dark—The medicinal plants production network in Uttarakhand, Singapore. *Journal of Tropical Geography, 34*, 229–243.
Pei, S., Alan, H., & Wang, Y. (2020). Vital roles for ethnobotany in conservation and sustainable development. *Plant Diversity, 42*(6), 399.
Phondani, P. C., Maikhuri, R. K., & Saxena, K. G. (2014). The efficacy of herbal system of medicine in the context of allopathic system in Indian central Himalaya. *Journal of Herbal Medicine, 4*(3), 147–158.
Phondani, P. C., Maikhuri, R. K., Rawat, L. S., Farooquee, N. A., Kala, C. P., et al. (2010). Ethnobotanical uses of plants among the Bhotiya tribal communities of Niti Valley in central Himalaya, India. *Ethnobiology Research and Applications, 8*, 233–244.
Rawat, Y. S., & Everson, C. S. (2012). Ecological status and uses of juniper species in the cold desert environment of the Lahaul Valley, north-western Himalaya India. *Journal of Mountain Science, 9*(5), 676–686.
Rawat, Y. S., Negi, V. S., Moussa, I. M., Zaman, W., & Elansary, H. O. (2023). Diversity, distribution and vegetation assessment of woody plant species in the cold desert environment, North-Western Himalaya India. *Sustainability, 15*(13), 10429.
Reyes-García, V., García-del-Amo, D., Benyei, P., Fernández-Llamazares, Á., Gravani, K., Junqueira, A. B., & Soleymani-Fard, R. (2019). A collaborative approach to bring insights from local observations of climate change impacts into global climate change research. *Current Opinion in Environmental Sustainability, 39*, 1–8.
Savo, V., Lepofsky, D., Benner, J. P., Kohfeld, K. E., Bailey, J., & Lertzman, K. (2016). Observations of climate change among subsistence-oriented communities around the world. *Nature Climate Change, 6*(5), 462–473.
Saxena, K.G., Liang, L., Xue, X. (2011). Global change, biodiversity, and livelihood in the cold desert region of Asia. Bishen Singh Mahendra Pal Singh. ISBN: 978–81–211-0780-8.
Schippmann, U. W. E., Leaman, D., & Cunningham, A. B. (2006). A comparison of cultivation and wild collection of medicinal and aromatic plants under sustainability aspects. *Frontis*, 75–95.
Sekar, K. C., Pandey, A., Giri, L., Joshi, B. C., Bhatt, D., Bhojak, P., Dey, D., Thapliyal, N., Bisht, K., Bisht, M., & Negi, V. S. (2022). Floristic diversity in cold desert regions of Uttarakhand Himalaya India. *Phytotaxa, 537*(1), 1–62.
Sharma, R. K., & Shrestha, D. G. (2016). Climate perceptions of local communities validated through scientific signals in Sikkim Himalaya India. *Environmental Monitoring and Assessment, 188*(10), 578.

Shrestha, U. B., Gautam, S., & Bawa, K. S. (2012). Widespread climate change in the Himalayas and associated changes in local ecosystems. *PLoS One, 7*(5), e36741.

Singh, L., Bhatt, I. D., Negi, V. S., Nandi, S. K., Rawal, R. S., & Bisht, A. K. (2021). Population status, threats, and conservation options of the orchid *Dactylorhiza hatagirea* in Indian Western Himalaya. *Regional Environmental Change, 21*(2), 21–40.

Thakur, M., Asrani, R. K., Thakur, S., Sharma, P. K., Patil, R. D., Lal, B., & Parkash, O. (2016). Observations on traditional usage of ethnomedicinal plants in humans and animals of Kangra and Chamba districts of Himachal Pradesh in North-Western Himalaya, India. *Journal of Ethnopharmacology, 191*, 280–300.

Vashistha, R., Nautiyal, B. P., & Nautiyal, M. C. (2006). Conservation status and morphological variations between populations of *Angelica glauca* Edgew and *Angelica archangelica* Linn Garhwal Himalaya. *Current Science, 91*(11), 1537–1542.

WIPO. (2020). Traditional knowledge and cultural expressions. World intellectual property organization. https://www.wipo.int (Retrieved: 31 August 2025).

Biogeochemical Dynamics of Ladakh's High-Altitude Permafrost

Vasudha Agnihotri

1 Introduction

High-altitude permafrost is defined as the ground comprising soil, rocks, and other superficial deposits that remain at or below 0 °C for at least two consecutive years (Wani et al., 2021). The permafrost region in Ladakh is mainly controlled by altitude rather than latitude, which mostly occurs above 4000 masl (Li & Cheng, 1999). The region is characterized by a high-altitude cold desert climate and is situated in the Trans-Himalayan Region. Ladakh's high-altitude permafrost area is generally thinner, warmer, and mostly discontinuous. These attributes make it susceptible to climate change, which contributes to permafrost thawing and various related environmental impacts (Chevuturi et al., 2016; Guo & Wang, 2016). Ladakh receives limited precipitation, mostly in the form of snow, leading to arid conditions in the region. Minimal soil moisture and sparse vegetation cover also influence the thermal regime in the area (Zhang et al., 2021a). These conditions, further combined with intense solar radiation during the daytime and rapid cooling at night, create a suitable environment for permafrost development and the formation of a shallow active layer, which responds sensitively to these thermal variations (Yoshikawa et al., 2025). The active layer, found above the permafrost frozen ground (permafrost table), thaws due to extreme diurnal and seasonal temperature fluctuations (Zhang et al., 2021b). As a cold-arid desert with low precipitation, intense solar and ultraviolet radiation exposure, and low atmospheric pressure, the partial pressure of oxygen and carbon dioxide is low (Fu et al., 2025). During summer, in the cold desert ecosystem, an active layer of soil thaws, and waterlogging occurs due to the impermeable permafrost table (Hu et al., 2019). Due to the thawing process, microbial activities and biogeochemical cycling of nutrients occur in the active layer, while organic carbon is generally stored

V. Agnihotri (✉)
G.B. Pant National Institute of Himalayan Environment-Himachal Pradesh Regional Centre, Mohal, Kullu, Himachal Pradesh, India
e-mail: vasudha@gbpihed.nic.in

Y. S. Rawat and C. P. Kala (eds.), *Cold Deserts*,
https://doi.org/10.1007/978-3-032-29119-6_13

in the underlying permafrost layer (Perez-Mon et al., 2021; Reyes & Lougheed, 2015; Wu et al., 2021). Therefore, the distribution of permafrost in Ladakh depends on local microclimatic conditions, altitudes, and slope orientations. North-facing slopes and shaded valley areas tend to preserve permafrost more effectively because of reduced solar heating and moist conditions (Munkhjargal et al., 2020). Vegetation and rock debris function as insulation for the ground, affecting thermal stability and heat exchange processes (Dash et al., 2024; Kenner et al., 2019; Mahanta et al., 2024; Wang et al., 2016). Kenner et al. (2019) found that the talus slope with thick rock debris in Ladakh is having lower active layer thickness in comparison to bare soil areas. Areas with dense vegetation cover have shown reduced permafrost thaw rates due to shading and moisture retention effects (Dash et al., 2024), which also slow down the microbial decomposition process by maintaining lower soil temperatures, preserving organic carbon stocks within permafrost soils. Any type of disturbance (i.e., anthropogenic and natural) in these areas, including the removal of vegetation or surface cover, causes ground warming that can accelerate permafrost degradation, resulting in the release of stored carbon in any form (Mahanta et al., 2024). In this context, the present study deals with the biogeochemical dynamics of Ladakh's high-altitude cold desert permafrost processes and the impact of permafrost thawing on geomorphological and hydrological processes in the region.

2 Physical and Geochemical Structure of Permafrost Soil

2.1 Permafrost Profile

The permafrost structure is composed of several layers: an active layer, permafrost table, permafrost body, talik, and permafrost base (Fig. 1). The active layer, which is the uppermost section, thaws during summer and refreezes in winter, supporting plant growth and microbial activity in high-altitude cold desert ecosystems. The thickness of the active layer varies with the seasons and is influenced by temperature fluctuations owing to daily and seasonal changes in cold, arid environments (Humlum et al., 2003). The permafrost table is situated at the bottom of the active layer. A permafrost body is a layer that remains frozen for more than two consecutive years, with thicknesses ranging from a few meters to over 1500 m (Jones et al., 2016). A talik is an unfrozen ground layer or patch found within or beneath the permafrost (Connon et al., 2018). The permafrost base is the deepest layer, where ground temperatures exceed 0 °C owing to geothermal heat, marking the transition from frozen to unfrozen ground at depth (Langford et al., 2019). These layered characteristics of permafrost are observed in various cold regions, including alpine, arctic, and subarctic environments, underscoring the complexity and sensitivity of permafrost to climate change scenarios.

Permafrost profiles may vary from continuous, discontinuous, to degrading permafrost depending on climatic conditions and local geography. In areas with

Fig. 1 Permafrost profile: representative picture

continuous permafrost, the frozen ground forms a thick, uninterrupted layer beneath a very thin active layer. Discontinuous permafrost is characterized by a fragmented profile in which frozen ground is interrupted by taliks or unfrozen patches. It is frequently found beneath water bodies or on sun-exposed, south-facing slopes (Devoie et al., 2019; Ishikawa et al., 2005; Roy-Leveillee & Burn, 2017). In the degrading permafrost areas, the active layer thickens significantly, leading to the formation of a residual thaw layer, an active layer that thickens to the level where winter cold cannot reach the bottom to refreeze it, creating an unfrozen ground between the seasonal frost and the permafrost table (Chen et al., 2012; Zhang et al., 2008). Trumpet curves serve as a tool for comprehending permafrost stability, forecasting the depth of the active layer, and modeling the effects of climate change, such as increased thawing and potential ground subsidence (Thermokrast) (Holloway et al., 2020).

2.2 *Soil Geochemistry*

Chemical weathering and nutrient cycling processes occur mainly in the active layer of the permafrost, while the permafrost body has served as a geochemical archive of carbon, salts, and trace elements since the Pleistocene and Holocene epochs (Fouché et al., 2020). The permafrost table is impermeable, which prevents the vertical leaching of solutes, leading to a concentration of organic matter and ions in the transition zone (Prokushkin et al., 2007). With changing climatic conditions toward warming, the active layer thickness is increasing due to climate change and global warming, which further triggers the decomposition of previously frozen soil organic matter, shifting the redox potential of soil and facilitating the release of greenhouse gases (Aminitabrizi et al., 2020). Oxidation of carbonates and sulfides is further altering the ionic composition of surrounding watersheds in the cold desert ecosystems (Herndon et al., 2015).

The active layer soil permafrost region in Ladakh is mainly driven by an extremely cold-arid climate and is reported to be alkaline with low moisture content and a distant vertical stratification of organic and mineral components (Shao et al., 2022). The soil in the Changthang area is alkaline (7.8–8.8 pH) and is dominated by exchangeable calcium ions. The Soil Organic Carbon (SOC) content was reported to be low, ranging from 0.1–0.6%, and higher Soil Organic Carbon (SOC) was found at a depth of 10 cm (Mu et al., 2024). Acharya et al. (2012) reported that Soil Organic Carbon (SOC) distribution in the top 30 cm soil of cropland and grassland areas in Nubra Valley of Ladakh was around 43.94 t ha^{-1} and 46.9 t ha^{-1}, respectively. Due to the global regional warming impacts, permafrost thaws, which causes the mobilization of nutrients such as nitrogen and phosphorus along with carbon. This causes leaching of solutes into nearby headwater catchments, which can alter the hydro-geochemistry of high-altitude ecosystems, along with the release of greenhouse gases (Fouché et al., 2020; Skierszkan et al., 2024). Nutrient cycling in permafrost regions is driven by seasonal soil dynamics and the metabolic states of microbial communities. Organic matter originating from dead plants and animals is trapped by the permafrost body due to the inactivity of microbes at subzero levels (Mackelprang et al., 2011). Therefore, these organic matters are generally in a frozen state. Thawing occurs in an active layer during summer, which causes the release of Dissolved Organic Nitrogen and greenhouse gas nitrous oxide (N_2O) (Lacroix et al., 2022; Patzner et al., 2022). After obtaining a suitable environment, microbial degradation starts and causes the mineralization of organic nitrogen and phosphorus, which allows plants to grow. Increased nitrogen and phosphorus content causes greening of the area, which helps absorb more CO_2 but can stimulate microbial activities and accelerate the decomposition of ancient carbon (Bouranis et al., 2025).

2.3 *Hydrological Influences and Redox Gradients*

In permafrost areas, hydrological processes are characterized by seasonal thawing and freezing cycles (Jeelani et al., 2024). After thawing, the active layer allows water infiltration and lateral flow, whereas vertical drainage is restricted by frozen layers. These lead to variable water saturation zones, which create special heterogeneity in oxygen diffusion and redox gradients (Ebrahimi & Or, 2017). In anaerobic zones, microbial communities utilize nitrite, iron, and sulfate as electron receptors in place of oxygen in waterlogged conditions during denitrification, iron reduction, and sulfate reduction, whereas aerobic conditions favor oxidative processes (Bouranis et al., 2025; Yang et al., 2021). Therefore, there is an interplay between hydrology and redox reactions, which influences organic matter decomposition, greenhouse gas emissions (e.g., methane production under reducing conditions), and nutrient cycling (Aminitabrizi et al., 2020; Chang et al., 2019). In the Tso Moriri basin of Ladakh, permafrost and its active layer control supra-permafrost water flow, which sustains dry-season streams amid low precipitation. In the Upper Ganglass catchment, ground ice melt helps sustain low snow year hydrology (Wani et al., 2021).

3 Microbial Ecology in Ladakh

Biogeochemical cycles are primarily driven by microbes present in cold desert areas, such as Ladakh (Fig. 2). Microbial ecology is of critical relevance to permafrost regions and is dependent on variables such as intense UV radiation, low oxygen levels, and severe arid conditions. Changing climatic conditions are causing the active layer to thaw, which is awakening dormant microbial communities and degrading carbon. Soil properties, such as soil organic carbon, pH, and electrical conductivity, are prominent factors affecting microbial diversity (Ali et al., 2025). Their metabolic activities cause the release of greenhouse gases, such as carbon dioxide and methane, through which the microbes contribute to altering biogeochemical cycles. Microbial processes are important for maintaining soil health through nitrogen mineralization and carbon fixation. Alpha diversity of microbes found dominant in the active layer of Ladakh permafrost in comparison to that of Arctic or Russian permafrost, which might be due to localized thawing conditions (Ali et al., 2024).

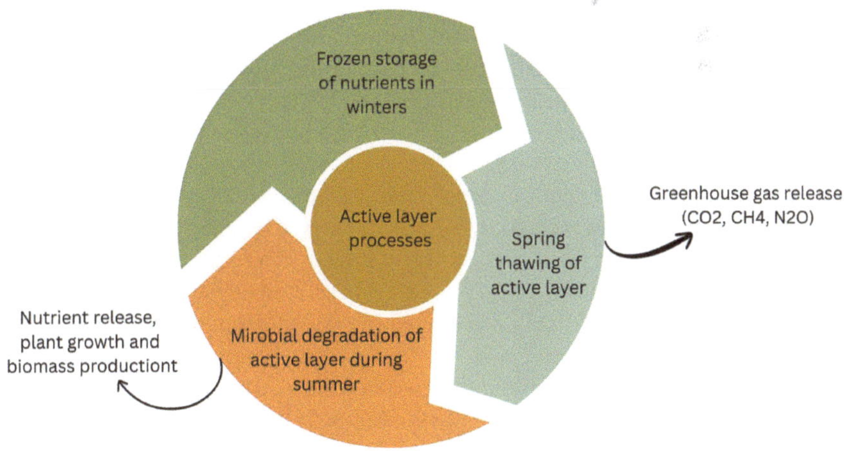

Fig. 2 Seasonal processes in permafrost active layer

Specialized bacteria (such as *Actinobacteria*, *Firmicutes*, *Bacteriodetes*, and *Proteobacteria*) are reported in the Changthang region, which are involved in nitrogen mineralization and ammonia oxidation processes, and support the alpine vegetation. These bacteria have shown specific pathways for stress tolerance and energy metabolism, which help them survive in extreme conditions, such as high salinity, intense UV radiation, and severe temperature fluctuations (Yadav et al., 2024). Psychrophiles, such as *Cryobacterium*, have been reported in regions that are susceptible to moderate warming conditions in the region. Choudhary et al. (2024) reported *Thermococcus* sp. in the thermal alkaline springs of Ladakh, which are hyperthermophilic archaea generally found in deep-sea hydrothermal vents along with the Euryarchaeota group of archaea, which includes halophiles, methanogens, and extreme thermophiles. Fungi such as Basidiomycota have also been reported in this region (Ali et al., 2024). The acidic sites limit methanogens, whereas alkaline soil in the presence of Archaea-rich clusters enhances the fermentation to methane production process (Waldrop et al., 2023). These microbes have shown biotechnological potential in the form of industrial enzymes, plant growth-promoting substances, especially for low-temperature agriculture, and for producing novel bioactive compounds and antibiotics (Ali et al., 2024; Choudhary et al., 2024).

4 Elemental Cycling and Related Greenhouse Gas Emissions

The permafrost region contains a significant quantity of organic carbon, estimated to be between 1460 and 1600 Pg, which is vulnerable to microbial decomposition during thawing. This decomposition process results in the release of Dissolved Organic Carbon (DOC), Dissolved Organic Nitrogen (DON), Dissolved Inorganic Carbon (DIN), and greenhouse gases, especially under aerobic and anaerobic conditions (Fig. 3) (Ernakovich & Wallenstein, 2015; Park et al., 2025). In the Ladakh permafrost region, greenhouse gas emissions are affected by a mix of thermal insulation, seasonal microbial activity, and hydrological transport (Aminitabrizi et al., 2020). The active layer of rocky glaciers serves as an insulator in a process known as thermal decoupling, which reduces the rate of greenhouse gas release (Domine et al., 2022). Thawing can increase the release of stored gases. Carbon is generally released as Dissolved Organic Carbon (DOC) rather than directly into the atmosphere, and it seeps into nearby water bodies, where it undergoes further microbial breakdown (Knutson et al., 2025). Microbial activity persists during winter owing to stable subsurface soil temperatures in the area, allowing greenhouse gas emissions to continue even under cold conditions (Jansson & Taş, 2014).

Nitrogen mineralization occurs in the thawed active layer of permafrost, where organic nitrogen from microbial biomass or plant matter is transformed into ammonium ions. In Ladakh, this process is sluggish during winter because of the extremely low temperatures but accelerates in the summer months. In alkaline soils with oxygen

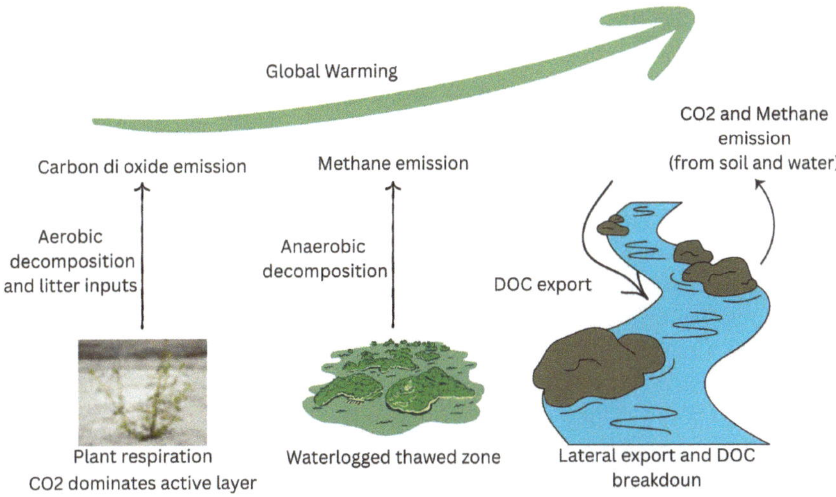

Fig. 3 Permafrost active layer greenhouse gas emission with global warming

present, ammonia-oxidizing bacteria such as *Nitrosomonas* and Archaea convert ammonium ions into nitrate ions, which often seep into nearby water bodies. Under anaerobic conditions, particularly in waterlogged sediments such as those in Tso Kar, facultative bacteria such as *Pseudomonas* convert nitrate ions into nitrogen gas or nitrous oxide (N_2O). New nitrogen sources are also developed by free-living diazotrophs, and cyanobacterial crusts fix atmospheric nitrogen into biologically available forms (Mao et al., 2020). In the permafrost areas of Ladakh, iron is plentiful, mainly existing as ferric ions (Fe^{+3}) in a frozen form. When permafrost thaws, especially in waterlogged environments lacking oxygen, bacteria that consume iron utilize these ferric ions while decomposing organic material, transforming iron into its ferrous form (Fe^{+2}) (Kwon et al., 2016). Similarly, in the saline-rich regions of Ladakh, sulfate-reducing bacteria convert sulfate ions into hydrogen sulfide under anaerobic conditions (Wunder et al., 2021). These processes are linked by redox coupling. When oxygen-rich meltwater infiltrates the active layer of soil in permafrost regions, it oxidizes pyrite (Fe_2S), reintroducing sulfate into the system. This generates acidity, which causes localized chemical weathering of nearby carbonate rocks (Pan et al., 2019). With an increase in active layer thawing, water can interact with deeper mineral horizons. This causes a shift in the pH of that area and can mobilize ions such as mercury, arsenic, lead, and cadmium (Ji et al., 2021), which can enter the food web in high-altitude areas by accumulating in fish and macroinvertebrates. These processes can collectively modify the geochemical landscape of high-altitude cold desert regions through biogeochemical and hydrological changes associated with permafrost thawing (Aschenbach et al., 2013; Hansel et al., 2015).

5 Geomorphological and Hydrological Implications

The high-altitude cold desert of Ladakh acts as a geological binder, securing loose debris and stabilizing the steep slopes. However, as temperatures rise and the permafrost region deteriorates, this area becomes prone to landslides and rockfalls. In high-altitude permafrost zones, moraines are held together by core ice. As the temperature increased, the ice began to melt, creating voids within the moraine particles. These voids allow water from glacial lakes to seep through, eventually forming pipes and tunnels. The development of these pipes or tunnels can lead to a sudden release of water without prior overtopping (Majeed et al., 2021; Schmidt et al., 2020). Thawing permafrost also affects the hydrological cycles of the region by altering groundwater storage. While examining the ground thermal regimes at the Chang La and South Pullu sites, Wani et al. (2020) observed the formation of taliks in the permafrost areas of Ladakh, which act as subsurface drainage channels that facilitate groundwater movement to deeper aquifers. This situation is exacerbated by the dual impact of climate stressors: erratic precipitation and decreased snowfall lead to reduced surface water availability, whereas the deepening permafrost table may affect the supply of perennial springs (Archer et al., 2010; Chen et al., 2019; Shayanmehr et al., 2022). Thawing of permafrost also challenges critical transit routes, such as Khardungla and Chang La, where the soil does not behave as a solid foundation for road construction. The hydration heat released during concrete curing can melt the surrounding ice, which causes the structure to sink before it is set. The black asphalt used for road construction also absorbs solar radiation, which conducts heat downwards and increases the thawing process. This phenomenon causes the development of wavy characteristics in the Himalayan region. The Border Road Organization in India is working toward a solution by adopting different strategies, such as vented rock embankments, employment of high-strength geo-grids, and utilization of bitumen mixed with plastic waste (PET/LDPE) to increase the flexibility of roads under extreme temperature gradients. Further technological modifications are required to address the complex challenges caused by permafrost degradation (De Grandpré et al., 2012; Hjort et al., 2022; Wu et al., 2020).

6 Conclusion

The Ladakh permafrost region represents a distinct high-altitude cold desert ecosystem that is approaching a critical threshold owing to climate change. Traditionally, this area has operated as a closed nitrogen cycle and has served as a significant atmospheric methane sink. However, the rapid increase in the depth of the active layer, now reaching up to 4.2 m, may activate dormant microbial communities, releasing elements such as carbon, nitrogen, and toxic trace metals. Additionally, the transition to anaerobic waterlogged conditions in thawed areas alters the biogeochemical landscape, turning it from a methane sink into a potential source. These changes lead to regional geohazards, including slope instability, drying of high-altitude springs, and mobilization of dissolved carbon, nitrogen, and other elements, along with heavy metal ions, into nearby rivers and other water bodies. Therefore, future research should address the gaps in monitoring deep-layer carbon and other elemental stocks, as well as trace metal flux through field interventions and remote sensing techniques. The stability of permafrost in Ladakh is crucial for regional security, and resource management must focus on developing permafrost-resilient infrastructure and climate-smart conservation strategies to mitigate its loss and ensure the long-term stability of water and structures in the region.

Acknowledgements The author expresses gratitude to the Director, G.B. Pant National Institute of Himalayan Environment, Almora, Uttarakhand, for the support provided by the institution. National Mission on Himalayan Studies (NMHS) is also duly acknowledged for project funding.

References

Acharya, S., Charan, G., Singh, N., & Srivastava, R. B. (2012). Soil organic carbon sequestration of cold desert Ladakh. *Range Management and Agroforestry, 33*(1), 79–82.

Ali, A., Dolma, P., Vishnivetskaya, T. A., et al. (2025). Exploring prokaryotic diversity in permafrost-affected soils of Ladakh's Changthang region and its geochemical drivers. *Scientific Reports, 15*, 15388. https://doi.org/10.1038/s41598-025-94542-9

Ali, A., Vishnivetskaya, T. A., & Chauhan, A. (2024). Comparative analysis of prokaryotic microbiomes in high-altitude active layer soils: Insights from Ladakh and global analogues using In-Silico approaches. *Brazilian Journal of Microbiology, 55*, 2437–2452. https://doi.org/10.1007/s42770-024-01365-3

Aminitabrizi, R., Tfaily, M. M., Chanton, J. P., Heyman, H. M., Rich, V. I., Fudyma, J. D., Saleska, S. R., Wilson, R. M., & Hodgkins, S. B. (2020). Controls on soil organic matter degradation and subsequent greenhouse gas emissions across a permafrost thaw gradient in Northern Sweden. *Frontiers in Earth Science, 8*, 557961. https://doi.org/10.3389/feart.2020.557961

Archer, D. R., Shah, S. M., Forsythe, N., & Fowler, H. J. (2010). Sustainability of water resources management in the Indus Basin under changing climatic and socio economic conditions. *Hydrology and Earth System Sciences, 14*(8), 1669–1680. https://doi.org/10.5194/hess-14-1669-2010

Aschenbach, K., Doležal, J., Řeháková, K., Angel, R., Janatková, K., & Conrad, R. (2013). Methanogens at the top of the world: Occurrence and potential activity of methanogens in newly deglaciated soils in high-altitude cold deserts in the Western Himalayas. *Frontiers in Microbiology, 4*, 359. https://doi.org/10.3389/fmicb.2013.00359

Bouranis, J. A., Mcgivern, B. B., Makke, G., Jurgensen, S. K., Bosman, S. H., Stemple, B., Chanton, J. P., Wrighton, K. C., & Tfaily, M. M. (2025). Metabolic redox coupling controls methane production in permafrost-affected peatlands through organic matter quality-dependent energy allocation. *Global Change Biology, 31*(8). https://doi.org/10.1111/gcb.70390

Chang, K.-Y., Crill, P. M., Rich, V. I., Grant, R. F., Riley, W. J., & Saleska, S. R. (2019). Large carbon cycle sensitivities to climate across a permafrost thaw gradient in subarctic Sweden. *The Cryosphere, 13*(2), 647–663. https://doi.org/10.5194/tc-13-647-2019

Chen, L., Wu, T., Xie, C., Li, R., Wu, X., Hu, G., Qiao, Y., & Wang, W. (2012). Formation and distribution of residual thaw layers in the permafrost regions of the Qinghai-Tibet Plateau. *Theoretical and Applied Climatology, 111*(1–2), 131–139. https://doi.org/10.1007/s00704-012-0643-9

Chen, Y., Sun, C., Fan, Y., Fang, G., & Li, B. (2019). Hydrological and water cycle processes of inland river basins in the arid region of Northwest China. *Journal of Arid Land, 11*(2), 161–179. https://doi.org/10.1007/s40333-019-0050-5

Chevuturi, A., Thayyen, R. J., & Dimri, A. P. (2016). Climate change over Leh (Ladakh), India. *Theoretical and Applied Climatology, 131*(1–2), 531–545. https://doi.org/10.1007/s00704-016-1989-1

Choudhary, G., Kumari, S., Anu, K., & Devi, S. (2024). Deciphering the microbial communities of alkaline hot spring in Panamik, Ladakh, India using a high-throughput sequencing approach. *Brazilian Journal of Microbiology, 55*, 1465–1476. https://doi.org/10.1007/s42770-024-01346-6

Connon, R., Hayashi, M., Quinton, W., Veness, T., & Devoie, É. (2018). The influence of shallow Taliks on permafrost thaw and active layer dynamics in Subarctic Canada. *Journal of Geophysical Research: Earth Surface, 123*(2), 281–297. https://doi.org/10.1002/2017jf004469

Dash, S. S., Dash, J. P., Mahanta, P., Pradhan, S. K., & Roy, P. S. (2024). Vegetation-permafrost interactions in the Himalayan cold desert: Implications for carbon cycling and thermal insulation. *Journal of Arid Environments, 221*, Article 105154.

De Grandpré, I., Stephani, E., & Fortier, D. (2012). Degradation of permafrost beneath a road embankment enhanced by heat advected in groundwater. *Canadian Journal of Earth Sciences, 49*(8), 953–962. https://doi.org/10.1139/e2012-018

Devoie, É. G., Connon, R. F., Craig, J. R., & Quinton, W. L. (2019). Taliks: A tipping point in discontinuous permafrost degradation in peatlands. *Water Resources Research, 55*(11), 9838–9857. https://doi.org/10.1029/2018wr024488

Domine, F., Fourteau, K., Picard, G., Lackner, G., Sarrazin, D., & Poirier, M. (2022). Permafrost cooled in winter by thermal bridging through snow-covered shrub branches. *Nature Geoscience, 15*(7), 554–560. https://doi.org/10.1038/s41561-022-00979-2

Ebrahimi, A., & Or, D. (2017). Mechanistic modeling of microbial interactions at pore to profile scale resolve methane emission dynamics from permafrost soil. *Journal of Geophysical Research: Biogeosciences, 122*(5), 1216–1238. https://doi.org/10.1002/2016jg003674

Ernakovich, J. G., & Wallenstein, M. D. (2015). Permafrost microbial community traits and functional diversity indicate low activity at in situ thaw temperatures. *Soil Biology and Biochemistry, 87*, 78–89.

Fouché, J., Christiansen, C. T., Lafrenière, M. J., Grogan, P., & Lamoureux, S. F. (2020). Canadian permafrost stores large pools of ammonium and optically distinct dissolved organic matter. *Nature Communications, 11*(1). https://doi.org/10.1038/s41467-020-18331-w

Fu, Z., Wu, Q., Chen, A., Wang, L., Jiang, G., Gao, S., Yun, H., & Chen, J. (2025). Non-temperature environmental drivers modulate warming-induced 21st-century permafrost degradation on the Tibetan Plateau. *Nature Communications, 16*(1). https://doi.org/10.1038/s41467-025-63032-x

Guo, D., & Wang, H. (2016). CMIP5 permafrost degradation projection: A comparison among different regions. *Journal of Geophysical Research: Atmospheres, 121*(9), 4499–4517. https://doi.org/10.1002/2015jd024108

Hansel, C. M., Lentini, C. J., Jardine, P. M., Tang, Y., Johnston, D. T., & Wankel, S. D. (2015). Dominance of sulfur-fueled iron oxide reduction in low-sulfate freshwater sediments. *The ISME Journal, 9*(11), 2400–2412. https://doi.org/10.1038/ismej.2015.50

Herndon, E. M., Yang, Z., Janot, N., Regier, T. Z., Gu, B., Wullschleger, S. D., Bargar, J., Liang, L., & Graham, D. E. (2015). Geochemical drivers of organic matter decomposition in arctic tundra soils. *Biogeochemistry, 126*(3), 397–414. https://doi.org/10.1007/s10533-015-0165-5

Hjort, J., Doré, G., Bjella, K., Streletskiy, D., Wu, Q., & Luoto, M. (2022). Impacts of permafrost degradation on infrastructure. *Nature Reviews Earth & Environment, 3*(1), 24–38. https://doi.org/10.1038/s43017-021-00247-8

Holloway, J. E., Jin, H., Lewkowicz, A. G., Douglas, T. A., Li, X., Turetsky, M. R., & Baltzer, J. L. (2020). Impact of wildfire on permafrost landscapes: A review of recent advances and future prospects. *Permafrost and Periglacial Processes, 31*(3), 371–382. https://doi.org/10.1002/ppp.2048

Hu, G., Hao, J., Zhu, X., Wu, T., Li, R., Wu, X., Xie, C., & Zhao, L. (2019). Thermal properties of active layer in permafrost regions with different vegetation types on the Qinghai-Tibetan Plateau. *Theoretical and Applied Climatology, 139*(3–4), 983–993. https://doi.org/10.1007/s00704-019-03008-2

Humlum, O., Instanes, A., & Sollid, J. L. (2003). Permafrost in Svalbard: A review of research history, climatic background and engineering challenges. *Polar Research, 22*(2), 191–215. https://doi.org/10.3402/polar.v22i2.6455

Ishikawa, M., Zhang, Y., Ohata, T., Kadota, T., & Sharkhuu, N. (2005). Ground thermal and moisture conditions at the southern boundary of discontinuous permafrost, Mongolia. *Permafrost and Periglacial Processes, 16*(2), 209–216. https://doi.org/10.1002/ppp.483

Jansson, J. K., & Taş, N. (2014). The microbial ecology of permafrost. *Nature Reviews Microbiology, 12*(6), 414–425. https://doi.org/10.1038/nrmicro3262.1

Jeelani, G., Hassan, W., Padhya, V., Deshpande, R. D., Dimri, A. P., & Lone, S. A. (2024). Significant role of permafrost in regional hydrology of the Upper Indus Basin, India. *Science of the Total Environment, 919*, 1702863.

Ji, X., Abakumov, E., Xie, X., & Polyakov, V. (2021). Mobilization of geochemical elements to surface water in the active layer of permafrost in the Russian Arctic. *Water Resources Research, 57*(1). https://doi.org/10.1029/2020wr028269

Jones, B. M., Parsekian, A. D., Baughman, C. A., Stephani, E., Jones, M. C., Berg, E. E., Grosse, G., Babcock, E. L., & Romanovsky, V. E. (2016). Presence of rapidly degrading permafrost plateaus in south-central Alaska. *The Cryosphere, 10*(6), 2673–2692. https://doi.org/10.5194/tc-10-2673-2016

Kenner, R., Noetzli, J., Hoelzle, M., Raetzo, H., & Phillips, M. (2019). Distinguishing physical and thermal causes of rockfall in a high-mountain permafrost face. *Permafrost and Periglacial Processes, 30*(3), 227–241. https://doi.org/10.1002/ppp.2010

Knutson, J. K., Westermann, S., De Wit, H. A., Dörsch, P., & Clayer, F. (2025). Water chemistry and greenhouse gas concentrations in waterbodies of a thawing permafrost peatland complex in northern Norway. *Biogeosciences, 22*(15), 3899–3914. https://doi.org/10.5194/bg-22-3899-2025

Kwon, M. J., O'Loughlin, E. J., Boyanov, M. I., Brulc, J. M., Johnston, E. R., Kemner, K. M., & Antonopoulos, D. A. (2016). Impact of organic carbon electron donors on microbial community development under iron- and sulfate-reducing conditions. *PLoS One, 11*(1), Article e0146689. https://doi.org/10.1371/journal.pone.0146689

Lacroix, F., Stimmler, P., Schaller, J., Göckede, M., Zaehle, S., Kutzbach, L., Caldararu, S., & Holl, D. (2022). Mismatch of N release from the permafrost and vegetative uptake opens pathways of increasing nitrous oxide emissions in the high Arctic. *Global Change Biology, 28*(20), 5973–5990. https://doi.org/10.1111/gcb.16345

Langford, J. E., Nagare, R. M., Mohammed, A. A., Schincariol, R. A., & Quinton, W. L. (2019). Transient and transition factors in modeling permafrost thaw and groundwater flow. *Groundwater, 58*(2), 258–268. https://doi.org/10.1111/gwat.12903

Li, X., & Cheng, G. (1999). A GIS-aided response model of high-altitude permafrost to global change. *Science in China Series D: Earth Sciences, 42*(1), 72–79. https://doi.org/10.1007/bf02878500

Mackelprang, R., David, M. M., Chavarria, K. L., Jansson, J. K., Waldrop, M. P., Rubin, E. M., Deangelis, K. M., & Blazewicz, S. J. (2011). Metagenomic analysis of a permafrost microbial community reveals a rapid response to thaw. *Nature, 480*(7377), 368–371. https://doi.org/10.1038/nature10576

Mahanta, P., Dash, S. S., Singh, A. K., & Thayyen, R. J. (2024). Role of blocky rock debris in cooling high-altitude permafrost: Field observations from the Ladakh Range, India. *Cold Regions Science and Technology, 218*, Article 104082. https://doi.org/10.1016/j.coldregions.2023.104082

Majeed, U., Rashid, I., Sattar, A., Nüsser, M., & Schmidt, S. (2021). Recession of the Gya Glacier and reconstruction of the 2014 glacial lake outburst flood (GLOF), Ladakh, India. *Science of the Total Environment, 796*, Article 148919. https://doi.org/10.1016/j.scitotenv.2021.148919

Mao, C., Kou, D., Chen, L., Qin, S., Zhang, D., Peng, Y., & Yang, Y. (2020). Permafrost nitrogen status and its determinants on the Tibetan Plateau. *Global Change Biology, 26*(9), 5290–5302.

Mu, M., Zhu, Y., Liu, H., Peng, X., Jia, Y., Zhang, G., Fan, C., Mu, C., & Qiao, Y. (2024). Topographic drivers of permafrost organic carbon accumulation on the Northern Qinghai-Tibet Plateau. *Permafrost and Periglacial Processes, 35*(3), 373–383. https://doi.org/10.1002/ppp.2221

Munkhjargal, M., Yadamsuren, G., Yamkhin, J., & Menzel, L. (2020). Ground surface temperature variability and permafrost distribution over mountainous terrain in northern Mongolia. *Arctic, Antarctic, and Alpine Research, 52*(1), 13–26. https://doi.org/10.1080/15230430.2019.1704347

Pan, F., Fu, Y., Cai, Y., Guo, Z., Wang, B., Liu, H., Gao, A., & Wu, J. (2019). Kinetic exchange of remobilized phosphorus related to phosphorus-iron-sulfur biogeochemical coupling in coastal sediment. *Water Resources Research, 55*(12), 10494–10517. https://doi.org/10.1029/2019wr025941

Park, S. W., Mun, J. H., Lee, H., Steinert, N. J., An, S. I., Shin, J., & Kug, J. S. (2025). Continued permafrost ecosystem carbon loss under net-zero and negative emissions. *Science Advances, 11*(7), eadn8819.

Patzner, M. S., Straub, D., Borch, T., Joss, H., Bryce, C., Mckenna, A. M., Kleindienst, S., Hoeschen, C., Logan, M., Young, R. B., Scholten, T., Zhou, Z., Kappler, A., & Mueller, C. W. (2022). Microbial iron cycling during palsa hillslope collapse promotes greenhouse gas emissions before complete permafrost thaw. *Communications Earth & Environment, 3*(1). https://doi.org/10.1038/s43247-022-00407-8

Perez-Mon, C., Qi, W., Vikram, S., Frossard, A., Makhalanyane, T., Cowan, D., & Frey, B. (2021). Shotgun metagenomics reveals distinct functional diversity and metabolic capabilities between 12 000-year-old permafrost and active layers on Muot da Barba Peider (Swiss Alps). *Microbial Genomics, 7*(4). https://doi.org/10.1099/mgen.0.000558

Prokushkin, A. S., Ruehlow, S., Mcdowell, W. H., Schulze, E.-D., & Gleixner, G. (2007). Source- and substrate-specific export of dissolved organic matter from permafrost-dominated forested watershed in central Siberia. *Global Biogeochemical Cycles, 21*(4). https://doi.org/10.1029/2007gb002938

Reyes, F. R., & Lougheed, V. L. (2015). Rapid nutrient release from permafrost thaw in Arctic aquatic ecosystems. *Arctic, Antarctic, and Alpine Research, 47*(1), 35–48. https://doi.org/10.1657/aaar0013-099

Roy-Leveillee, P., & Burn, C. R. (2017). Near-shore talik development beneath shallow water in expanding thermokarst lakes, Old Crow Flats, Yukon. *Journal of Geophysical Research: Earth Surface, 122*(5), 1070–1089. https://doi.org/10.1002/2016jf004022

Schmidt, S., Nüsser, M., Baghel, R., & Dame, J. (2020). Cryosphere hazards in Ladakh: The 2014 Gya glacial lake outburst flood and its implications for risk assessment. *Natural Hazards, 104*(3), 2071–2095. https://doi.org/10.1007/s11069-020-04262-8

Shao, M., Zhang, S., Song, S., Niu, B., Yun, H., Lei, T., & Pei, Y. (2022). Soil texture influences soil bacterial biomass in the permafrost-affected alpine desert of the Tibetan plateau. *Frontiers in Microbiology, 13*, 1007194. https://doi.org/10.3389/fmicb.2022.1007194

Shayanmehr, S., Babošová, M., Shahnoushi Foroushani, N., Rastegari Henneberry, S., Porhajašová, J. I., Sabouhi Sabouni, M., & Mohammadi, H. (2022). The impacts of climate change on water resources and crop production in an Arid Region. *Agriculture, 12*(7), 1056. https://doi.org/10.3390/agriculture12071056

Skierszkan, E. K., Glas, B., Thorson, J., Welch, L. A., James, M., McAlister, J., van der Velden, A., Bond, J., Kokelj, S. V., St-Arnault, M., & Mayer, B. (2024). Stable isotopes and major ions reveal how permafrost thaw and groundwater influence the geochemistry of Arctic surface waters. *Scientific Reports, 14*, 12948. https://doi.org/10.1038/s41598-024-63301-9

Waldrop, M. P., Chabot, C. L., Liebner, S., et al. (2023). Permafrost microbial communities and functional genes are structured by latitudinal and soil geochemical gradients. *The ISME Journal, 17*, 1224–1235. https://doi.org/10.1038/s41396-023-01429-6

Wang, T., Yang, D., Yang, Y., Piao, S., Li, X., Cheng, G., & Fu, B. (2016). Permafrost thawing puts the frozen carbon at risk over the Tibetan Plateau. *Science Advances, 2*(8), Article e1600319. https://doi.org/10.1126/sciadv.1600319

Wani, J. M., Thayyen, R. J., Gruber, S., Ojha, C. S. P., & Stumm, D. (2020). Single-year thermal regime and inferred permafrost occurrence in the upper Ganglass catchment of the cold-arid Himalaya, Ladakh, India. *Science of the Total Environment, 703*, Article 134631.

Wani, J. M., Thayyen, R. J., Ojha, C. S. P., & Gruber, S. (2021). The surface energy balance in a cold and arid permafrost environment, Ladakh, Himalayas, India. *The Cryosphere, 15*(5), 2273–2293.

Wu, M. H., Wang, Y. F., Chen, J. W., Kang, S. C., Thies, J. E., Chen, S. L., Bardgett, R. D., Chen, S. Y., Xue, K., Wang, X. M., Chen, T., & Rui, J. P. (2021). Reduced microbial stability in the active layer is associated with carbon loss under alpine permafrost degradation. *Proceedings of the National Academy of Sciences, 118*(25). https://doi.org/10.1073/pnas.2025321118

Wu, Q., Ma, W., Yu, Q., Sheng, Y., & Chen, J. (2020). Engineering in the rugged permafrost terrain on the roof of the world under a warming climate. *Permafrost and Periglacial Processes, 31*(3), 417–428. https://doi.org/10.1002/ppp.2059

Wunder, L. C., Aromokeye, D. A., Yin, X., Otersen, C., Kasten, S., Schnakenberg, A., Willis-Poratti, G., Dohrmann, I., Römer, M., Bohrmann, G., Richter-Heitmann, T., & Friedrich, M. W. (2021). Iron and sulfate reduction structure microbial communities in (sub-)Antarctic sediments. *The ISME Journal, 15*(12), 3587–3604. https://doi.org/10.1038/s41396-021-01014-9

Yadav, P., Das, J., Sundharam, S. S., & Krishnamurthi, S. (2024). Analysis of culturable bacterial diversity of Pangong Tso Lake via a 16S rRNA tag sequencing approach. *Microorganisms, 12*(2), 397. https://doi.org/10.3390/microorganisms12020397

Yang, S., Knoblauch, C., Walz, J., In 'T Zandt, M. H., Wagner, D., Jetten, M. S. M., Probst, A. J., Liebner, S., & Bornemann, T. L. V. (2021). Effects of a long-term anoxic warming scenario on microbial community structure and functional potential of permafrost-affected soil. *Permafrost and Periglacial Processes, 32*(4), 641–656. https://doi.org/10.1002/ppp.2131

Yoshikawa, K., Casassa, G., Azócar, G., Iribarren Anacona, P., Masías, P., Miyata, T., Takahashi, N., Schorghofer, N., Ramirez, E., & Cáceres Correa, B. (2025). Permafrost and active layer characteristics in the high arid Andes between the Altiplano and high Atacama Desert. *Arctic, Antarctic, and Alpine Research, 57*(1). https://doi.org/10.1080/15230430.2025.2532202

Zhang, G., Nan, Z., Zhao, L., Liang, Y., & Cheng, G. (2021a). Qinghai-Tibet Plateau wetting reduces permafrost thermal responses to climate warming. *Earth and Planetary Science Letters, 562*, Article 116858. https://doi.org/10.1016/j.epsl.2021.116858

Zhang, G., Nan, Z., Yin, Z., & Zhao, L. (2021b). Isolating the contributions of seasonal climate warming to permafrost thermal responses over the Qinghai-Tibet Plateau. *Journal of Geophysical Research: Atmospheres, 126*(21). https://doi.org/10.1029/2021jd035218

Zhang, T., Frauenfeld, O. W., Serreze, M. C., Ettinger, A., Hodge, C., Gilichinsky, D., Barry, R. G., Ye, H., Saunders, F., & Zhang, X. (2008). Spatial and temporal variability in active layer thickness over the Russian Arctic drainage basin. *Journal of Geophysical Research: Atmospheres, 110*(D16). https://doi.org/10.1029/2004jd005442

Hydrology and Water Governance in Cold Deserts: Glacial Dynamics, Resource Management, and Natural Disaster Resilience

Shivali Sharma, Harinder Singh Banyal, and Nitesh Kumar

1 Introduction

Cold deserts are unique and fragile ecosystems, characterized by high altitudes, low temperature, limited precipitation, and extreme aridity (Laity, 2009). These are typically situated in high-altitude regions, having low temperature with low precipitation most of the year. The Gobi Desert (Mongolia and China), Atacama Desert (Peru and Chile), Patagonian Desert (Argentina), Taklamakan Desert (China), the Great Basin Desert (USA), Kyzylkum Desert (Central Asia), Ladakh (India), and Spiti Valley (Himachal Pradesh, India) are some of the cold deserts present worldwide. There is huge seasonal variation occurring in weather in cold deserts with short, dry, arid, and cloudless summer with temperature reaching up to 36 °C and winters are long, chilled, and windy with freezing temperature of −32 °C. They are one of the most fragile ecosystems with limited natural resources supporting endemic biotic communities (Banerji & Basu, 2010). It is well known that the cold deserts have limited water resources, streams supplied by glaciers, unproductive soil, and a short growth season due to their severe climate (Banerji & Basu, 2010).

These cold deserts despite of having harsh environment help in sustaining downstream populations by providing water to large number of downstream rivers. Therefore, studying the hydrology of these unique ecosystem is quite important with respect to availability, distribution, and governance of water resources for their sustainability (Shen & Chen, 2010). Hydrology in cold deserts is primarily governed by glacial and snowmelt dynamics, which in turn influence water availability, ecosystem health, and human livelihoods. Understanding eco-hydrological dynamics in cold and arid areas is very important due to limited and unpredictable presence of water resources

S. Sharma · H. S. Banyal (✉) · N. Kumar
Department of Biosciences, Himachal Pradesh University, Shimla, H.P., India
e-mail: dr.harinderbanyal@gmail.com

Y. S. Rawat and C. P. Kala (eds.), *Cold Deserts*,
https://doi.org/10.1007/978-3-032-29119-6_14

(Rodriguez-Iturbe & Porporato, 2005). Hydrological processes' knowledge is essential for controlling the hydrological system, identifying the source of water, and enhancing predictive precipitation-runoff models (Andrew, 2009; Kevin et al., 2010).

Water-related natural disasters such as cloudbursts, glacial lake outburst floods (GLOF), landslide dam outburst floods (LDOF), and glacial surge dam outburst floods occur frequently in the cold desert areas (Thayyen, 2020). Cryospheric and hydrological alterations coupled with socio-economic growth are increasingly impending the downstream water security, highlighting the urgent need for efficient management strategies (Aggarwal et al., 2021). Due to increasing anthropogenic pressure and climate change-induced glacial retreat, effective water governance is crucial in these areas. The topographical limitations in this area are a major management challenge for water resources. Particularly weather conditions and discharge monitoring along with limited studies in the cold deserts hinder our understanding about the hydrological processes occurring in the region.

Therefore, this study deals with the physical hydrology of cold deserts; the constraints and dependencies on water resources; and the critical role of governance in mediating access, equity, and sustainability of these resources. The study critically examines the interplay between glacial hydrology, water resource management, and disaster resilience in cold deserts, globally.

2 The Physical Hydrology of Cold Deserts

Hydrological processes in cold deserts are predominantly influenced by snow or glacier melt, minimal rainfall, and seasonal variations in temperature regulating river flows by melt water discharge (Lone et al., 2021). Moreover, the presence of permafrost, freeze–thaw cycles, and presence of sparse vegetation further complicate the hydrological balance, often resulting in flash floods or droughts (Sharifnezhadazizi, 2022). Glacier melt and freeze thawing are the primary thermal processes, which contribute to intimate and complex coupling between water and heat transfer (Mou et al., 2008). Cold Region Hydrological Model (CRHM) given by Pomeroy et al. (2007) incorporates advanced physical-based representations of cold deserts such as snow relocation, snowmelt, sublimation, infiltration into frozen soil, and tundra runoff dynamics. It represents a robust and adaptable modeling platform that enhances our ability to predict and understand complex hydrological processes in cold deserts, making it a valuable asset for research, water resources planning, and climate adaptation efforts.

2.1 Water Resources and Their Use

Cold deserts, especially those in Trans-Himalayan and Central Highlands, are the home to some of the biggest and most intricate glacier systems outside the Polar

Regions. There are large number of huge sized and high-density glaciers present in Karakoram Range due to its steep terrain and intense avalanche activity. For instance, the Biafo Glacier, second longest and third largest glacier of Karakoram Range, comprising 63% of glacier basin of the area, with 75% of perennial snow or ice cover. The Baltoro Glacier, despite having large basin size (400 km^2), has less glacial ice coverage (42%) (Gao et al., 2020). Perennial snow and ice cover are similar to the Biafo Glacier (72%), but its accumulation zone is much smaller, i.e., on 12% of the total basin area. Another glacier, the Toltar-Baltar Glacier, represents extremely steep glacier basin, highlighting characteristics common in Karakoram but rarely seen elsewhere. The glacier comprises remarkable feature of "ablation zone" but there is absence of accumulation zone. It has a watershed elevation of 7,750 m and total fall of 4,760 m making it steeper than Biafo, despite its shorter length (Hewitt, 2014). The Batura Glacier, a major valley-type glacier in the Hunza Basin of Northwest Karakoram, with an extensive debris ablation zone, around 27% of the glacier is covered with debris. Also the glacier shows the decrease in the area from 285 km^2 (1974) to 243.5 km^2 (2020) (Gao et al., 2020). The Inylchek Glacier system, comprising of Southern and Northern Inylchek glaciers in Tian Shan mountains in China/Kyrgyzstan, drains an area of over 800 km^2 and extends up to 60 km in length, with an average width of approximately 1.9 km near Merzbacher meadow (Mayer et al., 2008). The Siachen Glacier, located in the eastern Karakoram Range of the India–Pakistan border, is the largest glacier in the Karakoram and in all of High Mountain Asia. It stretches about 74 km in length and varies from 1 to 8 km in width, covering an area of approximately 936 km^2, with an average ice thickness of ~**300 m** (Frey et al., 2014).

These glaciers are vital resources of water in cold deserts, storing precipitation as ice and releasing it in the form of water during warmer months (National Research Council, 2012). Water from glaciers melting, permafrost, and snowfields is the major source of irrigational water in agriculture in cold-arid zones, globally (Nüsser et al., 2019). However, global warming has significantly altered glacial mass balance, leading to accelerated retreat and the formation of unstable glacial lakes. Key issues of concern in cold deserts include declining of glaciers reducing long-term water availability, melting of glaciers because of climate change leads to moraine-dammed lakes prone to sudden breaching and early and snowmelt affects the normal water flow in the region. Local inhabitants of the cold deserts face a number of hydrological constraints such as low and erratic rainfall, deep and saline groundwater, poor socio-economic condition and education systems, sandy terrain with high permeability, and development of water resources with particular focus on target (Tundup et al., 2017). Varied adaptive technologies are designed for a proper water management and distribution in diverse region on the basis of regional requirements (Carey et al., 2017; Mark et al., 2017). Local communities which are highly reliant on glaciers and snowmelt for water have established distinctive water management strategies and are now seeking alternative solutions because of continuous climate change and lifestyle alterations (Pandey et al., 2022). Monitoring glacial changes using remote sensing, ground surveys, and hydrological models is crucial for assessing water supply sustainability.

2.2 Water Governance Framework

Freshwater resources sustainable management and ecosystem services provided by them remains a critical challenge of the twenty-first century. Many water-related issues arise from the failures in governance than from the physical scarcity of the water (Pahl-Wostl, 2017). Governance is defined as "the process of decision-making and the process by which decisions are implemented or not" (UNWWAP, 2006). "Water governance refers to the range of political, social, economic and administrative systems that are in place to regulate development and management of water resources and provisions of water services at different levels of society" (Rogers & Hall, 2003). Key recommendations for enhancing water governance comprises distinguishing water as an economic good; enacting new, workable water laws and policies; setting up efficient participatory structures and procedures; adopting integrated water management approaches at basin level (Tortajada, 2010). An interdisciplinary three-step approach is projected by Van Rijswick et al. (2014) to address water scarcity, water quality, and flood risks. It integrates water system analysis, law, economics, and public administration to build a shared understanding through new concepts. Three steps include: (1) developing knowledge of the water system and related values and policies; (2) ensuring stakeholder engagement, clarifying trade-offs and responsibilities; and (3) implementing agreed service level via appropriate infrastructure, implementation, and conflict resolving mechanisms.

Governance of water resources in cold deserts requires integrated and adaptive approaches for proper management and distribution of water. For example, in order to address Ladakh's increasing water scarcity, Sonam Wangchuck and his team from SEMCOL (Students' Educational and Cultural Movement of Ladakh) launched the Ice Stupa program in 2013. Inspired by traditional methods and improved upon earlier designs by Chewang Norphel, these structures require less maintenance and are built at lower altitudes. They create vertical, cone-shaped artificial glaciers that store winter meltwater and release it in the spring (Anonymous, 2025). Strategies must include: empowerment of local panchayats and user associations (Madhav, 2025), trans-boundary water sharing arrangements and policies (Ward, 2013), and investments in hydrological monitoring and modeling and data-driven decision-making (Cantor et al., 2021).

2.3 Climate Change and Glacial Retreat

Cold desert's hydrology (Fig. 1) is characterized by cold winters, precipitation, significant water storage by seasonal snow cover, glacial geomorphology, seasonally or permanently frozen ground, frequent episodic runoff events, and correlation between mass and energy balance through phase variations (Pomeroy et al., 2007). Increase in the emission of greenhouse gases, including CO_2 and other trace gases, has changed the balance of energy on the earth and will endure in future, causing a global shift in

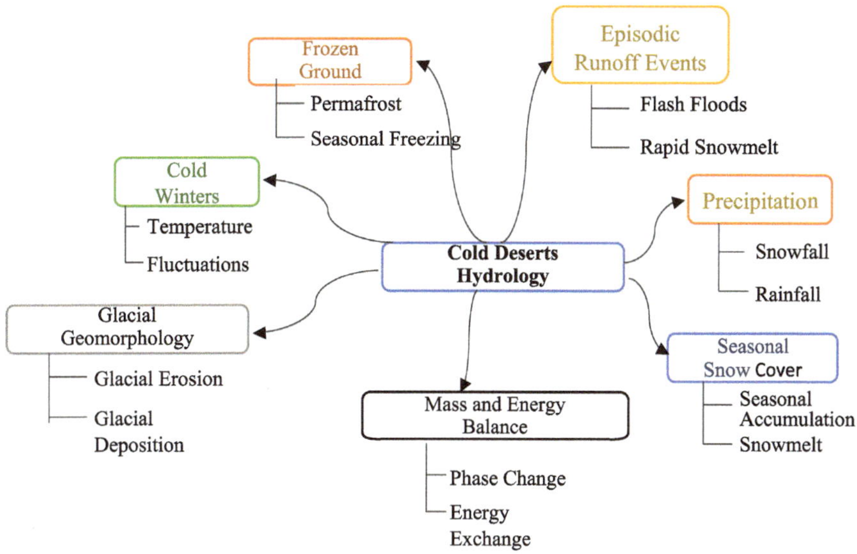

Fig. 1 Hydrological characteristics of cold deserts

the climate (IPCC, 2013). The relationship between temperature and rain, the ensuing effects on the hydrological cycle, and how these factors translate into various reactions occurring in aquatic and terrestrial ecosystems will determine the ramifications of these climatic changes on cold desert ecosystems (Snyder et al., 2019).

In arid regions such as in cold deserts, climate plays an important role in controlling hydrological processes along with topography and geology (Houston, 2006). Climate change linked to development, resource stress, and disaster risk has significant impacts on high-altitude cold deserts. Climate change has become a global challenge, creating varying levels of vulnerability across different communities, sectors, and regions (IPCC, 2014). According to IPCC (Intergovernmental Panel on Climate Change) (1996, 2001) assessment reports, the trend of global warming would be more noticeable at high altitudes. The majority of glaciers throughout the world are retreating more quickly as an impact of climate change (Hu et al., 2024; Sommer et al., 2020), which has significant effects on the universal water cycle. The hydrological processes of high-altitude alpine watersheds have undergone substantial modifications as a result of climate change (Singh et al., 2021; Wang et al., 2021). The positive correlation between the level of development and climate change impacts was seen by Banerji and Basu (2010) in the Himalayan cold deserts, indicating that more developed areas face greater hydrological pressures, as there is decrease in snow cover, water availability, and natural vegetation. Furthermore, a positive correlation was also found between climate change and rising frequency of natural disasters including floods, droughts, landslides, avalanches, etc., posing a serious threat to local livelihoods.

Natural resources management in the context of climate change demands a flexible and adaptive approach, enabling adjustments to management strategies as environmental conditions continue to evolve. One conceptual framework adopted by the U.S. Fish and Wildlife Service (USFWS, 2010) and the U.S. Forest Service (USFS, 2011) categorizes adaptation strategies into three types, viz., resistance, resilience, and response (Millar et al., 2007).

- **Resistance strategies** focus to strengthen the ability of ecosystems to sustain their core structure, functions, and processes in the face of short-term climate-related stressors including drought, wildfire, insect outbreaks, and disease.
- **Resilience strategies** focus on reducing ecosystem vulnerability and intensifying the capacity of ecosystem components to absorb disturbances while continuing to function, thereby lessening the harshness of climate change effects.
- **Response strategies** are designed to facilitate ecological transitions in anticipation of or in reaction to long-term environmental alterations, supporting adaptive shifts in ecosystem structure and composition.

It is important to note that a particular management strategy may fall under one or more of these categories, depending on the objectives and the specific environmental context (Snyder et al., 2019). Local people's conception of climate change is a key factor in understanding and developing mitigation strategies to cope up with the impact of climate change (Hasan & Kumar, 2019). This perception plays a vital role in determining how communities respond to climate-related hazards and junctures (Arora et al., 2021; Debela et al., 2015). Moreover, understanding local perceptions not only helps in the formulation of effective climate policies but also acts as an indicator of their willingness to adopt behavioral changes aimed at reducing the negative effects of climate change (Arora et al., 2021; Niles & Mueller, 2016).

3 Disaster Risk and Resilience in Cold Desert Ecosystem

According to the Pressure and Release (PAR) model, disasters comprised of three interconnected progression levels of vulnerability, viz., fundamental causes, unsafe conditions, and dynamic pressures (Blaikie et al., 2004). There are two types of disasters: those that occur slowly, like desertification, and those that happen quickly, like flash floods and other events. The two biggest threats to the ecology in the cold deserts are climate change and the likelihood of disasters (Banerji & Basu, 2010). Climate change, natural disasters, and human needs are the highest contributor to environmental threat. Climate change-induced disasters including GLOFs, flash floods, and landslides are more prevalent in cold deserts (Ramya et al., 2023). Increase in anthropogenic pressure due to development, resource use practices, and increasing demand supply also poses significant threat to these natural ecosystems. The first hydrological level at extremely high altitude faces significant threat (Banerji & Basu, 2010). Unusual snowfall and untimely melting of glaciers is another big threat to the local economy of cold deserts, so introducing artificial glaciers (a novel innovation) will minimize the water shortage in the region (Namtak & Sharma, 2018).

3.1 Key Resilience Strategies for Hydrological Management in Cold Deserts

- In order to maintain ecological balance, the cold desert inhabitants employ and manage resources using traditional methods like organic farming.
- Adoption of more advanced methods, like lift irrigation, sprinklers, and drip irrigation, would increase water resource productivity and lessen stress.
- In the watersheds that are severely water-stressed, drought-resistant agricultural cultivars must also be introduced.
- Alternatives would need to be sought in order to encounter the desires of the local community, given the increasing depletion of natural resources accessible for home and professional usage.
- Restoration of alpine wetlands and vegetation to buffer hydrological shocks.
- A thorough disaster management plan must be developed, outlining the roles of communities, state agencies, and non-state organizations in terms of preparedness, response, and risk reduction strategies.
- Numerous conservation and preservation initiatives can be implemented, such as structural and vegetative measures such as check dams, storage measures, plants, and floodplain development.
- Cold desert village communities have received training in conservation, management, budgeting, and water sharing. To ameliorate the access and management of water as well as to meet the village's sanitation needs, women-led Domestic Water and Sanitation Committees (DWSC) have been established (Banerji & Basu, 2010). For instance, in Ladakh, local youth collaborated to construct Ice-Stupas to store winter glacial meltwater for spring use, which demonstrated indigenous innovation rooted in local stewardship. Also, traditional Farmers Managed Irrigation System (FMIS) in Miyar Valley of Lahaul-Spiti District, Himachal Pradesh offer another valuable example. The Kuhls (open channel irrigation system) made by local people to capture and divert snowmelt water from nearby glaciers to agricultural fields (Padigala, 2014).

4 Integrated Water Resource Management (IWRM) Practices in Cold Deserts

In cold deserts, IWRM refers to a coordinated and sustainable approach to manage limited water resources of high-altitude, arid regions where water availability is scarce and climatic conditions are extreme (Agarwal et al., 2000). These areas receiving low annual precipitation are mainly dependent on glacial melt water to fulfill their needs. In spite of having large number of glaciers and perennial rivers such as Indus River, Shyok River, Nubra River, and Suru River in the cold deserts, the region suffers from a severe water shortage for domestic purposes and irrigation, because of the unavailability of glacier water and even the rivers and their tributaries flow deep in the gorges, making water extraction difficult. The rivers draining the region also carry a large amount of sediment load during snowmelt periods (Sharma, 2006). Moreover, glaciers, main water resource of the region, are also shrinking due to global warming (Romshoo et al., 2022).

So, the implementation of IWRM practices addressing the complex relations between hydrology, climate, and local communities is critical in such type of ecosystems (Fig. 2). IWRM is defined as "a process which promotes the coordinated development and management of water, land, and related resources, in order to maximize the resultant economic and social welfare in an equitable manner without compromising the sustainability of vital ecosystems" (GWP, 2000). A holistic approach for IWRM includes inclusion of local inhabitants, linking water management plan with agriculture, energy and biodiversity, and adoption of adaptive management strategies in order to combat evolving climate change and socio-economic conditions. Furthermore, traditional practices play a critical role in the management of natural water resources effectively.

For example, in Ladakh (cold desert region in India), several water management practices have been undertaken by local communities living there (Saxena et al., 2021), such as

- Glacier melt water is collected in depressions known as Zings on the entry of the village and is brought to agricultural fields by Kuhls (water channels).
- Ex situ water conservation system includes pressurized irrigation, such as drip irrigation system, which reduces the irrigation water losses up to 25%.

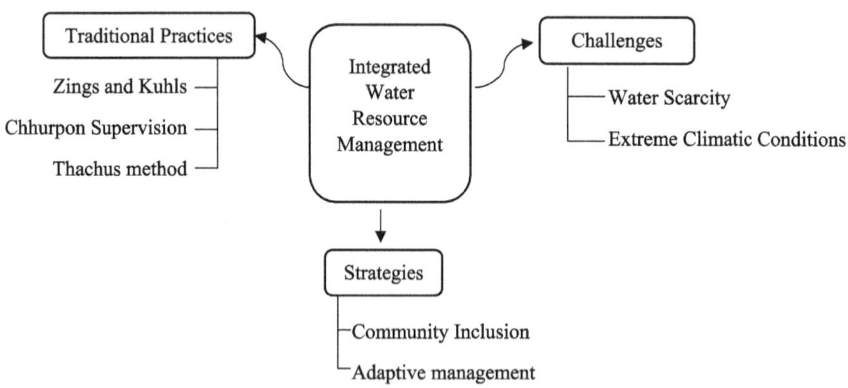

Fig. 2 Integrated water resource management in cold deserts (Zings: water storage ponds for irrigation in Ladakh; Kuhls: gravity-fed irrigation channels; Chhurpon: traditional water distributor (village official) in Ladakh; Thachus: traditional water harvesting system in cold deserts where winter runoff is frozen using retaining walls, then melted snow was used for irrigation)

- Since micro-irrigation technologies have the capacity to provide water and conserve water in more sustainable way in cold, arid high desert agro-ecosystems, their application is becoming more and more common.
- Additionally, Chhurpon (water lord) appointed by local peoples in these areas having knowledge of water flow, field conditions, and requirements supervises distribution of water to households.
- A method known as "Thachus" (traditional ice-based irrigation system) is adopted in Ladakh, in which an irrigation channel is made in the month of November and December, diverting runoff water to shady regions of mountains, slowing down of water, and let it to freeze. Then melted snow during April was used for irrigation.

5 Recommendations

So as to address the growing challenges arises due to climate change along with glacial retreat and increasing water scarcity in the cold desert ecosystems, a set of strategic recommendations is essential, which will promote sustainable management, integrate local knowledge with modern technologies, and reduce disaster impacts. The following recommendations provide a roadmap for improving water security in the cold deserts:

- Local communities must adopt an IWRM approach to ensure sustainable use of glacial melt water present in the region (Flügel, 2011).
- Real-time monitoring stations should be established for tracking of glacial processes using remote sensing (Kimothi et al., 2022).
- Disaster risk reduction (DRR) programs (Barua et al., 2022) must be established for early warning of natural disasters such as the GLOFs, flash floods, and avalanches.
- Proper policy reform and improvement of coordination between various water management institutes and agencies can help to enhance community engagement and sustainability of water resources.
- Moreover, water conservation infrastructure such as formation of artificial glaciers (Angmo et al., 2021), micro-irrigation system (Patel et al., 2023; Saxena et al., 2021), and rainwater harvesting further optimize water use in cold deserts.
- Designing of resilient infrastructure that can withstand hydrological variability and extreme weather events linked to climate change.
- Incorporation of traditional knowledge with advanced technologies of water governance further enhances hydrological conditions of the area (Saxena et al., 2021).
- Additionally, local stakeholders and youth must be trained in hydrological science, glacial behavior, and emergency response strategies.

6 Future Perspectives

Looking forward, various future perspectives such as technological innovations, traditional knowledge integration, policy reforms, and climate awareness will address the long-term challenges of water scarcity, glacial retreat, and climate-induced disasters in cold deserts. With a special focus on sustainability, innovation, and collaboration, the following future perspectives must be undertaken to strengthen the water resilience, which ensure equal water governance and enhance ecological balance across cold desert ecosystems:

- Nature-based eco-solutions such as afforestation, restoration of wetlands, and creation of buffer zone help in mitigation of the GLOFs and also help in aquifer recharge.
- Shift toward decentralized, community-managed water governance models empowered by legal backing and technological support.
- Adoption of Artificial Intelligence (AI), remote sensing, machine learning, and various other innovative technologies improve hydrology of the region by glacial melt forecasting, climate change, and disaster prediction.
- Collaboration among climatologists, hydrologists, sociologists, and policymakers to address complex water governance issues will further help in formulation of sustainable water management practices.
- Foster trans-boundary agreements and treaties for shared glacial and river systems in cold desert regions.

7 Conclusion

Cold deserts, a fragile ecosystem, having extreme climatic conditions, limited water availability, and low temperature pose unique challenges for hydrological management, especially in the context of climate change. It is crucial to adopt holistic approach such as integrating traditional water harvesting systems, promoting glacier and snowmelt conservation techniques, enhancing watershed management, implementing decentralized water storage techniques (Zings and Ice-Stupas) and climate adaptive strategies aligning with ecological sustainability, to address the increasing water challenges in the cold deserts. Ensuring water security in cold deserts requires integrated water resource management (IWRM), combining traditional knowledge with modern technologies (such as remote sensing and GIS for monitoring water resources, drip and sprinkler irrigation system for efficient water use, and artificial glaciers), community participation, resilient infrastructure, and adaptive governance. Strengthening local capacities, enhancing hydrological monitoring, and promoting trans-boundary cooperation are essential for building climate-resilient and sustainable water management systems in these ecologically sensitive landscapes. Investments in hydrological infrastructure, disaster risk reduction programs, and inclusive policy reforms are pivotal to augment the adaptive capacity of both ecosystems and local communities. Looking ahead, a multi-disciplinary approach that combines traditional knowledge systems, cutting-edge technologies like AI and remote sensing, and robust governance mechanisms is essential for building climate-resilient hydrological systems in cold desert ecosystem. Through collaborative efforts and sustained innovation, the cold deserts of the world can transform from water-stressed zones into models of resilience and sustainable water governance.

References

Agarwal, A., de los Angeles, M. S., Bhatia, R., et al. (2000). *Integrated water resources management* (pp. 1–67). Global Water Partnership.

Aggarwal, A., Frey, H., McDowell, G., Drenkhan, F., Nüsser, M., Racoviteanu, A., & Hoelzle, M. (2021). Adaptation to climate change induced water stress in major glacierized mountain regions. *Climate and Development, 14*(7), 665–677. https://doi.org/10.1080/17565529.2021.1971059

Andrew, J. L. (2009). Hydrograph separation for karst watersheds using a two-domain rainfall–discharge model. *Journal of Hydrology, 364*, 249–256. https://doi.org/10.1016/j.jhydrol.2008.11.001

Angmo, P., Angmo, D., & Namgail, D. (2021). Water crisis: A big constraints for agricultural practices in high Himalaya of Ladakh and affordable strategies to mitigate the crisis. In V. Fupta & F. D. Sheikh (Eds.), *Good agriculture practices in cold arid regions* (pp. 134–147). AkiNik Publications.

Anonymous. (2025). *How Sonam Wangchuk's ice stupa tackle Ladakh's water scarcity.* Retrieved August 29, 2025, from https://www.ceew.in/sustainable-agriculture-initiatives/ice-stupa-artificial-glaciers

Arora, P., Ali, S. N., & Morthekai, P. (2021). Local perceptions and trends of climate change in the Sikkim Himalaya, North-East India. *Journal of Climate Change, 7*(2), 1–26. https://doi.org/10.3233/JCC210008

Banerji, G., & Basu, S. (2010). Adapting to climate change in Himalayan cold deserts. *International Journal of Climate Change Strategies and Management, 2*(4), 426–448. https://doi.org/10.1108/17568691011089945

Barua, P., Mitra, A., & Eslamian, S. (2022). Disaster risk reduction: Detecting Himalayan Glacial lake outburst floods. In S. Eslamian & F. Eslamian (Eds.), *Disaster risk reduction for resilience: Disaster and social aspects* (pp. 393–403). Springer. https://doi.org/10.1007/978-3-030-99063-3_16

Blaikie, P., Cannon, T., Davis, I., & Wisner, B. (2004). *At Risk, natural hazards, people's vulnerability and disasters* (2nd edn). Routledge. https://doi.org/10.4324/9780203714775

Cantor, A., Kiparsky, M., Hubbard, S. S., Kennedy, R., Pecharroman, L. C., Guivetchi, K., Darling, G., McCready, C., & Bales, R. (2021). Making a water data system responsive to information needs of decision makers. *Frontiers in Climate, 3*, Article 761444. https://doi.org/10.3389/fclim.2021.761444

Carey, M., Molden, O. C., Rasmussen, M. B., Jackson, M., Nolin, A. W., & Mark, B. G. (2017). Impacts of glacier recession and declining meltwater on mountain societies. *Annals of American Association and Geographers, 107*, 350–359. https://doi.org/10.1080/24694452.2016.1243039

Debela, N., Mohammed, C., Bridle, K., Corkrey, R., & McNeil, D. (2015). *Perception of climate change and its impact by smallholders in pastoral/agropastoral systems of Borana, South Ethiopia* (Vol. 4, p. 236). Springer Plus. https://doi.org/10.1186/s40064-015-1012-9

Flügel, W. A. (2011). Development of adaptive IWRM options for climate change mitigation and adaptation. *Advances in Science and Research, 7*(1), 91–100. https://doi.org/10.5194/asr-7-91-2011

Frey, H., Machquth, H., Huss, M., Huggel, C., Bajracharya, S., Bolch, T., Kulkarni, A., Linsbauer, A., Sazmann, N., & Stoffel, M. (2014). Estimating the volume of glaciers in the Himalayan-Karakoram region using different methods. *The Cryosphere, 8*, 2313–2333. https://doi.org/10.5194/tc-8-2313-2014

Gao, H., Zou, X., Wu, J., Zhang, Y., Deng, X., Hussain, S., Wazir, M. A., & Zhu, G. (2020). Post-20th century near-steady state of Batura Glacier: Observational evidence of Karakoram Anomaly. *Scientific Reports, 10*(1), 1–10.

GWP (Global Water Partnership) (2000). Integrated Water Resources Management, 22 Stockholm: GWP Secretariat. TAC Background Papers No. 4.

Hasan, Md. K., & Kumar, L. (2019). Comparison between meteorological data and farmer perceptions of climate change and vulnerability in relation to adaptation. *Journal of Environmental Management, 237*, 54–62. https://doi.org/10.1016/j.jenvman.2019.02.028

Hewitt, K. (2014). Karakoram glaciers: Types and terrain. In *Glaciers of the Karakoram Himalaya. Advances in Asian human-environmental research* (pp. 61–86). Springer. https://doi.org/10.1007/978-94-007-6311-1_3

Houston, J. (2006). Variability of precipitation in the Atacama Desert: Its causes and hydrological impact. *International Journal of Climatology, 26*(15), 2181–2198. https://doi.org/10.1002/joc.1359

Hu, J., Yao, X., Zhang, C., Zhang, Y., & Ma, Y. (2024). Spatial and temporal changes of glaciers and glacial lakes in the Northern Tianshan mountains over the past 30 years. *Journal of Geographical Sciences, 34*, 1857–1880. https://doi.org/10.1007/s11442-024-2274-3

IPCC. (1996). *Climate change 1995: Impacts, adaptations, and mitigation of climate change scientific-technical analyses*. Cambridge University Press.

IPCC. (2001). *Climate change 2001: Impacts, adaptation, and vulnerability—Contribution of Working Group II to the third assessment report of the Intergovernmental Panel on Climate Change—Summary for policymakers* (p. 1032). IPCC Cambridge University Press.

IPCC. (2013). Summary for policymakers. In T. F. Stocker, D. Qin, G. K. Plattner, M. Tignore, S. K. Allen, J. Boschung, A. Nauels, Y. Xia, V. Bex, & P. M. Midgley (Eds.), *Climate change 2013, the physical science basis. Contribution of Working Group I to the fifth assessment report of the Intergovernmental Panel on Climate Change* (pp. 3–29). Cambridge University Press.

IPCC. (2014). Climate Change: Synthesis Report Contribution of Working Groups I, II and III to the Fifth Assessment Report of the Intergovernmental Panel on Climate Change; Technical Report; IPCC.

Kevin, W. T., Brent, B. W., & Thomas, W. D. E. (2010). Characterizing the role of hydrological processes on lake water balances in the Old Crow Flats, Yukon Territory, using water isotope tracers. *Journal of Hydrology, 386*(4), 103–117. https://doi.org/10.1016/j.jhydrol.2010.03.012

Kimothi, S., Singh, R., Gehlot, A., Akram, S. V., Malik, P. K., Gupta, A., & Bilandi, N. (2022). Intelligent energy and ecosystem for real-time monitoring of glaciers. *Computers and Electrical Engineering, 102*, Article 108163. https://doi.org/10.1016/j.compeleceng.2022.108163

Laity, J. J. (2009). *Deserts and desert environments* (Vol. 3). Wiley.

Lone, S. A., Jeelani, G., Deshpande, R., Mukherjee, A., Jasechko, S., & Lone, A. (2021). Meltwaters dominate groundwater recharge in cold arid desert of Upper Indus River Basin (UIRB), western Himalayas. *Science of Total Environment, 786*, Article 147514. https://doi.org/10.1016/j.scitotenv.2021.147514

Madhav, M. V. (2025). Catalysing water governance: Empowering communities for sustainable water management. *Indian Journal of Public Administration, 71*(1), 119–136. https://doi.org/10.1177/00195561241309140

Mark, B. G., French, A., Baraer, M., Carey, M., Bury, J., Young, K. R., Polk, M. H., Wigmore, O., Lagos, P., Crumley, R., McKenzie, J. M., & Lautz, L. (2017). Glacier loss and hydro-social risks in the Peruvian Andes. *Global and Planetary Change, 159*, 61–76. https://doi.org/10.1016/j.gloplacha.2017.10.003

Mayer, C., Lambrecht, A., Hagg, W., Helm, A., & Scharrer, K. (2008). Post-drainage ice dam response at lake Merzbacher, Inylchek Glacier, Kyrgyzstan. *Geografiska Annaler: Series A, Physical Geography, 90*(1), 87–96. https://doi.org/10.1111/j.1468-0459.2008.00336.x

Millar, C. I., Stephenson, N. L., & Stephens, S. L. (2007). Climate change and forests of the future: Managing in the face of uncertainty. *Ecological Applications, 17*, 2145–2151. https://doi.org/10.1890/06-1715.1

Mou, L., Tian, F., Hu, H., & Sivapalan, M. (2008). Extension of the representative elementary watershed approach for cold regions: Constitutive relationships and an application. *Hydrology and Earth System Sciences, 12*(2), 565–585. https://doi.org/10.5194/hess-12-565-2008

Namtak, S., & Sharma, R. C. (2018). Traditional knowledge and innovative practices for water management and conservation in Indian cold desert, Leh Ladakh. *Indian Journal of Ecology, 45*(4), 737–744.

National Research Council. (2012). *Himalayan glaciers: Climate change, water resources, and water security* (p. 143). The National Academies Press.

Niles, M. T., & Mueller, N. D. (2016). Farmer perceptions of climate change: Associations with observed temperature and precipitation trends, irrigation, and climate beliefs. *Global Environmental Change, 39*, 133–142. https://doi.org/10.1016/j.gloenvcha.2016.05.002

Nüsser, M., Dame, J., Kraus, B., Baghel, R., & Schmidt, S. (2019). Socio-hydrology of "artificial glaciers" in Ladakh, India: Assessing adaptive strategies in a changing cryosphere. *Regional Environmental Change, 19*, 1327–1337. https://doi.org/10.1007/s10113-018-1372-0

Padigala, B. (2014). Social capital and local institutions: A perspective to assess communities adaptation potential to climate change. *Handbook of climate change adaptation* (pp. 1–21). Springer.

Pahl-Wostl, C. (2017). An evolutionary perspective on water governance: From understanding to transformation. *Water Resources Management, 31*, 2917–2932. https://doi.org/10.1007/s11269-017-1727-1

Pandey, P., Ali, S. N., & Simon, A. (2022). Rock glacier Oasis: An alternative for agro-pastoralism in a changing environment in the Himalayan cold desert. *Geographical Journal, 188*(4), 585–590. https://doi.org/10.1111/geoj.12468

Patel, A., Kushwaha, N. L., Rajput, J., & Gautam, P. V. (2023). Advances in micro-irrigation practices for improving water use efficiency in dryland agriculture. In A. Naorem & D. Machiwal (Eds.), *Enhancing resilience of dryland agriculture under changing climate: Interdisciplinary and convergence approaches* (pp. 157–176). Springer Nature Singapore. https://doi.org/10.1007/978-981-19-9159-2_10

Pomeroy, J. W., Gray, D. M., Brown, T., Hedstorm, N. R., Quinton, W. L., Granger, J., & Carey, S. K. (2007). The cold regions hydrological model: A platform for basing process representation and model structure on physical evidence. *Hydrological Processes, 21*(19), 2650–2667. https://doi.org/10.1002/hyp.6787

Ramya, A., Poornima, R., Karthikeyan, G., Priyatharshini, S., Thanuja, K. G., & Dhevagi, P. (2023). Climate-induced and geophysical disasters and risk reduction management in mountains regions. In S. Sharma, J. C. Kuniyal, P. Chand, & P. Singh (Eds.), *Climate change adaptation, risk management and sustainable practices in the Himalaya*. Springer. https://doi.org/10.1007/978-3-031-24659-3_16

Rodriguez-Iturbe, I., & Porporato, A. (2005). *Ecohydrology of water-controlled ecosystems: Soil moisture and plant dynamics*. Cambridge University Press.

Rogers, P., & Hall, A.W. (2003). Effective water governance. TEC background papers Nr. 7. Global water partnership technical committee (TEC).

Romshoo, S. A., Murtaza, K. O., Shah, W., Ramzan, T., Ameen, U., & Bhat, M. H. (2022). Anthropogenic climate change drives melting of glaciers in the Himalaya. *Environmental Science and Pollution Research International, 29*(35), 52732–52751. https://doi.org/10.1007/s11356-022-19524-0

Saxena, A., Raghuvanshi, M. S., & Suna, T. (2021). Traditional water management of subsistence agriculture system in cold arid Ladakh: A review. *International Journal of Scientific and Engineering Research, 12*(4), 1019–1028. https://doi.org/10.13140/RG.2.2.16170.75207

Sharifnezhadazizi, Z. (2022). Data fusion and synergy of active and passive remote sensing; An application for freeze thaw detections. Doctoral dissertation, The City College of New York.

Sharma, P. D. (2006). Natural resource management in Indian cold arid region. *Journal of Indian Society of Soil Science, 54*(4), 408–417.

Shen, Y., & Chen, Y. (2010). Global perspective on hydrology, water balance, and water resources management in arid basins. *Hydrological Processes, 24*(2), 129–135. https://doi.org/10.1002/hyp.7428

Singh, V., Jain, S. K., & Shukla, S. (2021). Glacier change and glacier runoff variation in the Himalayan Baspa River Basin. *Journal of Hydrology, 593*, Article 125918. https://doi.org/10.1016/j.jhydrol.2020.125918

Snyder, K. A., Evers, L., Chambers, J. C., Dunham, J., Bradford, J. B., & Loik, M. E. (2019). Effects of changing climate on the hydrological cycle in cold desert ecosystems of the Great Basin and Columbia Plateau. *Rangeland Ecology and Management, 72*(1), 1–12. https://doi.org/10.1016/j.rama.2018.07.007

Sommer, C., Malz, P., Seehaus, T. C., Lippl, S., Zemp, M., & Braun, M. H. (2020). Rapid glacier retreat and downwasting throughout the European Alps in the early 21st century. *Nature Communications, 11*, 3209. https://doi.org/10.1038/s41467-020-16818-0

Thayyen, R.J. (2020). Hydrology of the cold-Arid Himalaya. In A. Dimri, B. Bookhagen, M. Stoffel, & T. Yasunari (Eds.), *Himalayan weather and climate and their impact on the environment*. Springer. https://doi.org/10.1007/978-3-030-29684-1_20

Tortajada, C. (2010). Water governance: Some critical issues. *International Journal of Water Resources Development, 26*(2), 297–307. https://doi.org/10.1080/07900621003683298

Tundup, P., Wani, M. A., Dawa, S., Hussain, S., & Laskit, J. (2017). Water harvesting and conservation under cold desert condition of Ladakh (J&K): Constraints and strategies. *International Journal of Current Microbiology and Applied Sciences, 6*(2), 1796–1800.

United States Fish and Wildlife Service (USFWS). (2010). *Rising to the urgent challenge: strategic plan for responding to accelerating climate change.* Retrieved April 23, 2025 from https://www.fws.gov/home/climatechange/pdf/CCStrategicPlan.pdf

United States Forest Service (USFS). (2011). *National roadmap for responding to climate change* (p. 32). FS-957b. US Department of Agriculture, Forest Service.

UNWWAP (United Nations World Water Assessment Programme). (2006). Water: A shared responsibility. The United Nations World Water Development Report 2; United Nations Educational, Scientific and Cultural Organization (UNESCO).

van Rijswick, M., Edelenbos, J., Hellegers, P., Kok, M., & Kuks, S. (2014). Ten building blocks for sustainable water governance: An integrated method to assess the governance of water. *Water International, 39*(5), 725–742. https://doi.org/10.1080/02508060.2014.951828

Wang, X., Luo, Y., Sun, L., & Shafeeque, M. (2021). Different climate factors contributing for runoff increases in the high glacierized tributaries of Tarim River Basin, China. *Journal of Hydrology: Regional Studies, 36*, Article 100845. https://doi.org/10.1016/j.ejrh.2021.100845

Ward, F. A. (2013). Forging sustainable transboundary water-sharing agreements: Barriers and opportunities. *Water Policy, 15*, 386–417. https://doi.org/10.2166/wp.2012.003

Transformation in Pastoral Practices and Changing Climate in Cold Desert Areas of India

Chandra Prakash Kala

1 Introduction

Pastoralism, a traditional way of management and utilization of grazing resources through domestication of animals, has been practiced by various ethnic communities in the cold desert of Indian Himalaya, including the Changpa in Ladakh and Bhotiya in Uttarakhand State of India (Kala, 2004; Namgail et al., 2007). It is primarily practiced by communities for meeting their livelihoods because the cold desert areas sustain limited resources for crop farming. There are different forms of pastoralism, which include nomadic (constant movement in search of pastures), transhumant (regular seasonal migration), and agro-pastoralism (cultivate crops combining herding with farming) depending on the ecosystems, availability of resources, and cultural milieu (Koocheki & Gliessman, 2005; Ladon et al., 2023). The pastoral communities derive most of their food and economic resources from their animals, using them for milk, meat, wool, transport, and trade (Kala, 2002, 2021; Namgail et al., 2007). Seasonal migrations practiced by pastoralists help prevent overgrazing in one location and encourage the health of grassland ecosystems (Dong et al., 2023).

Although the way of livings of pastoral communities have several resemblances, there are differences in the size of their herds, the composition of their livestock, and how they use pasturelands (Kala, 1998; Mishra, 2001). Furthermore, these pastoral communities operate in specific geographical areas within the Himalayan cold desert areas and follow specific migratory routes, all under the protection of well-established traditional institutions and customs (Kala, 2004, 2021). In the Indian cold deserts, the pastoralism is deeply tied to indigenous knowledge, traditional practices, and the rich cultural heritage of ethnic communities such as the Changpa, Gaddi, and Bhotiya (Bhatnagar et al., 2006; Kala & Maikhuri, 2011). Studying pastoralism in this area is

C. P. Kala (✉)
Indian Institute of Forest Management, Bhopal, Madhya Pradesh, India
e-mail: cpkala@yahoo.co.uk

Y. S. Rawat and C. P. Kala (eds.), *Cold Deserts*,
https://doi.org/10.1007/978-3-032-29119-6_15

vital for understanding the region's unique cultural, ecological, social, and economic systems, as well as the challenges the pastoral communities face due to environmental and socioeconomic changes. Besides, reviewing the pastoral practices offers insights into adaptive strategies developed over centuries to survive in challenging terrains and climates. The present chapter therefore aims to review the historical livelihood strategies of pastoral communities, transformation of their pastoral practices, and vulnerability of pastoralism to the ongoing anthropogenic climate change.

2 History of Pastoralism in Indian Cold Desert

It is believed that the nomadic pastoralists from the Steppes of Central Asia migrated to inhabit the Ladakh regions during the Bronze Age (approximately the fourth millennium BCE). Evidence of ancient art further indicates that the initial settlers in this area were pastoralists drawn to the valleys and high mountains of the cold desert for summer transhumance (UNESCO, 2015). Archaeological findings suggest that pastoral activities were present in Ladakh approximately 4000 years ago (Aldenderfer, 2011). It is also reported in literature that the inhabitants of Kham Province of Tibet were also migrated to Ladakh in the eighth century, and by the early nineteenth-century Ladakh had established itself as a transit emporium for trade and commerce (Joshi et al., 2024; Warikoo, 2003). Though the cold deserts of Ladakh and Lahual-Spiti have been socially intertwined throughout history, they possess distinct historical backgrounds due to their remote geographic positions and different access routes from both Indian and Tibetan territories. While Ladakh was situated along the trade routes connecting Punjab to Kashmir, and extending further to Baltistan (Skardo), Kashgar, Yarqand, Khotan (Eastern Central Asia or Xinjiang), Gartok, and Lhasa (in Tibet), with Leh serving as an important trade center (Warikoo, 2020), Spiti valley remained more secluded, divided into eastern and western valleys (Kala, 2000). The eastern side was linked to Ladakh and Tibet, while the western side connected with Kinnaur and Kullu via high mountain passes (UNESCO, 2015).

Buddhism was introduced in Ladakh during the second century (Jina, 1996; Kala, 2005), and during the Namgyal Dynasty (1460 to 1842), several prominent monasteries emerged as key regional traders in Ladakh including Hemis and Hanle, largely due to Tibet's significant political and cultural influence (Rizvi, 1983). The Namgyal rulers were patrons of Buddhism and granted land and resources to monasteries. This provided the monastic estates with a stable economic base from which they could engage in commercial activities (Jina, 1996). Monasteries owned and managed pasture lands, using local institutions to regulate grazing seasons, water usage, and livestock movement. Monastic leaders supervised systems like lurhupa, where villagers used to protect monastery plantations from grazing animals in exchange for access to resources, thus mediating between pastoralists and community needs (Ladon & Garkoti, 2024).

The colonial and post-colonial periods imposed restrictions on grazing lands and shrinking pastures, leading to influence the traditional pastoralism and forcing communities to diversify their livelihoods or settle permanently. Despite these pressures, pastoralism remains integral to the identity and survival of many groups in the Indian cold desert (UNESCO, 2015). Over the period of time, pastoralists in the cold deserts established strong social networks, oral traditions, and distinctive customs to address resource scarcity, negotiate grazing access, and ensure the health of their herds. Their movements were influenced not only by environmental factors such as snow cover or monsoons but also by agreements with sedentary farmers and adjustments to changing state policies (Bhasin, 2011).

3 Pastoral Communities of the Indian Cold Deserts

Pastoralism has been practiced all along the borderline of Indian Himalayan highlands of cold desert areas. The pastoral practices in the highlands of northern part of Sikkim, including Lhonak Valley, are shaped by the continuous interaction between Dokpas and yaks. Dokpas originally belong to the southern part of the Tibetan plateau, and yak rearing is the central part of their livelihood and culture (Rai & Lal, 2025). In Niti Valley of Chamoli district in Uttarakhand state of India, which forms a part of cold desert, Tolchha and Marcha sub-groups within the Bhotiya tribal community have been practicing transhumance for their livelihood since historical past (Phondani et al., 2010; Kala & Maikhuri, 2011). Niti Valley remains geographically isolated, hence the people's cultural traditions and the interconnected production subsystems for agriculture, animal husbandry, and rangeland-based products have remained intact. The inhospitable climatic conditions are made of suitable for living by moving from one settlement to other by its inhabitants. One of the settlement is at the lower elevations in the valley where people reside in winter due to heavy snowfall on higher reaches, where they stay during summer (Kala & Maikhuri, 2011). Bampa, Gamsali, and Niti villages in Niti Valley are the migratory villages inhabited by the Bhotiya tribal community during summer season. Along with small-scale farming, these villagers have historically engaged in cleaning, spinning, and weaving woolen goods like jackets, shawls, pullovers, carpets, and blankets.

In Ladakh, the Changpa is a major group of nomadic pastoralists who migrated from Tibet in the eighth century A.D. (Jina, 1996), and they utilize the cold desert pasturelands of Changthang for grazing. They practice Buddhism and possess cultural and linguistic connections to Tibet (Rizvi, 1983). In Changthang Plateau, they have been living on move with their herds of yak, sheep, and pashmina goats (Ahmed, 2002). In the Zanskar Valley of Ladakh, the Tibetan Buddhist herder communities who migrate to highland pastures during the summer months for rearing yak, cow, and zhomo use to prepare Doksa. These herders, mainly women, manage their livestock and produce dairy products like milk, butter, and churpey (dried cheese) in the Doksa, which serves as temporary settlements for herders and their cattle during the summer for the pastoralist lifestyle, away from the villages (Raj et al., 2020). The cold desert areas of Pin Valley in Lahaul-Spiti district of Himachal Pradesh have been utilized by Gaddi pastoral community for sheep and goat grazing during the summer season. The traditional lifestyle of this transhumance pastoral community is deeply rooted into seasonal movements between lower Shivalik ranges and high-altitude Himalayan pasturelands (Kala, 2021).

4 Trans-Border Conflicts and Reorganization of Pastoralism

The trans-border conflicts between India and China in 1962 had impacted the communities across the border line from Uttarakhand to Sikkim and Lahual-Spiti to Ladakh cold deserts of India. This war resulted in the ban on frequent movement of pastoral communities between Tibet and India across the borderline. The closure of the borderline has severely impacted the trade and commercial activities of the pastoral communities across the borderline in the Indian cold desert areas. Prior to 1962, the trade enabled the Bhotiya community in Uttarakhand to acquire precious jewelry, clothing, and household goods (Kala & Maikhuri, 2011). Approximately two-thirds of the Bhotiya population depended almost exclusively on trade for their sustenance (Negi & Maikhuri, 2002). The rearing of sheep and goats contributed to trade by aiding in the transportation of goods between different locations and nurturing connections with various parties involved in the trade. The Dokpas in Sikkim who were engaged in transboundary migration to reach and utilize pastures in Tibet were territorially trapped (Rai & Lal, 2025).

In eastern part of Ladakh, the Changpa pastoral community was affected in two primary ways—the constraints of grazing land resources due to ban on trans-border movement and high influx of Tibetan refugees due to Chinese occupation of Tibet (Bhatnagar et al., 2006; Dolkar, 2022). Prior to the conflict, Changpas had access to a variety of traditional pastures in western Tibet, which became inaccessible due to the increased Chinese presence and control in Tibet. Numerous lakes dispersed throughout the pasturelands were vital for Changpas pastoral practices (Namgail et al., 2007). Numerous pastures, including Skagzung, which serve as the main winter grazing area for the Rupshu Changpas, had become inaccessible, thereby significantly increasing the stocking density in eastern Ladakh (Bhatnagar et al., 2006). Simultaneously, there was a heavy influx of Tibetan refugees in 1959 after the Chinese occupation of Tibet (Malik & Wani, 2017; Dolkar, 2022). These refugees settled mainly on the eastern part of Ladakh (called as Changthang) along with their livestock, situated in the southwestern part of the larger Changthang Plateau that extends into Tibet. This resulted in a significant decrease in available pasturelands to the Changpa, thereby putting pressure on grass and water resources. As a result, tensions emerged among different pastoral groups in Changthang, eastern Ladakh, leading to alterations in their movement patterns (Dollfus, 2013).

The entire livelihood of the pastoral communities was suddenly disrupted due to complete prohibition of cross-border trade. The pastoral communities either had to explore new sources for their livelihood or had to find out the possibilities to enhance the existing livelihood sources. The end of cross-border movement meant pastoralists had to adapt to local, more limited grazing areas, increasing pressure on local resources and often reducing herd sizes (Farooquee & Nautiyal, 1996; Kala, 2002). For instance, Bhotiya community in Uttarakhand restricted their income sources to agricultural activities and raising of sheep and goats (Kala & Maikhuri, 2011). In addition, the seasonal rains determine the marginal agriculture in cold deserts of Uttarakhand. As they no longer maintained large population of livestock, they acquired the necessary raw materials for producing woolen goods from transhumant herders (Kala, 2002). In Sikkim, Dokpas gradually assimilated into the local culture and became "Sikkimized," while still preserving elements of their own culture (Chakravarti, 1997). They have also modified their herding strategies, following the closure of the border. In winter, all the herds of entire community are assigned to the hired herders who remain with the yaks, performing tasks such as feeding, caring for them, and safeguarding them against wildlife attacks. Dokpas have established a permanent homestead, at present, where they reside after their seasonal migrations. Unlike in the past, they have started to adopt a more sedentary lifestyle instead of relocating their homestead (Rai & Lal, 2025).

In Changthang, Changpa community has been continuously struggling to adapt to the changing situations and circumstances around them. Their patterns of living have been evolved on the tested experience of centuries for rearing the specialized domestic animals in the harsh but known geographical conditions with time-bound pastoral activities (Kala, 2021). A report claims that in 2007, a total of 124,530 sheep and goats, along with 10,390 yaks, were directly impacted by the scarcity of pastures due to the desert locust invasion. The pastures, which were already overgrazed, suffered a locust attack that resulted in a degradation of pasture (Bhasin, 2011). The loss of conventional pastoral pathways and practices has led to an identity crisis for the pastoral communities across the borderline, as transhumance was fundamental to their social frameworks and cultural institutions.

Bhotiya community also explored new avenues in the forms of associating with tourism and seeking jobs in the plains of India (Kala & Maikhuri, 2011), whereas Changpa doubled their livestock population from approximately 212,500 in 1977 to 487,000 in 1999 (of which 94% sheep and goats), and engaged in growing commercialization of cashmere (pashmina) production (Bhatnagar et al., 2006). Pashmina, also known as cashmere, is a premium luxury fiber obtained from the underwool of the indigenous goat, which has been promoted by the government. Changpas rarely engage with pashmina, as they consider it challenging to weave, opting instead to sell it to the prospective traders or the Pashmina Cooperative in Leh, the headquarters of Ladakh. However, the pastoral communities keep the wool and hair derived from sheep and yaks, respectively, and utilize it to create several textile products ranging from coverings and containers to fabric for their robes, hats, and boots (Ahmed, 2002).

5 Protected Areas and Pastoralism

The cold desert of India is home to a several unique and remarkable species of plants and animals, prompting the Government to create several national parks and wildlife sanctuaries aimed at their protection. Hemis National Park in Ladakh named after the nearby Hemis Monastery is known for its diverse high-altitude wildlife, including the elusive snow leopard. Pin Valley National Park in Lahaul-Spiti district of Himachal Pradesh obtains a good population of Ibex and snow leopard. A part of the cold desert in Sikkim toward north is protected within the Khangchendzonga National Park. The cold desert Lhonak Valley situated partially within Khangchendzonga National Park serves as the sole known breeding ground for the Black-necked or Tibetan Crane in the Eastern Himalayas and is also a significant stopover point for migratory waterbirds (Chettri et al., 2006). The flora and fauna of cold desert areas of Niti Valley in Uttarakhand are conserved as an extension of buffer zone of the Nanda Devi Biosphere Reserve. Besides in 2009, the cold desert of Lahual-Spiti was recognized as a Biosphere Reserve, which includes Pin Valley National Park and its surrounding areas, along with Chandratal, Sarchu, and the Kibber Wildlife Sanctuary in Himachal Pradesh. Other protected environments for the cold deserts flora and fauna are the Changthang Wildlife Sanctuary, situated in the Leh district, and Karakoram Wildlife Sanctuary located in the easternmost part of the Karakoram Range is known for its high-altitude terrain and home to migratory populations of the Tibetan antelope or Chiru (UNESCO, 2015).

Protected areas in the Indian cold desert, such as Pin Valley National Park and Hemis National Park, have influenced pastoralism, shaping community practices, landscapes, and rural livelihoods. The national park status of a protected area led to increased regulation as per the Wildlife Protection Act (1972), such as restrictions on herd movement or blanket bans on livestock grazing in certain areas or designated zones, which impact herd movements and the intensity of pastoral grazing (Kala, 2004). Besides, the depredation of livestock by the wild predators in and around the protected areas creates conflicts between the pastoral communities and the protected area managers. For instance, the presence of large predators like snow leopards and wolves in cold desert areas of Ladakh and Lahual-Spiti results in livestock losses, which subsequently influences economic decisions and causes tensions between conservation goals and herder livelihoods (Sonam et al., 2022).

A study reports significant losses of livestock, mainly domestic goats and sheep, to Tibetan wolf, followed by snow leopard in Gya-Miru Wildlife Sanctuary of Ladakh (Namgail et al., 2007). In addition, the wolves and snow leopard are reported to kill horses significantly in this wildlife sanctuary. The pastoral communities in the cold desert areas are said to carry a strong adverse opinion on the presence of wolves in comparison to other large predators that coexist in the same habitat (Bhatia et al., 2020), which may be attributed to their greater visibility and social structure (Eriksson et al., 2015; Sonam et al., 2022). In Changthang Wildlife Sanctuary of Ladakh during the 1980s, herders expressed concerns about the level of forage consumption by kiang and the subsequent competition with their livestock, leading to calls for the government to relocate the kiang back to Tibet, which was believed to be migrated from Tibet (Fox et al., 1991). It is also perceived by the pastoral communities that due to establishment of protected areas the number of wildlife species has increased significantly, which has either created competition for forage consumptions in pastures or they are direct threats as carnivores to the life of their livestock.

On the other hand, a study reports that goats and sheep create resource constraints for ibex in the Pin Valley National Park by effectively excluding them from specific pastures. The presence of other livestock, including yaks, donkeys, and cattle, impose a minimal impact on Asiatic ibex, and in suitable habitats, Asiatic ibex coexists with horses for foraging (Bagchi et al., 2004). Nevertheless, a significant amount of forage is consumed by most of the livestock, which consequently diminishes the forage availability for Asiatic ibex (Bagchi et al., 2004). This study further states that the domestic livestock reduce both the density and diversity of wild herbivores in the cold deserts of the Indian trans-Himalaya by creating resource constraints. In addition, competitive exploitation of grazing resources in the cold desert areas by blue sheep and local livestock demonstrates that the survival rate of juvenile blue sheep diminishes (Suryawanshi et al., 2010). With similar understanding, prior to the establishment of the Protected Area network, policymakers believed that pastoral practices were primarily accountable for the depletion of natural resources in the Himalayas (Chakravarty-Kaul, 1998), leading to the conclusion that such practices should be halted within the protected areas. The conservation of cold desert biodiversity is crucial for sustainability of cold desert ecosystems; nevertheless, the prohibition of pastoral practices in these protected areas has resulted in tensions between pastoral communities and the managers of the protected areas (Kala, 2004).

6 Transformation of Traditional Knowledge of Pastoral Communities

The pastoral communities in cold desert areas of India have accumulated a great deal of knowledge over the centuries of experimentations and learnings on pastoral grazing practices. The knowledge as accumulated is passed down to the next generations through which the grazing resources in the pastureland are utilized during summer and also how to meet the nutritional needs of livestock species during stall feeding in the winters is practiced. The rotational grazing system is followed as per the customary practices through which the pastoral communities move from one pastures to others in search of good quantity and quality of forage to their livestock (Kala, 1998). However, over the time, there are clear indications of changes in the livestock production system and decline in traditional knowledge systems associated with pastoralism. A study in cold desert landscape of Spiti reports that the evolving aspirations of local residents are primary factors responsible for the loss of traditional ecological knowledge, which are subsequently associated with employment opportunities in more profitable alternatives such as cash crops and tourism (Singh et al., 2020). In addition, the out-migration of the younger generation for higher education and the arrival of immigrant laborers who have become new actors in livestock management are the other factors responsible for major social and ecological transformations in the traditional pastoral management systems (Singh et al., 2015, 2020). The sheep raring of Dokpa community in the cold desert of Sikkim has been drastically declined, as the Tibetan sheep population reduced from 30,000 to 235 from 1982, due to several reasons, including shrinking of quality pastures and spread of pathogenic diseases (Banerjee, 2009; Kumar et al., 2016). The wool of Tibetan sheep was utilized for making coat (bhakku) and blankets and meat for consumption, besides the beads were collected and transported to lower elevations for farming of vegetables and millets (Kumar et al., 2016). Since there is a drastic fall in population of Tibetan sheep such traditional practices almost no longer exist.

7 Effects of Climate Change on Pastoral Practices

The phenomenon of climate change, characterized by rising temperatures, reduced snowfall, and altered precipitation patterns, has resulted in the degradation of pastures in cold desert regions, thereby jeopardizing livestock productivity and the livelihoods of pastoral communities (Singh et al., 2020). The accelerated melting of glaciers and increasingly unpredictable precipitation have affected water availability and traditional water utilization and storage systems, such as kuhls (diverting water from streams or rivers by gravity through a network of open canals) and zings (small storage ponds or tanks constructed to collect and store glacial meltwater or streamflow), compelling pastoralists to travel considerable distances to access these resources (Pandey et al., 2022). Moreover, the intensity, frequency, and scale of droughts, floods, cloudbursts, and snowstorms have intensified, leading to crop failures, livestock losses, and the degradation of grazing lands (Kala, 2014). Variations in rainfall and snowfall negatively influence grass regeneration in pasturelands, potentially escalating competition for limited forage between livestock and wildlife, as well as increasing the risks of disease transmissions (Khanyari et al., 2022). Climate change is one of the major drivers of cold desert pastures degradation, endangering both ecological integrity and the sustainability of pastoral livelihoods in these delicate ecosystems.

The extreme climatic events and variations in climate do not only degrade the pastures but they are the major causes of livestock mortality in the cold desert areas. For instance, studies on livestock mortality in cold deserts of Mongolia report that livestock mortality is predominantly associated with winter temperatures (November–February) and mass mortality with cold winters, apart from the summer droughts (July–September) and deficits in precipitation serve as significant catalysts for mortality (Sternberg et al., 2011; Rao et al., 2015). In cold deserts of India, the mortality rate of livestock including yaks has been rising in recent years due to snowstorms in the high altitudes, consequently discouraging the herders to such professions (Joshi et al., 2020; Rai et al., 2024). In cold desert areas of Spiti, the extreme climatic events are held responsible for shortage of palatable grasses in the pastures, which subsequently decline the number of livestock and their composition (Singh et al., 2015).

A study in Lachen and Lachung Cold Deserts of North Sikkim reports that pastoral yak rearing is coping up with the climate change by adapting to the prevailing climatic conditions. For instance, the yak herders have advanced their upward migration by 15 to 30 days, which is attributed to rising temperatures in March, and the duration of migration has now increased by 45 days due to decline and change in snowfall (Feroze et al., 2019). It is further reported that their downward migration is limited to mid-elevations (2750–3000 m), compared to the previous limit of around 2000 m, while the upward migration has ascended approximately 500 m due to higher temperatures and a lack of quality pastures (Feroze et al., 2019). Besides adapting and mitigating the impacts of climate change through reorganization of their customary pastoral practices, the pastoral communities may diversify their sources of livelihoods by aligning their pastoral activities with tourism in order to attract their young generations for continuing pastoralism.

8 Conclusion

Pastoralism is one of the unique economic adaptation strategies, which is influenced by ecological factors, and is highly specialized for food production in challenging environments such as the cold desert areas. The pastoral system is quite dynamic, as the planned actions of pastoralists are consistently responsive to changing conditions. The ethnic communities inhabited in the cold desert areas of India, such as Ladakh, Lahual-Spiti, northernmost parts of Sikkim, and Niti Valley of Uttarakhand, have developed a system for survival by utilizing the available potential resources through pastoralism, which is vulnerable at present due to various reasons, including anthropogenic climate change. Though the pastoral communities have been gradually adopting changes in their pastoral systems to adjust with the prevailing environmental conditions, it often does not provide the expected results, which subsequently discourages the communities to continue the traditions. Pastoralism in cold desert regions may be sustained by recognizing indigenous knowledge, promoting adaptive management strategies, and aligning pastoral livelihoods with wider conservation and development initiatives.

References

Ahmed, M. (2002). *Living fabric: Weaving among the nomads of Ladakh Himalaya* (p. 192). Orchid Press.

Aldenderfer, M. (2011). Peopling the Tibetan plateau: Insights from archaeology. *High Altitude Medicine and Biology, 12*(2), 141–147.

Bagchi, S., Mishra, C., & Bhatnagar, Y. V. (2004). Conflicts between traditional pastoralism and conservation of Himalayan ibex (*Capra sibirica*) in the Trans-Himalayan mountains. *Animal Conservation Forum, 7*(2), 121–128. https://doi.org/10.1017/S1367943003001148

Banerjee, S. (2009). Shift from transhumance and subtle livelihood patterns of the Bhotia community and its impact on Tibetan sheep population in Sikkim (India). *World Applied Sciences Journal, 7*, 1540–1546.

Bhasin, V. (2011). Pastoralists of Himalayas. *Journal of Human Ecology, 33*(3), 147–177.

Bhatia, S., Redpath, S. M., Suryawanshi, K., & Mishra, C. (2020). Beyond conflict: Exploring the spectrum of human–wildlife interactions and their underlying mechanisms. *Oryx, 54*(5), 621–628.

Bhatnagar, Y. V., Wangchuk, R., Prins, H. H., Van Wieren, S. E., Mishra, C. (2006). Perceived conflicts between pastoralism and conservation of the kiang Equus kiang in the Ladakh Trans-Himalaya, India. *Environmental Management, 38*(6), 934–941. https://doi.org/10.1007/s00267-005-0356-2

Chakravarti, P. K. (1997). Sikkim: A study in social geography. In: P. S. Jina (Ed.), *Recent Researches on the Himalaya* (pp. 21–25). New Delhi: Indus Publishing.

Chakravarty-Kaul, M. (1998). Transhumance and customary pastoral rights in Himachal Pradesh: Claiming for high pastures for Gaddis. *Mountain Research and Development, 18*, 5–17.

Chettri, S. K., Singh, K. K., & Krishna, A. P. (2006). Anthropogenic pressures on the natural resources in fringe areas of the Khangchendzonga biosphere reserve. *International Journal of Ecology and Environmental Sciences, 32*(3), 229–240.

Dolker, P. (2022). Nomadic pastoralism of Changthang, Ladakh, at a crossroads: Changing socioeconomic characteristics, livelihood and livestock composition. *SN Social Sciences, 2*, 2. https://doi.org/10.1007/s43545-021-00302-7

Dollfus, P. (2013). Transformation processes in nomadic pastoralism in Ladakh. *Himalaya, the Journal of the Association for Nepal and Himalayan Studies, 32*(1), 61–72.

Dong, S., Zhang, Y., Shen, H., Li, S., & Xu, Y. (2023). Grazing management and pastoral production. In: S. Dong, Y. Zhang, H. Shen, S. Li, Y. Xu (Eds.), *Grasslands on the Third Pole of the World*. Springer, Cham. https://doi.org/10.1007/978-3-031-39485-0_7

Eriksson, M., Sandström, C., & Ericsson, G. (2015). Direct experience and attitude change towards bears and wolves. *Wildlife Biology, 21*, 131–137. https://doi.org/10.3390/ani11061735

Farooquee, N. A., & Nautiyal, A. (1996). Livestock ownership patterns among transhumant in high-altitude villages of the central Himalayas. *Nomadic Peoples, 39*, 87–95.

Feroze, S. M., Ray, L. I., Singh, K. J., & Singh, R. (2019). Pastoral yak rearing system is changing with change in climate: An exploration of North Sikkim in Eastern Himalaya. *Climatic Change, 157*(3), 483–498. https://doi.org/10.1007/s10584-019-02551-1

Fox, J. L., Nurbu, C., & Chundawat, R. S. (1991). The mountain ungulates of Ladakh, India. *Biological Conservation, 58*, 167–190.

Jina, P. S. (1996). *Ladakh: The land and the people*. Indus Publishing Company.

Joshi, H., Sharma, S., & Joshi, P. (2024). Sustainability of agro-pastoralism in highlands of the Trans-Himalaya: Transformation in 200 years. *Current Agriculture Research Journal, 12*(3). https://doi.org/10.12944/CARJ.12.3.30

Joshi, S., Shrestha, L., Bisht, N., Wu, N., Ismail, M., Dorji, T., Dangol, G., & Long, R. (2020). Ethnic and cultural diversity amongst yak herding communities in the Asian highlands. *Sustainability, 12*(3), 957. https://doi.org/10.3390/su12030957

Kala, C. P. (1998). Ecology and conservation of alpine meadows in the Valley of Flowers National Park, Garhwal Himalaya. Ph.D. Thesis, Forest Research Institute (Deemed University), Dehradun. Uttarakhand.

Kala, C. P. (2002). Indigenous knowledge of Bhotiya tribal community on wool dyeing and its present status in the Garhwal Himalaya, India. *Current Science, 83*(7), 814–817.

Kala, C. P. (2004). Pastoralism, plant conservation, and conflicts on proliferation of Himalayan knotweed in high altitude protected areas of the Western Himalaya, India. *Biodiversity and Conservation, 13*(5), 985–995.
Kala, C. P. (2005). Health traditions of Buddhist community and role of amchis in trans-Himalayan region of India. *Current Science, 89*(8), 1331–1338.
Kala, C. P. (2014). Deluge, disaster and development in Uttarakhand Himalayan region of India: Challenges and lessons for disaster management. *International Journal of Disaster Risk Reduction, 8*, 143–152.
Kala, C. P. (2000). Status and conservation of rare and endangered medicinal plants in the Indian trans-Himalaya. *Biological conservation, 93*(3), 371–379. https://doi.org/10.1016/S0006-3207(99)00128-7
Kala, C. P. (2021). Ethnic food knowledge of highland pastoral communities in the Himalayas and prospects for its sustainability. *International Journal of Gastronomy and Food Science, 23*, Article 100309. https://doi.org/10.1016/j.ijgfs.2021.100309
Kala, C. P., & Maikhuri, R. K. (2011). Mitigating people-park conflicts on resource use through ecotourism: A case of the Nanda Devi Biosphere Reserve, Indian Himalaya. *Journal of Mountain Science, 8*(1), 87–95.
Khanyari, M., Robinson, S., Milner-Gulland, E. J., Morgan, E. R., Rana, R. S., & Suryawanshi, K. R. (2022). Pastoralism in the high Himalayas: Understanding changing practices and their implications for parasite transmission between livestock and wildlife. *Pastoralism, 12*(1), 44. https://doi.org/10.1186/s13570-022-00257-1
Koocheki, A., & Gliessman, S. R. (2005). Pastoral nomadism, a sustainable system for grazing land management in arid areas. *Journal of Sustainable Agriculture, 25*(4), 113–131.
Kumar, B., Avasthe, R. K., Islam, R., Bhutia, P., Mishra, A. K., & Tantia, M. S. (2016). Characterization and evaluation of Tibetan Sheep: A key source of livelihood in alpine ecosystem of Sikkim Himalaya. *Journal of Livestock Biodiversity, 6*(2), 51–55.
Ladon, P., Nüsser, M., & Garkoti, S. C. (2023). Mountain agropastoralism: Traditional practices, institutions and pressures in the Indian Trans-Himalaya of Ladakh. *Pastoralism, 13*(1), 30. https://doi.org/10.1186/s13570-023-00289-1
Ladon, P., & Garkoti, S. C. (2024). Local institutions in Ladakh, India: Community-based resource management in mountain agropastoralism. *Current Science, 127*(7), 812–817.
Malik, I., & Wani, F. (2017). Pashmina goat farming in cold arid desert of Ladakh: A geographical study of Changthang region. *International Journal of Advanced Scientific Research and Management, 2*(11), 110–119.
Mishra, C. (2001). *High altitude survival: Conflicts between pastoralism and wildlife in the trans-Himalaya* (Doctoral Thesis). Wageningen University, The Netherlands.
Namgail, T., Bhatnagar, Y. V., Mishra, C., & Bagchi, S. (2007). Pastoral nomads of the Indian Changthang: Production system, landuse and socioeconomic changes. *Human Ecology, 35*(4), 497–504.
Negi, C. S., & Maikhuri, R. K. (2002). Nanda Devi Biosphere Reserve-pro-poor tourism: An option. *The Botanica, 52*, 101–111.
Pandey, P., Ali, S. N., & Simon, A. (2022). Rock glacier Oasis: An alternative for agro-pastoralism in a changing environment in the Himalayan cold desert. *The Geographical Journal, 188*(4), 585–590.
Phondani, P. C., Maikhuri, R. K., Rawat, L. S., Farooquee, N. A., Kala, C. P., Vishvakarma, S. R., Rao, K. S., & Saxena, K. G. (2010). Ethnobotanical uses of plants among the Bhotiya tribal communities of Niti Valley in Central Himalaya, India. *Ethnobotany Research and Applications, 8*, 233–244.
Rai, N., & Lal, U. (2025). Yak pastoralism in transition: Changes and challenges in yak-rearing practices in Lhonak valley, Sikkim. *Nomadic Peoples, 29*(1), 35–57.
Rai, R. K., Joshi, S., Dorji, T., & Pant, B. (2024). Herders' preferences for maintaining yak herding in the Eastern Himalayas. *Rangeland Ecology & Management, 94*, 20–28. https://doi.org/10.1016/j.rama.2024.01.009

Raj, A., Angchuk, D., & Islam, M. A. (2020). Livelihood dependence on highland pastoralism (Doksa) in Trans-Himalayan region of Zanskar, Ladakh. *International Journal of Environment and Climate Change, 10*(12), 67–76.

Rao, M. P., Davi, N. K., D'Arrigo, R. D., Skees, J., Nachin, B., Leland, C., Lyon, B., Wang, S., & Byambasuren, O. (2015). Dzuds, droughts, and livestock mortality in Mongolia. *Environmental Research Letters, 10*, Article 074012. https://doi.org/10.1088/1748-9326/10/7/074012

Rizvi, J. (1983). *Ladakh: Crossroads of High Asia.* 3rd Walton Street. Oxford: Oxford University Press.

Singh, R., Sharma, R. K., & Babu, S. (2015). Pastoralism in transition: Livestock abundance and herd composition in Spiti, Trans-Himalaya. *Human Ecology, 43*(6), 799–810. https://doi.org/10.1007/s10745-015-9789-2

Singh, R., Sharma, R. K., Babu, S., & Bhatnagar, Y. V. (2020). Traditional ecological knowledge and contemporary changes in the agro-pastoral system of upper Spiti landscape, Indian trans-Himalayas. *Pastoralism, 10*(1), 15. https://doi.org/10.1186/s13570-020-00169-y

Sonam, K., Dorjay, R., Khanyari, M., Bijoor, A., Lobzang, S., Sharma, M., Suresh, S., Mishra, C., & Suryawanshi, K. R. (2022). A community-based conservation initiative for wolves in the Ladakh Trans-Himalaya, India. *Frontiers in Ecology and Evolution, 10*, Article 809817. https://doi.org/10.3389/fevo.2022.809817

Sternberg, T., Thomas, D., & Middleton, N. (2011). Drought dynamics on the Mongolian steppe, 1970–2006. *International Journal of Climatology, 31*(12), 1823–1830.

Suryawanshi, K. R., Bhatnagar, Y. V., & Mishra, C. (2010). Why should a grazer browse? Livestock impact on winter resource use by bharal Pseudois nayaur. *Oecologia, 162*(2), 453–462.

UNESCO. (2015). *Cold desert cultural landscape of India.* UNESCO. https://whc.unesco.org/en/tentativelists/6055/

Warikoo, K. (2003). India's gateway to Central Asia: Trans-Himalayan trade and cultural movements through Kashmir and Ladakh, 1846–1947. In M. Singh & V. Krassilchtchikov (Eds.), *Eurasion Vision* (pp. 163–175). Anamika Publishers and Distributors.

Warikoo, K. (2020). Ladakh: India's gateway to central Asia. *Strategic Analysis, 44*(3), 177–192.

Building Psychological Resilience for Sustainable Adaptation in Cold Desert Regions

Durgeshwary Kolhe and Arshad Bhat

1 Introduction

Cold deserts are among the toughest places for humans to live on our planet. These relentless landscapes—like the high-altitude plateaus of Ladakh and Tibet, the icy expanses of Antarctica, and the dry stretches of the Atacama Desert—really test our limits in incredible ways (Thomas & Goudie, 2016). As climate change speeds up, the people living in these areas are facing more and more challenges that require them to adapt both physically and mentally. In this article, we'll explore how communities can build psychological resilience to help them sustainably adapt to life in these cold desert environments. Life in these regions is undeniably tough; the severe weather puts immense pressure on human populations. Studies have shown that extreme cold, isolation, and scarce resources can lead to anxiety, depression, and difficulties with focus (Palinkas, 2003). Research conducted at Antarctic research stations has identified what's known as "winter-over syndrome," which includes mood disorders, sleep issues, and cognitive decline during long periods of winter isolation (Palinkas & Suedfeld, 2008). Similar mental health challenges have been observed among indigenous groups living in cold desert areas like Ladakh, where traditional coping strategies are being stretched thin due to the rapid impacts of climate change (Angchok & Singh, 2006).

The study by Nüsser et al. (2012) looks into irrigation and development in the Upper Indus Basin, particularly examining the socio-hydrological system in central Ladakh, India. These challenges are only getting worse as climate change accelerates. Cold desert ecosystems are incredibly sensitive to temperature changes, with many areas warming at rates that far surpass global averages (IPCC, 2019). Take Ladakh; for

D. Kolhe
School of Vedic Sciences, MIT-Arts Design and Technology University, Pune, Maharashtra, India

A. Bhat (✉)
Amity Institute of Liberal Arts, Amity University Mumbai, Bhatan, Maharashtra, India
e-mail: bhatarshad09@gmail.com

Y. S. Rawat and C. P. Kala (eds.), *Cold Deserts*,
https://doi.org/10.1007/978-3-032-29119-6_16

instance, researchers have observed glacier retreat, shifts in precipitation patterns, and a rise in water scarcity, all of which directly affect agricultural practices that have supported local communities for centuries (Nüsser et al., 2012). These environmental shifts bring about immediate stressors and long-term uncertainties that challenge psychological resilience. For generations, traditional knowledge systems have acted as psychological buffers in these tough environments. Studies in Tibet have highlighted how Buddhist mindfulness practices and community rituals offer crucial psychological support, helping residents navigate environmental challenges (Byg & Salick, 2009). Likewise, in the cold desert regions of the Andes, indigenous agricultural methods weave in spiritual elements that strengthen community bonds during tough times (Vidaurre de la Riva et al., 2013). These cultural practices are not just interesting; they represent vital psychological adaptation strategies that should be given more attention in climate adaptation planning.

Social connections play a crucial role in psychological resilience in extreme environments. Studies of isolated communities in Antarctica have consistently found that strong social bonds and effective communication are key factors in maintaining mental health during prolonged exposure to harsh conditions (Palinkas et al., 2004). In Ladakh, traditional systems of community labor sharing known as "bes" have historically provided both practical assistance and psychological support during agricultural seasons (Angchok & Singh, 2006). These social structures become even more important as environmental challenges intensify. Psychological flexibility appears to be a critical component of successful adaptation to cold desert living. Research with both indigenous communities and scientific personnel in these regions indicates that individuals who can adjust their expectations, find meaning in hardship, and maintain optimism, despite challenges show better psychological outcomes Norris et al. (2008). A study of Tibetan nomads facing climate-induced changes to their traditional lifestyles found that those who could incorporate new practices while maintaining cultural identity showed greater psychological wellbeing than those who resisted all change (Byg & Salick, 2009). The concept of "psychological hardiness" emerges repeatedly in the literature on cold desert adaptation and lifestyle. This trait—characterized by commitment to finding meaningful purpose, a belief in one's ability to influence surroundings, and viewing change as a challenge rather than a threat—correlates strongly with successful adaptation in extreme environments (Sandal et al., 2006). Interestingly, many traditional cultural practices seem designed to cultivate exactly these qualities, suggesting that indigenous knowledge contains important psychological wisdom in cold desert regions (Angchok & Singh, 2006).

Education and knowledge transfer play vital roles in building psychological resilience. Studies in Ladakh show that communities with strong intergenerational knowledge transfer about traditional agricultural techniques demonstrate greater confidence in facing changing environmental conditions (Angchok & Singh, 2006). A qualitative study by Tahernejad et al. (2024) explored factors influencing psychological resilience among farmers in Iran's drought-prone areas. The study identified both environmental factors (e.g., economic conditions, socio-cultural aspects, government policies) and intra-individual factors (e.g., personal characteristics, health status, psychological factors) as significant contributors to resilience. The research emphasized the importance of community support, traditional practices, and adaptive strategies in enhancing resilience. Similarly, research in Antarctic settings indicates that thorough preparation and training significantly reduce psychological distress during deployment (Palinkas & Suedfeld, 2008). These findings suggest that educational programs that combine traditional ecological knowledge (TEK) with scientific understanding of climate change could strengthen psychological resilience.

Gender considerations are also important in understanding psychological resilience in cold deserts. Research in the Andes and Himalayas indicates that women often bear different psychological burdens than men as climate change alters traditional livelihoods (Vidaurre de la Riva et al., 2013). In Ladakh, for example, as men increasingly seek employment outside traditional farming, women take on greater agricultural responsibilities while maintaining household duties, creating potential for increased stress (Goodall, 2004). Any comprehensive approach to psychological resilience must account for these gender-specific challenges. Developing sustainable adaptation strategies requires attention to both technological and psychological dimensions. Physical infrastructure improvements like better insulated housing, renewable energy systems, and water conservation technologies can reduce daily stressors in cold desert regions (Nüsser et al., 2012). However, research shows that such improvements are most effective when they align with local cultural values and involve community participation in planning and implementation, thus supporting psychological ownership and agency (Byg & Salick, 2009).

Mental health services tailored to the unique challenges of cold desert environments are increasingly recognized as essential components of adaptation strategies. Studies in Antarctica have demonstrated the effectiveness of regular psychological monitoring and support services for personnel working in Antarctica (Palinkas et al., 2004). However, such formalized support systems are often absent in indigenous cold desert communities. Models that integrate traditional healing practices with contemporary psychological approaches show promise for filling this gap (Wexler, 2014). Realistic optimism emerges as a particularly important psychological resource in cold desert adaptation. Research with both indigenous communities and scientific personnel indicates that maintaining hope while acknowledging difficulties is associated with better mental health outcomes than either blind optimism or fatalism Norris et al. (2008). Traditional cultural narratives in many cold desert regions contain themes of perseverance through hardship that cultivate this balanced perspective (Byg & Salick, 2009). Young people in cold desert communities face particular psychological challenges as they navigate between traditional knowledge systems and rapidly changing environmental and social realities. Studies in Ladakh have found that educational programs that validate TEK while providing scientific context for understanding climate change help young people develop psychological frameworks for addressing current challenges (Angchok & Singh, 2006).

Youth involvement in community adaptation planning also appears to reduce feelings of helplessness and build confidence. Looking forward, psychological resilience in cold desert regions will likely depend on maintaining cultural continuity while embracing necessary adaptations. Research in multiple settings suggests that communities that can preserve core cultural values and practices while strategically incorporating new approaches show greater collective psychological strength than those that either rigidly resist all change or abandon traditional systems entirely (Byg & Salick, 2009). Maintaining this equilibrium necessitates constant communication between generations as well as between scientific and traditional knowledge systems. In the end, creating psychological resilience for long-term adaptation in cold desert areas necessitates a comprehensive strategy that honors traditional wisdom, recognizes the significant obstacles posed by climate change, and creates plans that take into account both psychological and physical requirements.

The resilience of the people who live in these amazing environments may be just as crucial to building sustainable futures as technology advancements as these landscapes continue to change due to global climate change. Human survival and adaption in harsh conditions, such as cold deserts, depend heavily on psychological resilience. The ability to sustain positive mental functioning in the face of stress, trauma, and adversity is known as psychological resilience (Masten, 2001). This resilience is important in cold desert areas where people regularly deal with extreme weather conditions as well as growing uncertainty brought on by the effects of climate change. Studies have shown that psychological resilience directly influences people's ability to implement and maintain sustainable adaptation practices over time (Brown & Westaway, 2011). The connection between mental wellbeing, effective coping mechanisms, and sustainable adaptation forms a critical pathway for community survival in cold desert regions. Research in Ladakh has demonstrated that individuals with stronger psychological resources show greater willingness to adopt sustainable water conservation practices even when these require significant lifestyle adjustments (Angchok & Singh, 2006). Similarly, studies in Antarctic research stations reveal that personnel with better mental health outcomes engage more consistently in environmentally conscious behaviors, despite the harsh conditions (Palinkas et al., 2010). Psychological resilience encompasses several key components that support climate adaptation in cold desert regions. These include cognitive flexibility, emotional regulation, social connectedness, and meaning-making capacity (Bonanno, 2004). Research with Tibetan communities has shown how traditional mindfulness practices enhance these resilience components, helping residents maintain perspective while facing environmental challenges (Byg & Salick, 2009). These psychological resources enable individuals to respond adaptively rather than reactively to environmental stressors.

The relationship between psychological resilience and sustainable adaptation operates through several pathways. First, resilient individuals tend to maintain better decision-making capacity under stress, leading to more thoughtful resource management choices Norris et al. (2008). Second, psychological resilience supports the persistence needed to implement long-term adaptation strategies rather than seeking quick but unsustainable solutions (Brown & Westaway, 2011). Third, community-level resilience facilitates knowledge sharing and collective action that are essential for addressing large-scale environmental challenges (Adger, 2003).

2 Research Gap and Rationale

Despite growing recognition of climate change impacts in cold desert regions, research has disproportionately focused on physical and technological adaptation strategies while neglecting psychosocial dimensions. Pearce et al. (2015) found that less than 15% of climate adaptation studies in high-altitude desert regions addressed mental health or psychological resilience factors. This represents a significant oversight, as psychological factors often determine whether physical adaptation measures succeed or fail in practice (Berry et al., 2010). The research gap is particularly pronounced regarding empirical studies that examine specific resilience-building mechanisms in cold desert communities. While conceptual frameworks linking psychological resilience and climate adaptation exist, few studies have collected systematic data on how these processes unfold in the distinctive contexts of places like Ladakh, Tibet, or the Atacama (Berrang-Ford et al., 2011). This limits the evidence base for designing effective interventions that integrate psychological support with technological adaptation strategies. The rationale for addressing this research gap becomes increasingly urgent as climate change accelerates in cold desert regions.

These areas are experiencing warming at rates significantly above global averages, with some high-altitude desert regions showing temperature increases at twice the global rate (IPCC, 2019). As traditional coping mechanisms become less effective against these unprecedented changes, understanding psychological adaptation processes becomes critical for supporting vulnerable communities (Nüsser et al., 2012). Furthermore, cold desert regions often host indigenous communities with unique cultural knowledge systems that influence psychological resilience in ways not captured by Western psychological frameworks (Wexler, 2014). These knowledge systems contain valuable insights about maintaining psychological balance in harsh environments but remain understudied in climate adaptation literature. Exploring these indigenous perspectives could yield important lessons for building resilience in other contexts as well.

3 Objectives of the Study

The primary objective of this study is to explore the psychological resilience factors that influence sustainable adaptation in cold desert communities. This includes identifying both individual traits (such as psychological flexibility and hardiness) and social processes (such as community cohesion and knowledge transmission) that contribute to resilience (Sandal et al., 2006). The study aims to distinguish universal resilience mechanisms from culturally specific ones by examining the factors across different cold desert regions. A second key objective involves identifying mental health risks associated with climate change in cold desert environments and documenting effective coping strategies. Research indicates that climate-related stressors in cold deserts include not only direct impacts like water scarcity and temperature extremes but also secondary effects such as livelihood disruption and cultural loss (Berry et al., 2010). Understanding how these stressors affect mental health and which coping strategies effectively address them represents a crucial step toward comprehensive adaptation planning.

The third objective focuses on examining how community structures, cultural practices, and policy interventions either support or hinder psychological resilience in cold desert regions. This includes investigating TEK, community governance structures, and external support programs to determine how they interact to create enabling environments for resilience (Adger, 2003). The study pays particular attention to gender differences in resilience processes, as research suggests that men and women in cold desert communities often face different psychological challenges and utilize different coping resources (Vidaurre de la Riva et al., 2013). Therefore, the present study aims to develop a more comprehensive understanding of how psychological resilience operates within cold desert communities facing climate change. The findings of this study help to bridge the gap between physical adaptation strategies and psychological support mechanisms, potentially informing more holistic and effective approaches to climate change adaptation in these vulnerable regions. The experiences and perspectives of cold desert inhabitants themselves, the research is expected to contribute to more contextually appropriate and culturally sensitive adaptation frameworks.

4 Theoretical Framework

4.1 Psychological Resilience in Harsh Environments

The theoretical foundation for understanding psychological resilience in cold desert regions draws from several established frameworks that help explain human adaptation to extreme conditions. Bonanno's (2004) Resilience Theory provides a cornerstone for this research by distinguishing resilience from recovery. While recovery implies a period of dysfunction followed by gradual return to normal functioning, resilience represents the ability to maintain relatively stable, healthy psychological functioning despite exposure to significant adversity. Research in Antarctic stations has demonstrated this distinction, showing that approximately 35% of personnel maintain consistent psychological wellbeing throughout winter isolation periods, demonstrating resilience rather than merely recovering from distress (Palinkas & Suedfeld, 2008). Bonanno's (2004) framework identifies several pathways to resilience particularly relevant to cold desert inhabitants, including hardiness, self-enhancement, repressive coping, and positive emotion. Studies with Tibetan highland communities have shown how cultural practices often cultivate these precise qualities, with traditional rituals reinforcing hardiness and community celebrations maintaining positive emotion even during harsh winters (Byg & Salick, 2009). The theory also emphasizes that resilience is not rare but relatively common, which aligns with observations that most inhabitants of cold desert regions manage to maintain psychological functioning despite challenging conditions.

Bronfenbrenner's (1979) Ecological Systems Theory provides another crucial theoretical lens, emphasizing that human development and adaptation occur within nested environmental contexts. This framework is particularly valuable for understanding cold desert adaptation as it accounts for how multiple environmental levels—from immediate physical surroundings to broader cultural systems—simultaneously influence psychological resilience. Research in Ladakh demonstrates this ecological perspective, showing how individual psychological adaptation depends on family support systems, community water management practices, regional governmental policies, and global climate patterns (Angchok & Singh, 2006). The ecological perspective helps to explain why resilience factors may operate differently across cold desert regions depending on their specific environmental contexts. For example, strong community bonds appear universally important to resilience, but the specific mechanisms through which these bonds function vary between Andean communities (where reciprocal labor systems called "ayni" provide psychological security) and Ladakhi villages (where Buddhist monastic institutions offer spiritual guidance during hardship) (Angchok & Singh, 2006; Vidaurre de la Riva et al., 2013). This contextual understanding is essential for developing appropriately tailored interventions for sustainable development of the cold deserts.

Hobfoll's (1989) Conservation of Resources Theory adds another valuable dimension by conceptualizing psychological stress as resulting from threatened or actual loss of valued resources. According to this theory, people strive to protect, retain, and build resources that have survival value—whether material (i.e., water, food, shelter), conditional (i.e., security, stability), personal (i.e., self-efficacy, optimism), or social (i.e., community support, cultural identity). Research in cold desert regions demonstrates how this resource conservation perspective operates in practice, with communities developing sophisticated psychological strategies to manage scarce resources. Studies in the Atacama Desert, for example, documented how communities have developed elaborate psychological frameworks around water conservation that blend practical management strategies with cultural and spiritual significance, creating psychological resources that help members cope with water scarcity. Similarly, research with nomadic communities in Tibet shows how their psychological adaptation includes carefully managed expectations about resource availability that buffer against disappointment and conserve emotional energy during difficult periods (Byg & Salick, 2009). These examples illustrate how Hobfoll's (1989) theory helps explain the psychological dimensions of resource management in harsh environments.

4.2 Cold Desert-Specific Adaptation Frameworks

Beyond these broader theories, several frameworks address psychological adaptation specifically in cold desert contexts. Cognitive Adaptation Theory, developed in health psychology but increasingly applied to climate adaptation, emphasizes how individuals maintain psychological wellbeing through cognitive processes including finding meaning, gaining a sense of mastery, and enhancing self-esteem (Taylor, 1983). These cognitive processes prove particularly important in cold desert regions where inhabitants must continually reframe their relationship with challenging environments. Research in Ladakh illustrates this cognitive adaptation process, showing how residents interpret climate threats through cultural frameworks that provide meaning and maintain self-efficacy despite increasing environmental uncertainty (Angchok & Singh, 2006). Similarly, studies with indigenous communities in cold desert regions of the Andes demonstrate how traditional narratives about resilience in the face of environmental hardship provide cognitive frameworks that help community members interpret current climate challenges within meaningful historical contexts (Vidaurre de la Riva et al., 2013). These cognitive adaptations represent crucial psychological resources for sustainable adaptation.

Indigenous knowledge systems provide another important theoretical framework for understanding psychological adaptation in cold desert regions. TEK in these areas typically includes not just practical survival strategies but also psychological frameworks for maintaining emotional balance in extreme conditions. Research with Tibetan Buddhist communities demonstrates how their traditional knowledge incorporates sophisticated psychological practices for managing anxiety about environmental uncertainty, including mindfulness techniques that have been validated by contemporary psychological science (Byg & Salick, 2009). Similarly, studies in Ladakh document how TEK include intergenerational transmission of both practical skills and psychological attitudes conducive to resilience, such as acceptance of impermanence and community interdependence (Angchok & Singh, 2006). The persistence of these knowledge systems despite modernization pressures testifies to their psychological value. However, climate change now threatens traditional knowledge by rendering some historical patterns obsolete, creating a complex challenge for communities trying to maintain psychological coherence while adapting to new realities.

4.3 Integration with Climate Psychology

Recent developments in climate psychology provide additional theoretical perspectives for understanding psychological adaptation in cold desert regions. Research on climate anxiety and eco-distress has documented how awareness of environmental degradation can trigger profound psychological responses including grief, anxiety, and helplessness (Clayton et al., 2017). These responses appear particularly acute in regions experiencing visible climate impacts, including many cold desert areas where glacier retreat, changing precipitation patterns, and ecosystem shifts are readily observable. Studies in Ladakh have documented growing climate anxiety among younger generations who witness rapid glacial retreat while hearing elders' accounts of previously stable conditions (Nüsser et al., 2012). Similar findings from the Atacama region show how changing precipitation patterns create psychological distress by undermining traditional agricultural knowledge that previously provided certainty and structure. Understanding these psychological responses becomes crucial for developing effective adaptation strategies that address both practical and emotional needs. Theoretical frameworks for psychological interventions to enhance climate coping abilities have evolved significantly in recent years. These include approaches based on acceptance and commitment therapy, which helps individuals clarify values and take meaningful action despite distressing emotions about climate change (Hayes et al., 2006). Such approaches show promise for cold desert communities facing difficult transitions, as they support psychological flexibility while maintaining commitment to core values that may include environmental stewardship and community welfare.

Other intervention frameworks incorporate elements of positive psychology, focusing on cultivating hope, meaning, and post-traumatic growth in the face of climate challenges (Ojala, 2012). Research with youth in cold desert regions suggests that educational approaches incorporating these positive psychology elements—highlighting agency and meaningful action rather than just climate threats—support psychological resilience and sustained engagement with adaptation efforts (Angchok & Singh, 2006). These frameworks offer important guidance for designing interventions that build psychological capacity for long-term adaptation. The integration of these diverse theoretical perspectives creates a comprehensive framework for understanding psychological resilience in cold desert regions. This integrated approach recognizes that sustainable adaptation requires not just technological solutions but also psychological resources that help individuals and communities navigate profound environmental changes while maintaining mental wellbeing and cultural continuity. By applying this theoretical framework, researchers and practitioners can develop more holistic approaches to supporting cold desert communities facing unprecedented climate challenges.

5 Methodology (PRISMA Model)

5.1 Study Design

This research employs a systematic review using the Preferred Reporting Items for Systematic Reviews and Meta-Analyses (PRISMA) methodology to synthesize existing knowledge on psychological resilience in cold desert regions. The PRISMA approach provides a structured framework for conducting comprehensive literature reviews with transparency and reproducibility (Moher et al., 2009). This methodology is particularly appropriate for identifying psychological resilience strategies across diverse cold desert regions because it allows for systematic comparison of findings from different geographical and cultural contexts. The systematic review focuses specifically on identifying key psychological resilience strategies that support sustainable climate adaptation in cold desert environments. By using the PRISMA framework, the study ensures methodological rigor while examining both established psychological resilience factors documented in the literature and emerging adaptation approaches specific to cold desert contexts. This approach helps bridge theoretical understandings of resilience with practical applications in these extreme environments.

5.2 Eligibility Criteria

5.2.1 Inclusion Criteria

Recent available studies are included in the present study. This timeframe captures the evolution of research on psychological resilience in cold desert regions, spanning from early foundational work to contemporary studies addressing recent climate change impacts. The start date of 1990 coincides with increasing scientific attention to climate change impacts in high-altitude regions and the emergence of psychological resilience as a distinct research domain (Masten, 2001). The review includes research specifically focused on psychological resilience in cold desert regions, encompassing high-altitude deserts (e.g., Ladakh, Tibet), polar deserts (parts of Antarctica), and cold arid regions (e.g., portions of the Atacama). Studies must address psychological aspects of adaptation rather than merely physical or technological responses to environmental challenges. This criterion ensures the review remains focused on the psychological dimensions of adaptation rather than broader climate response literature. The review incorporates various research methodologies including empirical studies (both quantitative and qualitative), ethnographic research, participatory studies, and theoretical models with direct application to cold desert environments. This methodological diversity is essential for capturing the complex psychological processes involved in adaptation, which are often not fully represented through any single research approach (Berrang-Ford et al., 2011). Articles in English and other major languages relevant to cold desert communities are included to avoid geographical bias. Where necessary, professional translation services are utilized to ensure appropriate inclusion of research published in languages spoken in major cold desert regions, including Tibetan, Spanish (for Andean regions), and Russian (for Central Asian cold deserts). This approach helps address the common problem of English language bias in systematic reviews (Rasmussen & Montgomery, 2018).

5.2.2 Exclusion Criteria

Studies unrelated to climate adaptation or resilience are excluded from the review. While general psychological research conducted in cold desert regions may provide contextual information, this review focuses specifically on psychological processes related to climate change adaptation to maintain conceptual clarity and practical relevance. This exclusion helps concentrate the analysis on adaptation-specific psychological mechanisms. Non-peer-reviewed sources including news articles, opinion pieces, and non-academic reports are excluded to maintain scientific rigor. However, gray literature from established research institutions and governmental bodies (e.g., IPCC reports, national climate assessments) may be included when they contain original research or systematic analysis relevant to the review questions, as these often contain valuable information not yet published in academic journals (Adams et al., 2017). General psychological resilience research without specific focus on cold deserts is excluded. While broader resilience literature provides important theoretical frameworks, this review specifically examines how resilience processes operate in the unique contexts of cold desert environments. This criterion helps ensure the findings are ecologically valid for the specific regions under study.

5.3 Data Sources and Search Strategy

The review searches multiple academic databases to ensure comprehensive coverage of relevant literature. Databases searched include Scopus and Web of Science for broad multidisciplinary coverage, Google Scholar to capture articles from regional journals that may not be indexed in major databases, PubMed for research on health-related aspects of adaptation, and PsycINFO for specialized psychological literature. Additional databases specific to regional studies are also consulted, including the Himalayan database and the Andean Information Network, to ensure inclusion of regionally-specific research. Keywords used in the search strategy are carefully selected to capture the intersection of psychological resilience and cold desert adaptation. Primary terms include "psychological resilience," "cold desert adaptation," "climate change coping," "mental health in extreme environments," and "community resilience in cold regions." Additional region-specific terms such as "Ladakh," "Tibet," "Atacama," "Antarctic," and "high-altitude desert" are included to capture research on specific cold desert environments. Boolean operators are employed to create comprehensive search strings that capture relevant literature while excluding unrelated studies. The primary search string used is: ("psychological resilience" OR "mental health" OR "coping" OR "psychosocial adaptation") AND ("cold desert" OR "extreme environments" OR "high-altitude desert" OR "Ladakh" OR "Tibet" OR "Atacama" OR "Antarctic") AND ("climate change adaptation" OR "sustainability" OR "environmental change" OR "climate stress"). These search terms are adjusted for specific databases as needed to optimize results.

5.4 Study Selection and Screening

The study selection process follows the PRISMA (Fig. 1) four-phase flow diagram to ensure transparent and systematic inclusion of relevant literature. In Phase 1 (identification), initial database searches identify potential studies, and duplicate records are removed using reference. This phase establishes the total number of unique records to be screened. During Phase 2 (screening), two independent reviewers assess study abstracts against the inclusion and exclusion criteria to determine relevance. A standardized screening form is used to ensure consistent application of criteria across reviewers. Discrepancies between reviewers are resolved through discussion, with a third reviewer consulted when consensus cannot be reached. This approach minimizes selection bias and enhances reliability of the screening process (Higgins & Green, 2011). In Phase 3 (full-text review), complete articles that passed the initial screening undergo detailed assessment for eligibility. Two reviewers independently evaluate each full-text article against the inclusion criteria and methodological quality standards. Reasons for exclusion at this stage are documented and reported in the final review following PRISMA guidelines. This transparent reporting allows readers to evaluate the comprehensiveness of the review process (Moher et al., 2009). Throughout the screening process, snowball sampling (checking reference lists of included studies) is employed to identify additional relevant publications that may have been missed in the database searches. This technique is particularly important for capturing region-specific studies that may be published in journals not fully indexed in major databases.

5.5 Data Extraction and Synthesis

5.5.1 Key Variables Extracted

The data extraction process focuses on three primary categories of variables crucial for understanding psychological resilience in cold desert regions. First, psychological resilience factors are systematically documented, including coping mechanisms (e.g., cognitive reframing, traditional practices), community support structures (e.g., social networks, collective decision-making), and cognitive adaptability (e.g., flexibility in problem-solving, tolerance for uncertainty). These factors are coded using established resilience frameworks while remaining open to region-specific manifestations (Bonanno, 2004; Ungar, 2011). Second, climate change impacts on mental health are extracted, including documented cases of stress, eco-anxiety, trauma, and loss of place attachment resulting from environmental changes in cold desert regions. This category captures both acute psychological responses to specific climate events (e.g., drought, unusual weather patterns) and chronic stress related to ongoing environmental transformation (Berry et al., 2010). Third, adaptation strategies are

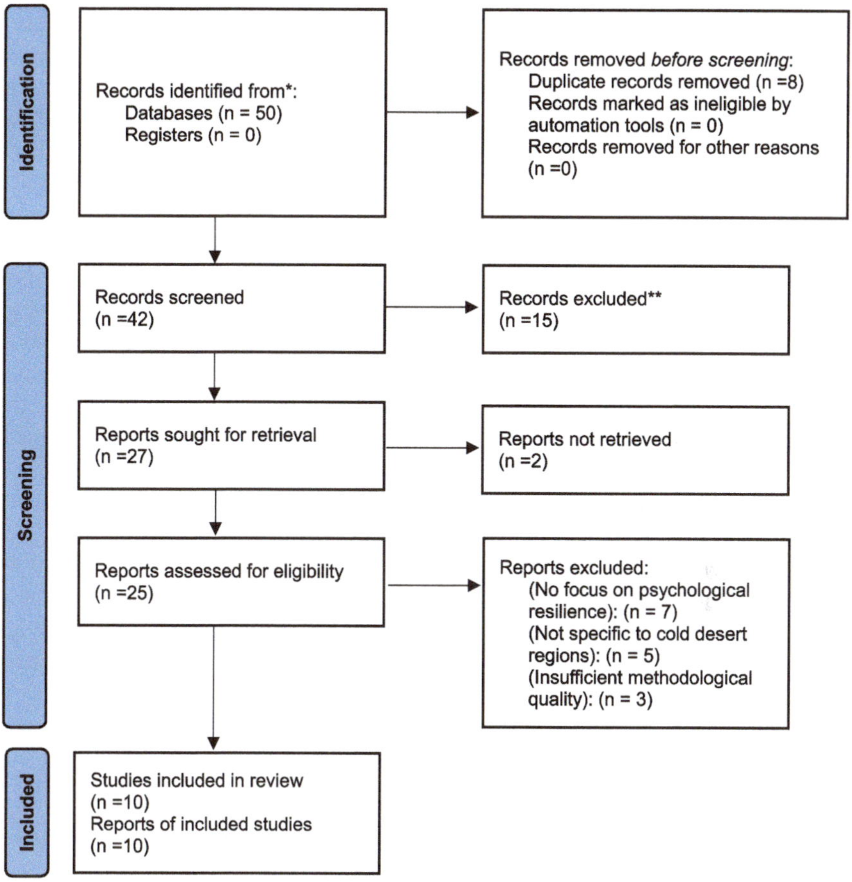

Fig. 1 PRISMA model

recorded, including behavioral adjustments (e.g., changes in daily routines, livelihood diversification), community-based initiatives (e.g., collective resource management, knowledge sharing practices), and policy-driven interventions (e.g., mental health services, educational programs). This category examines how psychological resilience translates into practical adaptation actions at different social scales.

5.5.2 Data Analysis Approach

The review employs narrative synthesis as the primary approach for analyzing qualitative data from diverse studies. This approach is appropriate given the heterogeneity of research designs, cultural contexts, and specific psychological processes examined across the literature on cold desert adaptation (Popay et al., 2006). The narrative synthesis follows a systematic process of organizing findings into conceptual categories while preserving contextual nuances from different regional studies. Thematic coding is used to identify patterns in psychological resilience factors across different cold desert regions. Initial coding categories are derived from established resilience frameworks, but the coding scheme evolves inductively as region-specific resilience factors emerge from the literature. This approach balances theoretical consistency with openness to contextual variations in how resilience manifests across cultural settings (Braun & Clarke, 2006). Two researchers independently code a subset of studies to establish inter-rater reliability before completing the coding process. For studies reporting quantitative results, descriptive statistics are compiled when possible to identify patterns in resilience factors and their relationships to adaptation outcomes. However, given the anticipated heterogeneity in measurement approaches and outcome variables across studies, meta-analysis is not planned as part of the current review. Instead, the synthesis focuses on identifying consistent psychological patterns while acknowledging methodological differences across studies.

The final synthesis integrates findings across studies to develop a comprehensive framework of psychological resilience factors in cold desert adaptation. This framework identifies both universal resilience processes that appear consistent across regions and context-specific factors that may operate differently in particular cultural or environmental settings. The analysis also highlights gaps in current knowledge and promising directions for future research on psychological dimensions of climate adaptation in cold desert regions.

6 Results and Discussion

6.1 Psychological Challenges in Cold Desert Regions

The systematic review revealed several significant psychological challenges faced by inhabitants of cold desert regions, with climate-induced stressors emerging as particularly prominent concerns. These challenges appear to be intensifying as climate change accelerates environmental transformation in these sensitive ecosystems.

6.1.1 Climate-Induced Stress and Mental Health Issues

The review identified a consistent pattern of climate-induced psychological distress across multiple cold desert communities. The nature and severity of these mental health impacts varied based on several factors including the speed of environmental changes, adaptive capacity, and cultural context. In Ladakh, rapid glacial retreat has been linked to increased anxiety and depression among farmers whose livelihoods depend on predictable meltwater (Nüsser et al., 2012). Similarly, research in Tibetan plateau communities documented significant psychological distress associated with changing precipitation patterns that undermine traditional agricultural knowledge (Byg & Salick, 2009). Several studies reported symptoms consistent with climate anxiety and ecological grief, and few studies found that 63% of their respondents in the Qinghai–Tibet plateau region reported moderate to severe worry about climate change impacts, with 27% showing clinical levels of anxiety related to environmental degradation. The psychological impact appears particularly pronounced among those with strong place attachment and environmental identity, which characterizes many traditional cold desert inhabitants (Cunsolo & Ellis, 2018). Post-traumatic stress disorder (PTSD) symptoms were documented in communities experiencing sudden climate-related disasters such as glacial lake outburst floods (GLOFs) in high-altitude regions. Research with survivors of the 2015 GLOF event in Ladakh found that 32% exhibited PTSD symptoms six months after the disaster, with those losing homes or family members most severely affected (Kumar & Shah, 2019). This suggests that while cold desert communities demonstrate remarkable resilience to gradual environmental stressors, acute climate disasters can overwhelm adaptive capacity and result in significant psychological trauma.

6.1.2 Isolation and Loneliness

Geographic isolation and extreme weather conditions create significant psychological challenges for cold desert inhabitants. Multiple studies documented how prolonged winters and limited mobility contribute to seasonal patterns of loneliness and depression. Research in remote Himalayan villages found that winter isolation was associated with increased symptoms of depression, particularly among elderly community members (Angchok & Singh, 2006). This isolation effect appears compounded by the out-migration of younger generations seeking economic opportunities, leaving behind increasingly vulnerable elderly populations. The psychological impact of isolation varies significantly based on cultural context and community structures. Studies with nomadic herders in Mongolia's Gobi region found that traditional collective coping strategies mitigated isolation effects, with seasonal gatherings providing crucial psychological support (Upton, 2020). In contrast, more sedentary communities experiencing fragmentation of traditional social structures reported higher rates of isolation-related psychological distress (Angchok & Singh, 2006) Modern communication technologies have emerged as both mitigating and complicating factors for isolation-related psychological challenges. While telecommunications reduce physical isolation, several studies noted the psychological tension created when community members gain increased awareness of lifestyles outside their remote regions. This awareness sometimes creates new sources of relative deprivation and identity conflict, particularly among younger community members (Nüsser et al., 2012).

6.1.3 Cultural Shifts and Identity Loss

Among the most profound psychological challenges identified in the review was the threat to cultural identity and traditional knowledge systems posed by rapid environmental change. Indigenous and traditional communities in cold desert regions have developed identities and knowledge systems intimately connected to their unique environments over centuries. As these environments transform, psychological distress related to identity disruption has emerged as a significant concern. Research with Andean communities documented how changing precipitation patterns undermine traditional agricultural calendars that previously structured not only economic activities but also cultural ceremonies and social life (Vidaurre de la Riva et al., 2013). The resulting "epistemological distress" manifests as community members question the validity of traditional knowledge that no longer accurately predicts environmental conditions. This creates profound psychological challenges as individuals navigate between preserving cultural heritage and adapting to new realities. Several studies documented intergenerational tensions emerging from these cultural shifts. Elder community members often report psychological distress related to inability to transfer traditional environmental knowledge to younger generations as this knowledge loses practical relevance (Byg & Salick, 2009). Meanwhile, younger community members experience psychological conflict between respecting traditional knowledge systems and pursuing adaptive strategies that may diverge from traditional practices.

This generational disconnect threatens social cohesion that previously served as a crucial psychological buffer during environmental challenges. Language loss compounds these cultural identity challenges, as many cold desert regions host endangered languages that contain sophisticated environmental knowledge systems. Prominent study found that communities experiencing rapid language loss reported higher rates of identity-related psychological distress and reduced collective efficacy in addressing environmental challenges. This suggests that cultural and linguistic preservation represents not merely a heritage concern but a crucial component of psychological resilience in the face of climate change. The review findings suggest that psychological adaptation in cold desert regions cannot be separated from broader questions of cultural continuity and transformation. The most effective approaches appear to be those supporting communities in maintaining cultural identity while selectively incorporating new knowledge and practices—a delicate balance requiring both psychological flexibility and cultural preservation. These three interconnected challenges—climate-induced stresses, isolation, and cultural shifts—create a complex psychological landscape for cold desert inhabitants. The subsequent sections will explore how communities navigate these challenges through various resilience strategies, drawing on both traditional knowledge systems and contemporary psychological approaches.

6.2 *Key Resilience-Building Strategies*

The systematic review identified three interconnected levels of psychological resilience strategies that have proven effective in cold desert communities facing climate change: individual-level approaches, community-based interventions, and policy/institutional frameworks. These strategies frequently overlap and reinforce each other, creating layered resilience systems that support adaptation to extreme environments.

6.2.1 Cognitive Reframing

Cognitive reframing emerged as a powerful individual resilience strategy across multiple studies. This process involves reconceptualizing climate-related challenges in ways that facilitate adaptation rather than despair. Research in Ladakh documented how individuals who successfully reframe unpredictable weather patterns as opportunities for innovation rather than threats to tradition demonstrate greater psychological wellbeing (Angchok & Singh, 2006). The effectiveness of cognitive reframing appears closely linked to cultural and spiritual frameworks. Studies with Buddhist communities in high-altitude regions reveal how traditional concepts like impermanence provide cognitive tools for reframing environmental change as part of natural cycles rather than catastrophic endings (Byg & Salick, 2009). These cultural frameworks don't eliminate distress but transform it into more adaptive forms that motivate sustainable action rather than avoidance or fatalism.

6.3 *Emotion Regulation Strategies*

The review identified several emotion regulation strategies particularly effective in cold desert contexts. Mindfulness practices, often embedded within traditional spiritual systems, help individuals tolerate environmental uncertainty and extreme conditions. Research with Tibetan communities documented how mindfulness practices reduced rumination about environmental threats and facilitated present-moment focus during challenging climate events. Acceptance-based coping strategies also featured prominently across studies. Unlike resignation, adaptive acceptance involves acknowledging environmental realities while maintaining agency for possible responses. Kumar and Shah (2019) found that communities practicing acceptance-based coping recovered more rapidly from glacial lake outburst floods compared to those employing denial or avoidance strategies. This suggests that psychological flexibility—rather than rigid optimism or pessimism—supports sustainable adaptation to climate uncertainty. Traditional knowledge often incorporates sophisticated emotion regulation techniques tailored to extreme environments.

For example, studies in Andean communities documented seasonal rituals that function as emotional processing mechanisms during environmental transitions, helping community members regulate anticipatory anxiety about water availability (Vidaurre de la Riva et al., 2013). These culturally embedded practices represent valuable psychological resources that merit preservation alongside technological adaptation strategies.

6.4 *Survival Mindset*

Cold desert inhabitants often develop distinctive psychological traits that collectively constitute a "survival mindset" supporting adaptation to extreme conditions. This mindset typically includes high tolerance for uncertainty, resourcefulness in resource-limited settings, and psychological endurance during prolonged environmental stress. Research with herding communities in Mongolia's Gobi region demonstrated how this survival mindset is deliberately cultivated through childrearing practices emphasizing delayed gratification and comfort with solitude (Upton, 2020). The survival mindset appears partially heritable through both genetic and cultural transmission. Several studies noted that the role of transgenerational narratives in cultivating psychological toughness, with community elders using historical examples of survival to instill resilience in younger generations (Byg & Salick, 2009). However, rapid environmental change sometimes outpaces these traditional psychological adaptations, creating new mental health challenges that require innovative responses.

6.5 *Social Support Networks*

Strong social support networks consistently emerged as powerful buffers against climate-related psychological distress. Communities maintaining cohesive social structures demonstrated significantly lower rates of depression and anxiety despite facing similar environmental challenges as more fragmented communities (Kumar & Shah, 2019). These networks provide emotional support, practical assistance during climate-related hardships, and shared meaning-making that helps individuals contextualize personal experiences. Traditional reciprocity systems in cold desert regions often serve dual economic and psychological functions. Research in Andean communities documented how traditional labor exchange systems (ayni) not only distribute agricultural work but also create psychological security through mutual obligation (Vidaurre de la Riva et al., 2013). Similarly, studies in Ladakh identified how resource-sharing traditions reduce psychological distress during periods of scarcity by establishing cultural expectations of community support during hardship (Angchok & Singh, 2006). The review highlighted how climate change sometimes strengthens rather than weakens these social support networks by necessitating collective adaptation efforts. Communities that respond to environmental challenges through collaborative problem-solving often report enhanced social cohesion and reduced psychological distress compared to pre-crisis periods (Nüsser et al., 2012). This suggests that facilitated community adaptation processes can simultaneously address practical needs and build psychological resilience.

6.6 Traditional Ecological Knowledge (TEK)

Traditional ecological knowledge (TEK) systems provide psychological as well as practical resources for climate adaptation. These knowledge systems typically blend environmental observations, cultural practices, and spiritual frameworks into coherent worldviews that support psychological stability despite environmental uncertainty Houston (2006). Research in Tibet documented how traditional environmental knowledge reduced climate anxiety by providing cognitive frameworks for understanding changing weather patterns, even when these frameworks required adaptation to new conditions (Byg & Salick, 2009). The intergenerational transmission of TEK creates psychological continuity that supports identity maintenance despite environmental change. Studies with Ladakhi communities found that young people engaging with traditional knowledge systems reported stronger environmental identity and lower rates of psychological distress as compared to peers disconnected from these traditions (Nüsser et al., 2012).This suggests that cultural transmission represents an important psychological resource rather than merely a heritage concern. Communities successfully integrating traditional knowledge with scientific information demonstrate particularly strong psychological resilience. Research across multiple cold desert regions found that this knowledge integration supports both practical adaptation and psychological wellbeing by affirming the value of cultural heritage while incorporating new information. Programs facilitating respectful dialogue between traditional knowledge holders and scientific experts show promise for enhancing both adaptation effectiveness and psychological resilience.

6.7 Psychological Preparedness Training

Formal and informal psychological preparedness training has emerged as an effective community-based resilience strategy. Several studies documented community-led education initiatives addressing climate change impacts, uncertainty management, and collective coping strategies. Research in Ladakh evaluated a community-based climate resilience program incorporating psychological components, finding significant improvements in reported self-efficacy and reduced climate anxiety among participants (Houston, 2006; Kumar & Shah, 2019). Intergenerational knowledge transfer often includes explicit psychological preparation for environmental challenges. Studies with nomadic communities documented how traditional education practices deliberately build psychological skills for managing isolation, resource scarcity, and environmental uncertainty (Upton, 2020). These practices suggest models for more formal psychological preparedness programs that respect cultural contexts while addressing emerging climate challenges. The most effective psychological preparedness approaches appear to blend traditional wisdom with contemporary psychological insights. Programs incorporating mindfulness-based stress reduction techniques alongside traditional coping strategies demonstrated particularly strong outcomes in multiple studies. An exploratory study by Saini et al. (2017) investigated cognitive deficits due to thermal stress among soldiers deployed in desert regions. The findings indicated that prolonged exposure to high temperatures adversely affected cognitive functions, emphasizing the need for interventions to mitigate thermal stress and enhance resilience among personnel in hot desert environments. This integration allows communities to maintain cultural continuity while accessing evidence-based approaches for addressing novel psychological challenges posed by accelerating climate change.

6.8 Mental Health Programs

Formal mental health programs tailored to cold desert contexts show promise for enhancing climate resilience, though their implementation remains limited in many regions. Research in Ladakh documented positive outcomes from a pilot program integrating mental health services into existing climate adaptation initiatives, with participants reporting improved coping skills and reduced climate anxiety (Kumar & Shah, 2019). The most effective programs employ culturally adapted approaches rather than imposing Western psychological models without modification. Several studies highlighted the importance of addressing structural barriers to mental health services in remote regions. Tele-health interventions show particular promise for overcoming geographic isolation, though implementation challenges include limited connectivity and cultural acceptability. Training local community members as mental health first responders offers another approach for expanding service availability while maintaining cultural relevance. The review identified significant gaps in trauma-informed approaches to climate disasters in cold desert regions. Communities experiencing acute climate events like glacial lake outburst floods often lack access to trauma-specific mental health support, potentially prolonging psychological impacts (Kumar & Shah, 2019). This highlights the need for disaster mental health planning specifically addressing the unique contexts of cold desert communities.

6.9 Resilience-Focused Education

Harrison et al. (2021) conducted a longitudinal mixed-methods study examining psychological resilience among participants undertaking a 25-day endurance challenge in an extreme environment. The study identified a "cluster effect" of stressors—such as environmental challenges, interpersonal dynamics, and logistical issues—that impacted participants' resilience over time. Key coping strategies included adopting a challenge mindset, leveraging social support, and maintaining focus on immediate goals. The research highlighted the dynamic and individualized nature of resilience in extreme settings educational initiatives explicitly addressing psychological dimensions of climate adaptation show growing evidence of effectiveness. School-based programs teaching both practical adaptation skills and psychological coping strategies demonstrate promising outcomes in several studies, with students reporting increased self-efficacy and reduced helplessness regarding climate change impacts (Nüsser et al., 2012). The most successful programs integrate local ecological knowledge with contemporary resilience concepts, affirming cultural heritage while building new skills. Intergenerational knowledge exchange programs serve both educational and psychological functions.

Research in Tibetan communities documented how structured dialogues between elders and youth regarding environmental change reduced psychological distress in both groups—giving elders a sense of continued value for their knowledge and youth access to traditional wisdom for contextualizing current changes (Byg & Salick, 2009). These exchanges support identity continuity while facilitating adaptive knowledge evolution. Educational initiatives fostering psychological connection to changing landscapes show particular promise. Programs helping community members develop meaningful relationships with transformed environments—rather than solely mourning lost landscapes—support psychological adaptation through renewed place attachment. This approach recognizes that sustainable adaptation requires not just practical knowledge but also emotional relationships with changing environments.

6.10 Psychological Incentives for Sustainable Practices

Policies incorporating psychological incentives for sustainable behaviors demonstrate growing effectiveness in cold desert regions. Research in Mongolia's Gobi region evaluated ecological payment systems designed to recognize nomadic herders' traditional conservation practices, finding these programs enhanced both environmental outcomes and psychological wellbeing by affirming cultural identity and environmental stewardship (Upton, 2020). These findings suggest economic incentives work best when aligned with existing cultural values rather than imposing external motivations. Community-based tourism initiatives show particular promise for creating psychological as well as economic benefits. Studies in Ladakh found that eco-tourism programs honoring traditional ecological knowledge enhanced community members' sense of cultural value and environmental efficacy (Angchok & Singh, 2006). These programs appear most successful when communities maintain substantial control over implementation, ensuring alignment with local values and priorities. Institutional recognition of traditional knowledge systems provides significant psychological benefits alongside practical advantages. Policies formally incorporating indigenous expertise into environmental management not only improve adaptation outcomes but also enhance community psychological wellbeing through cultural affirmation (Vidaurre de la Riva et al., 2013). This suggests that knowledge diversity should be valued not just for its practical contributions but also for its role in supporting psychological resilience.

7 Conclusion and Future Directions

7.1 *Summary of Key Findings*

This systematic review demonstrates that psychological resilience represents a critical dimension of sustainable climate adaptation in cold desert regions, functioning not merely as a co-benefit but as a fundamental prerequisite for effective adaptation. The findings reveal how climate stressors impact multiple dimensions of psychological wellbeing, from individual mental health to cultural identity and community stability. Climate-induced psychological challenges include increased anxiety and depression, isolation-related distress, and profound cultural identity disruption as traditional environmental knowledge loses predictive accuracy in rapidly changing conditions. Despite these challenges, the review identifies numerous resilience factors operating at individual, community, and institutional levels. Indigenous knowledge systems embody sophisticated psychological adaptation strategies developed through centuries of experience with environmental extremes. Social cohesion provides crucial psychological buffering against climate stressors, with strong community networks consistently associated with reduced mental health impacts. Institutional support—particularly when aligned with community values and priorities—enhances psychological resilience by providing resources and recognition for adaptation efforts. The research reveals that psychological adaptation is inseparable from cultural continuity in cold desert regions. Communities able to maintain core cultural values while flexibly incorporating new knowledge demonstrate the strongest psychological resilience. This suggests that adaptation approaches focused solely on technological or economic dimensions while neglecting cultural integrity may undermine rather than enhance long-term resilience. The most successful interventions respect cultural frameworks while providing additional resources for navigating unprecedented environmental changes.

7.2 Policy Recommendations

Based on these findings, several policy recommendations emerge for enhancing psychological resilience in cold desert regions facing climate change: First, mental health services should be systematically integrated into climate adaptation frameworks rather than treated as separate concerns. This integration should include both preventive approaches building general psychological resilience and targeted interventions addressing specific climate-related mental health impacts such as ecological grief and climate anxiety. Mental health support should be culturally appropriate and accessible despite geographic isolation, potentially through community-based models and tele-health options where infrastructure permits. Second, adaptation planning should prioritize community-led models that enhance psychological empowerment alongside physical resilience. Top-down adaptation approaches, even when technically sound, often undermine local agency and traditional knowledge systems that provide psychological stability. Participatory planning processes incorporating diverse knowledge sources and respecting community priorities yield more psychologically sustainable outcomes. This approach requires institutional humility and genuine power-sharing rather than merely consultative processes.

Third, educational and awareness programs should explicitly address eco-psychology and resilience-building alongside practical adaptation strategies. Climate education focusing solely on threats without building psychological capacity may inadvertently increase distress and learned helplessness. Curricula should incorporate both TEK and contemporary psychological insights regarding resilience, helping community members develop robust coping strategies for navigating environmental uncertainty. Fourth, cultural preservation initiatives should be recognized as climate adaptation strategies due to their psychological benefits. Language preservation, traditional knowledge documentation, and cultural practices maintenance support psychological continuity that enhances adaptive capacity. These cultural dimensions should receive funding and institutional support comparable to technological adaptation strategies given their critical role in maintaining psychological resilience. Finally, adaptation planning should incorporate longer timeframes acknowledging the psychological dimensions of change. Rapid transition strategies may achieve technical objectives while creating psychological disruption that undermines long-term sustainability. Phased approaches allowing for psychological adaptation alongside physical and economic changes better support community wellbeing during climate transitions.

7.3 Future Research Directions and Conclusion

The review identifies several promising directions for future research on psychological dimensions of climate adaptation in cold desert regions: Longitudinal studies tracking psychological resilience throughout climate adaptation processes would provide valuable insights currently missing from the predominantly cross-sectional research base. These studies should monitor both individual mental health outcomes and community-level psychological indicators such as social cohesion and cultural continuity over multi-year periods. Such longitudinal approaches would help identify critical intervention points for maintaining psychological wellbeing during climate transitions. Emerging technologies including machine learning and artificial intelligence offer promising tools for predicting mental health vulnerabilities in extreme climates. Research developing predictive models integrating environmental, social, and psychological variables could support more targeted and proactive mental health interventions. These approaches should incorporate both quantitative indicators and qualitative data from traditional knowledge systems to ensure comprehensive understanding of vulnerability factors.

Cross-cultural analysis comparing psychological adaptation strategies across geographically distant cold desert communities would help distinguish universal resilience factors from context-specific mechanisms. While this review identified commonalities across regions, systematic comparative research could provide deeper insights into transferable principles versus culturally-specific approaches. Such research should employ collaborative methodologies ensuring equitable knowledge co-production rather than extractive approaches. Finally, interdisciplinary research integrating psychological perspectives with climate science, anthropology, and indigenous knowledge systems represents a crucial frontier. The complex interactions between environmental change, cultural systems, and mental health require integrated approaches transcending traditional disciplinary boundaries. Research partnerships bringing together diverse expertise and knowledge systems would support more holistic understanding of psychological dimensions in climate adaptation. As climate change continues transforming cold desert regions at unprecedented rates, supporting psychological resilience becomes increasingly critical for sustainable adaptation. This review demonstrates that effective adaptation requires attention not just to physical infrastructure and economic systems but also to the psychological dimensions that enable communities to directing profound environmental change while maintaining wellbeing and cultural continuity. By integrating these psychological insights into climate adaptation planning, policymakers and communities can develop more holistic approaches that support both environmental sustainability and human flourishing in Earth's extreme environments.

References

Adams, R. J., Smart, P., & Huff, A. S. (2017). Shades of grey: Guidelines for working with the grey literature in systematic reviews for management and organizational studies. *International Journal of Management Reviews, 19*(4), 432–454. https://doi.org/10.1111/ijmr.12102

Adger, W. N. (2003). Social capital, collective action, and adaptation to climate change. *Economic Geography, 79*(4), 387–404. https://doi.org/10.1111/j.1944-8287.2003.tb00220.x

Angchok, D., & Singh, P. (2006). Traditional irrigation and water distribution system in Ladakh. *Indian Journal of Traditional Knowledge, 5*(3), 397–402. Retrieved May 01, 2025 from https://nopr.niscpr.res.in/handle/123456789/6745

Berrang-Ford, L., Ford, J. D., & Paterson, J. (2011). Are we adapting to climate change? *Global Environmental Change, 21*(1), 25–33. https://doi.org/10.1016/j.gloenvcha.2010.09.012

Berry, H. L., Bowen, K., & Kjellstrom, T. (2010). Climate change and mental health: A causal pathways framework. *International Journal of Public Health, 55*(2), 123–132. https://doi.org/10.1007/s00038-009-0112-0

Bonanno, G. A. (2004). Loss, trauma, and human resilience: Have we underestimated the human capacity to thrive after extremely aversive events? *American Psychologist, 59*(1), 20–28. https://doi.org/10.1037/0003-066X.59.1.20

Braun, V., & Clarke, V. (2006). Using thematic analysis in psychology. *Qualitative Research in Psychology, 3*(2), 77–101. https://doi.org/10.1191/1478088706qp063oa

Bronfenbrenner, U. (1979). *The ecology of human development: Experiments by nature and design.* Harvard University Press. Retrieved May 01, 2025 from https://khoerulanwarbk.wordpress.com/wp-content/uploads/2015/08/urie_bronfenbrenner_the_ecology_of_human_developbokos-z1.pdf.

Brown, K., & Westaway, E. (2011). Agency, capacity, and resilience to environmental change: Lessons from human development, well-being, and disasters. *Annual Review of Environment and Resources, 36*, 321–342.https://doi.org/10.1146/annurev-environ-052610-092905

Byg, A., & Salick, J. (2009). Local perspectives on a global phenomenon—climate change in Eastern Tibetan villages. *Global Environmental Change, 19*(2), 156–166. https://doi.org/10.1016/j.gloenvcha.2009.01.010

Clayton, S., Manning, C. M., Krygsman, K., & Speiser, M. (2017). *Mental health and our changing climate: Impacts, implications, and guidance.* American Psychological Association. Retrieved May 01, 2025 from https://www.apa.org/news/press/releases/2017/03/mental-health-climate.pdf.

Cunsolo, A., & Ellis, N. R. (2018). Ecological grief as a mental health response to climate change-related loss. *Nature Climate Change, 8*(4), 275–281. https://doi.org/10.1038/s41558-018-0092-2

Goodall, S. K. (2004). Rural-to-urban migration and urbanization in Leh, Ladakh. *Mountain Research and Development, 24*(3), 220–227. Retrieved May 01, 2025 from http://www.bioone.org/doi/pdf/10.1659/0276-4741%282004%29024%5B0220%3ARMAUIL%5D2.0.CO%3B2.

Harrison, D., Sarkar, M., Saward, C., & Sunderland, C. (2021). Exploration of psychological resilience during a 25-day endurance challenge in an extreme environment. *International Journal of Environmental Research and Public Health, 18*(23), 12707. https://doi.org/10.3390/ijerph182312707

Hayes, S. C., Luoma, J. B., Bond, F. W., Masuda, A., & Lillis, J. (2006). Acceptance and commitment therapy: Model, processes and outcomes. *Behaviour Research and Therapy, 44*(1), 1–25. https://doi.org/10.1016/j.brat.2005.06.006

Higgins, J. P., & Green, S. (Eds). (2011). *Cochrane handbook for systematic reviews of interventions* (Vol. 4). John Wiley & Sons. https://doi.org/10.1002/9780470712184

Hobfoll, S. E. (1989). Conservation of resources: A new attempt at conceptualizing stress. *American Psychologist, 44*(3), 513–524. https://doi.org/10.1037/0003-066X.44.3.513

Houston, J. (2006). Variability of precipitation in the Atacama Desert: Its causes and hydrological impact. *International Journal of Climatology, 26*(15), 2181–2198. https://doi.org/10.1002/joc.1359

IPCC. (2019). Special Report on the ocean and cryosphere in a changing climate. Intergovernmental panel on climate change. Retrieved 01 May, 2025, from https://www.ipcc.ch/srocc/.

Kumar, S., & Shah, A. (2019). Psychological impact of glacial lake outburst floods on affected communities in Ladakh, India. *International Journal of Disaster Risk Reduction, 37*, Article 101141. https://doi.org/10.1080/19475705.2024.2413893

Masten, A. S. (2001). Ordinary magic: Resilience processes in development. *American Psychologist, 56*(3), 227–238. https://doi.org/10.1037/0003-066X.56.3.227

Moher, D., Liberati, A., Tetzlaff, J., Altman, D. G., & PRISMA Group. (2009). Preferred reporting items for systematic reviews and meta-analyses: The PRISMA statement. *PLoS Medicine, 6*(7), e1000097. https://doi.org/10.1371/journal.pmed.1000097

Norris, F. H., Stevens, S. P., Pfefferbaum, B., Wyche, K. F., & Pfefferbaum, R. L. (2008). Community resilience as a metaphor, theory, set of capacities, and strategy for disaster readiness. *American Journal of Community Psychology, 41*(1–2), 127–150. https://doi.org/10.1007/s10464-007-9156-6

Nüsser, M., Schmidt, S., & Dame, J. (2012). Irrigation and development in the Upper Indus Basin: Characteristics and recent changes of a socio-hydrological system in central Ladakh, India. *Mountain Research and Development, 32*(1), 51–61. https://doi.org/10.1659/MRD-JOURNAL-D-11-00091.1

Ojala, M. (2012). Hope and climate change: The importance of hope for environmental engagement among young people. *Environmental Education Research, 18*(5), 625–642. https://doi.org/10.1080/13504622.2011.637157

Palinkas, L. A. (2003). The psychology of isolated and confined environments: Understanding human behavior in Antarctica. *American Psychologist, 58*(5), 353–363. https://doi.org/10.1037/0003-066X.58.5.353

Palinkas, L. A., & Suedfeld, P. (2008). Psychological effects of polar expeditions. *The Lancet, 371*(9607), 153–163. https://doi.org/10.1016/S0140-6736(07)61056-3

Palinkas, L. A., Johnson, J. C., Boster, J. S., & Houseal, M. (2004). Social support and depressed mood in isolated and confined environments. *Acta Astronautica, 54*(9), 639–647. https://doi.org/10.1016/S0094-5765(03)00236-4

Palinkas, L. A., Holloway, K. C., Rice, E., Fuentes, D., Wu, Q., & Chamberlain, P. (2010). Social networks and implementation of evidence-based practices in public youth-serving systems: A mixed-methods study. *Implementation Science, 5*(1), 1–11. https://doi.org/10.1186/1748-5908-6-113

Pearce, T., Ford, J., Willox, A. C., & Smit, B. (2015). Inuit traditional ecological knowledge (TEK), subsistence hunting and adaptation to climate change in the Canadian Arctic. *Arctic, 68*(2), 233–245. https://doi.org/10.14430/arctic4475

Popay, J., Roberts, H., Sowden, A., Petticrew, M., Arai, L., Rodgers, M., Britten, N. (2006). *Guidance on the conduct of narrative synthesis in systematic reviews. A product from the ESRC methods programme.* Lancaster University. https://doi.org/10.13140/2.1.1018.4643

Rasmussen, L. N., & Montgomery, P. (2018). The prevalence of and factors associated with inclusion of non-English language studies in Campbell systematic reviews: A survey and meta-epidemiological study. *Systematic Reviews, 7*(1), 1–12. https://doi.org/10.1186/s13643-018-0786-6

Saini, R., Srivastava, K., Agrawal, S., & Das, R. C. (2017). Cognitive deficits due to thermal stress: An exploratory study on soldiers in deserts. *Medical Journal Armed Forces India, 73*(4), 370–374. https://doi.org/10.1016/j.mjafi.2017.07.011

Sandal, G. M., Leon, G. R., & Palinkas, L. (2006). Human challenges in polar and space environments. *Reviews in Environmental Science and Bio/technology, 5*(2), 281–296. https://doi.org/10.1007/978-1-4020-6285-8_25

Tahernejad, A., Sohrabizadeh, S., & Mashhadi, A. (2024). Exploring factors affecting psychological resilience of farmers living in drought-affected regions in Iran: A qualitative study. *Frontiers in Psychology, 15*, 1418361. https://doi.org/10.3389/fpsyg.2024.1418361

Taylor, S. E. (1983). Adjustment to threatening events: A theory of cognitive adaptation. *American Psychologist, 38*(11), 1161–1173. https://doi.org/10.1037/0003-066X.38.11.1161

Thomas, D. S. G., & Goudie, A. S. (2016). The Dictionary of Physical Geography (3rd ed.). https://doi.org/10.1002/9781118782323.ch20

Ungar, M. (2011). The social ecology of resilience: Addressing contextual and cultural ambiguity of a nascent construct. *American Journal of Orthopsychiatry, 81*(1), 1–17. https://doi.org/10.1111/j.1939-0025.2010.01067.x

Upton, C. (2020). Conserving natures? Co-producing payments for ecosystem services in Mongolian rangelands. *Development and Change, 51*(1), 224–252. https://doi.org/10.1111/dech.12549

Vidaurre de la Riva, M., Lindner, A., Pretzsch, J. (2013). Assessing adaptation-climate change and indigenous livelihood in the Andes of Bolivia. *Journal of Agriculture and Rural Development in the Tropics and Subtropics, 114*(2), 109–122. Retrieved 01 May, 2025, from https://tu-dresden.de/bu/umwelt/forst/inter/tropen/ressourcen/dateien/forschung/INCA/publicationsinca/JARTS?lang=en.

Wexler, L. (2014). Looking across three generations of Alaska Natives to explore how culture fosters indigenous resilience. *Transcultural Psychiatry, 51*(1), 73–92. https://doi.org/10.1177/1363461513497417

Index

Y. S. Rawat and C. P. Kala (eds.), *Cold Deserts*,
https://doi.org/10.1007/978-3-032-29119-6

Kolibri 360 GmbH
Ferdinand-Jühlke-Straße 7
99095 Erfurt, Deutschland
produktsicherheit@kolibri360.de